D1592106

Industrial Firefighting for Municipal Firefighters

Industrial Firefighting for Municipal Firefighters

Craig H. Shelley, CFO, EFO, MIFireE

Anthony R. Cole, CFPS, CFEI, MIFireE

Timothy E. Markley, Fire Officer IV, CFEI, CFII

Disclaimer

The recommendations, advice, descriptions, and methods in this book are presented solely for educational purposes. The authors and publisher assume no liability whatsoever for any loss or damage that results from the use of any of the material in this book. Use of material in this book is solely at the risk of the user.

PennWell Corporation
1421 South Sheridan Road
Tulsa, Oklahoma 74112-6600 USA

800.752.9764
+1.918.831.9421
sales@pennwell.com
www.Fire EngineeringBooks.com
www.pennwellbooks.com
www.pennwell.com

Marketing Manager: Julie Simmons
National Account Executive: Francie Halcomb
Director: Mary McGee
Managing Editor: Jerry Naylis
Production / Operations Manager: Traci Huntsman
Production Editor: Tony Quinn
Cover Designer: Clark Bell
Book Designer: Kelly Cook
Book Layout: Susan Ormston

Library of Congress Cataloging in Publication Data

Shelley, Craig H., Cole, Anthony R., and Markley, Timothy E.
Industrial firefighting for municipal firefighters / Craig H. Shelley, Anthony R. Cole, and Timothy E. Markley.
p. cm.
Includes bibliographical references and index.
ISBN 978-1-59370-081-2
1. Fire prevention. 2. Fire extinction. 3. Industrial safety. I. Cole, Anthony R. II. Markley, Timothy E. III. Title.
TH9151.S44 2007
628.9'25--dc22

2007015290

Printed in the United States of America

1 2 3 4 5 11 10 09 08 07

CONTENTS

PREFACE

Having spent a combined 85 years in the municipal, industrial, and military fire service, as well as the industrial fire protection engineering and insurance-related consulting businesses, we came to realize the strong need to bridge the gap between the municipal fire department and the industrial process occupancies. Although there are many similarities, the differences are monumental. This gap has been recognized by professional organizations such as the Society of Fire Protection Engineers (SFPE), who have created a task force to study and evaluate this gap. We have tried to share both personal and professional knowledge and experiences from such organizations as the New York City Fire Department (FDNY), United States Air Force, Factory Mutual Engineering and Research (now FM Global), and countless visits to many of Fortune 500 companies across the globe.

After working in the industrial fire protection field for all these years, we looked back at some of these challenging situations and realized that what made them challenging was the fact that they were not the normal, bread-and-butter operations that a municipal fire department responds to on a daily basis. These operations consisted of large fuel fires such as those common to large atmospheric tanks or fires in large electric generation plants. These operations might include fires and leaks in large-diameter pipes that are typically found in refineries and petrochemical plants or explosion hazards associated with dust in woodworking, or food and beverage plants. At one of these incidents, a close friend was seriously burned by steam while operating at a large electrical power generation plant. At others, the fire department was unsure of the immediate and proper tactics to use until specialists were brought to the scene. As a result of some of these incidents, new standard operating procedures were developed, and new tools were purchased.

We began to research the available literature regarding fire tactics and procedures in use for municipal fire departments and found that much of it is written to address the commonplace operations concerning structural firefighting, but it was very limited in the amount of resources allotted to industrial firefighting available to the municipal firefighter. Yet in most communities, industrial occupancies are present, whether they are storage tanks, manufacturing plants, or pipelines. The time to understand the similarities and differences between municipal and industrial firefighting is prior to the incident, not during or after it.

And so the concept of this book was born. We wanted to find the best method of sharing their information with municipal fire departments, in an attempt to save lives and prevent injuries. Craig Shelley had a particular captain at one time during his career at the FDNY. The captain had a large sign over his desk in the company office declaring "Knowledge is Power." We truly believe that and hope this book imparts the knowledge to assist the municipal firefighters, fire inspectors, and fire marshals to perform their duties more efficiently and safely.

ACKNOWLEDGMENTS

We wish to thank the following people for their assistance with this book. Without their support, we would not have been able to pull it all together. Some people contributed suggestions, some donated photos, and others offered moral support.

Thanks go to Richard "Rick" Setzer, who helped the project get started and also contributed research to a number of chapters; Gregory Noll, who provided photos and over the years provided career advice and words of wisdom; Michael Reakes, whose photos made the authors look better than usual; and Lewis Bell, for his technical review of a number of chapters.

We also would be remiss if we didn't thank Jerry Naylis and the *Fire Engineering* staff for believing in the concept of this book and pushing us forward, trying to keep us on schedule; Dave Evans and Art Cote for allowing the authors' access and usage of photos and other pieces of artwork from NFPA; and also Eric LaVergne and Dave Cochran for supplying pictures and support. Dave McCoy, Jerry Smith, Bill Burkett, and Michael Barrett provided additional photographic support.

In addition, the following must also be recognized for contributed photographs and/or other behind-the-scenes support: Ansul, Val Pamboukes, Warren Stocker, Jeff Barlow, Jaime Baggett, Kevin Kupietz, David Bunzow, Daniel Levy, George Dodson, Chris Guillemette, and Ebbin A. Spellman. Victor Moussallem from National Foam made available several diagrams and photographs for the authors' use.

In memoriam, a special thank you goes to the late Tom Brennan, who coined the phrase "random thoughts" when highlighting various firefighting tactics and helpful tips in *Fire Engineering*. In our case studies, we have chosen to use this phrase when offering our thoughts on firefighting for a particular incident involving industrial firefighting.

And a special thank you to our wives and families, who put up with us during the past year and a half! Your patience has not gone unnoticed, and we appreciate it.

1

Purpose of This Book

Although many books have been written for firefighters regarding structural firefighting, and for industry workers on the subject of industrial process hazards, there are a limited number of books that specifically prepare the municipal firefighter for responses to a wide range of industrial fires.

Fig. 1–1. A large industrial fire may not only affect a business, but it also may affect the local or national economy. *(Courtesy of Williams Fire and Hazard Control)*

This book speaks to municipal firefighters, both career and volunteer, who may be called on to respond to an incident involving an industrial process or facility. It is also a pertinent guide to personnel at smaller industrial facilities that may have an in-house first-response capability but lack an experienced or full-time response. Larger industrial fire departments can use this book to check their response plans or prepare to respond to a fire at a neighboring plant or facility.

In today's business climate, industry has placed increased reliance on automation while reducing staffing levels at facilities. Advances in automation have resulted in decreased staff for operations personnel at industrial facilities. Subsequently, industrial fire brigades and departments are also shrinking. New products and processes bring new hazards to industry and concurrently to the fire service. Larger facilities are being constructed to meet greater production schedules, and their increasing value makes them more difficult to replace. A large industrial fire will not only affect the business where it occurred, but it also may well affect the local or national economy and many times it means the end to local jobs (fig. 1–1). How often do we see a spike in fuel prices after a petrochemical plant, refinery or pipeline has an emergency incident?

Although municipal firefighters respond on a daily basis to industrial fires and emergencies, even the largest fire departments often focus most of their training and attention to structural or wildland firefighting (fig. 1–2). Today's fire service training may be increasing its focus on current hot issues in the fire service such as rapid intervention teams (RITs), hazardous materials (HazMat), and weapons of mass destruction (WMD), to name a few. In reality, the chances of a municipal fire department responding to an industrial incident are much greater than the same department responding to a WMD incident. Yet, many of our efforts are directed toward those low-probability incidents. It has been said that industrial firefighting is a diverse subspecialty, requiring specialized knowledge and training. An improperly handled industrial incident can cause injuries and possibly death to civilians and responders. How can the fire service prepare those municipal firefighters who will respond to an industrial incident?

Fig. 1–2. Many fire departments focus their training mainly on structural firefighting responses. *(Courtesy of A. J. Morro)*

An *industrial incident* for the purpose of this book is defined as any incident that occurs or has the potential of occurring in a business or facility that manufactures, processes, produces, or stores products or goods. We have included in this definition those incidents that may occur in the marine or aircraft industry, including ship fires and air crashes due to their unusual response characteristics and products that may be involved.

It appears that the fire service is, at times, a reactive organization rather than a proactive one. Sometimes, we only focus and prepare for certain incidents after the incident has taken place. In New York City, a massive storage tank fire occurred in the early 1970s that highlighted the need for preplanning, especially the stockpiling of foam supplies, to combat such a fire. At the time, the fire department was ill-equipped to handle large hydrocarbon fires. Foam supplies had to be located and brought in from other states to combat this fire. Delivery systems to get the finished foam on to the fire were also lacking. How many departments today that have petroleum storage tanks in their districts have adequately preplanned, trained, conducted drills, and stockpiled adequate resources to mitigate an incident should it occur (fig. 1–3)? It was only after the incident mentioned earlier that the New York City Fire Department (FDNY) began a foam delivery system and foam training program in earnest. As a result, selected engine companies were designated as foam depots where the storage of bulk foam was kept in quarters. Even though it was a limited quantity, the policies and procedures were in place to transport the foam from these engine company's fire stations to the scene of the incident and be able to deliver it to the operational point where it was needed. Large-caliber foam nozzles and portable tanks were purchased, and foam coordinators were appointed. How will your department respond to a large industrial incident?

It is assumed by many firefighters that automatic detection and suppression systems will negate the need for a specialized or major response from the local fire department. Experience has proven this untrue. There are many incidents where the suppression systems were inoperative at the time of the fire, not installed, or were rendered inoperable because of the catastrophic event that caused the fire. At a recent incident in California, a municipal department was called to a large power plant containing fuel storage tanks, one of which was on fire. When interviewed after the fire, one of the fire officers stated that that they were not prepared for this magnitude of a fire because the tank in question was protected by a fixed fire suppression system. However, unknown to the fire department, this system had been removed because the storage tank was being demolished.

It is increasingly probable that firefighters will be called to an incident at one of these facilities due to an industrial accident, fire, or terrorist event, and municipal firefighters must be prepared. An educated and efficient response will assist with the mitigation of large-loss fires or other events, minimizing business interruption and the resultant dollar loss, as well as minimizing the impact to the community at large.

Fig. 1–3. A large petroleum storage tank fire will need adequate preincident response planning, training, and resources. *(Courtesy of Williams Fire and Hazard Control)*

The book is divided into two sections. The first section, chapters 1 through 11, contains general information pertaining to industrial firefighting. It describes the qualifications and skills required to successfully handle an industrial incident, highlights the functions and responsibilities of industrial emergency response teams, and discusses the incident command system and how it is structured for an industrial incident. This section also describes emergency response planning, including drills and exercises, and general firefighting tactics that may be used at various industrial incidents. Other areas that are examined are the fixed fire protection and detection systems that are in use and HazMat concerns. We also highlight homeland security issues that may affect the municipal fire department's response and/or planning for an industrial response. In today's world, just about every area of response has been affected by these homeland security issues. This section closely examines these issues including operational security issues, access issues, and restrictions to information sharing.

Section 2, chapters 12 through 34, targets specific occupancies, highlights their individual hazards, and presents a brief description of the process involved, common hazards, and firefighting tactics. Where applicable, case studies are presented that allow the reader to closely examine an incident and the tactics used to bring it to conclusion.

We, the authors, have vast experience in municipal, volunteer, and industrial fire departments, as well as experience in safety and process operations. It is our intention to impart our knowledge and experience to the reader so that, in understandable terms, the fire service can more effectively and safely respond to industrial incidents. It is not our intention to make the average firefighter a process engineer, but to provide a basic understanding of industrial hazards and firefighting techniques to the frontline fire officers and firefighters. This enhanced knowledge base will help the municipal firefighter interface with the industrial firefighting forces in a safe and effective way.

In researching this book and from personal experiences, we learned that there is a need for cooperation among municipal firefighters and industrial industries, brigades, and fire departments. We need to look over the fence that separates industry from the community and share experiences, training, resources, knowledge, and personnel so that industrial incidents can be eliminated, and if they do occur, the impact to our communities and our economy will be reduced.

2

Industrial Qualifications and Skills Required

INTRODUCTION

Some of the most important factors that must be considered when preparing fire-response forces are the qualifications and skills required for the responders. In addition to the local community's firefighting forces, the responders for an industrial facility may also include members of the *industrial fire brigade* for the industrial facility. The qualifications of the fire brigade members are critical considerations when planning your department's role in handling an industrial firefighting operation. The training and preparation requirements must be based on governmental, consensus, and industry standards. Some of the key reference documents concerning skills and qualifications are discussed in this chapter.

These qualification and skills references will help you to prepare your department for their response role and to help you to understand the key documents that industrial fire brigades use to prepare for their response. For example, knowing that the industrial brigade's capabilities are only to the exterior defensive level is essential information for the responding fire chief.

Since it is impossible to be prepared to handle all conceivable types of industrial incidents, your efforts must be geared toward planning for the most likely incidents that may occur. For example, if there is a hydrocarbon refining facility in your jurisdiction, and your department is required to respond to this facility, then you would make the necessary efforts to prepare your emergency forces for such a response (fig. 2–1). In your preparation, you will have to assess the qualifications and skills required for your firefighters as well as the skills and qualification of the industrial fire brigade or emergency response team (ERT). The industrial fire brigade could be an exceptional firefighting force with specialized delivery equipment and even large foam stores, or perhaps just the opposite. In any case you have to know their response capabilities before you respond.

Fig. 2–1. Preincident response planning and training are required by municipal fire departments for industrial facilities within their jurisdiction, and they must be prepared and trained for worst-case scenarios. *(Courtesy of Williams Fire and Hazard Control)*

If your municipal firefighting forces are located close to industrial facilities and are potentially required to respond to these facilities as part of a mutual aid agreement, it is essential, and probably legally required, for you to prepare for a safe and effective response to that facility. As will be discussed further, the level of preparedness can be moderated, depending on specific response expectations that have been defined in your department's organizational statement. Before you expend labor and resources on preparation, you will need the financial and political support from your local government. You must ensure that your personnel develop the proper skills and qualifications to meet the mission required by your local government.[1]

Without the full support from your local government, preparation for any type of response is not justified. A clear organizational statement defining exactly what level of response your local government expects is necessary to budget for personnel and equipment programs. One example that most departments are familiar with is emergency medical response. As you examine fire departments around the country, you will see that they perform various levels of emergency medical services, ranging from basic life support with automated external defibrillators (AEDs) to advanced cardiac life support. The skills and qualifications associated with these programs are substantial. The potential liability and ongoing maintenance of these programs require enormous efforts. In other words, if your city government determines that it is too costly to provide advanced cardiac life support capabilities out of the fire station, you should not plan for it. If you do provide emergency medical services, you should follow either your state's requirements or the national requirements for those services. Specifically if a first-responder emergency medical technician (EMT-Basic, EMT-1, or Advanced EMT-B) arrives on an emergency scene, you will know exactly what level of care the EMT should be able to provide. Industrial firefighting skills and proficiency levels are very much the same.

Just as your emergency medical services are defined by your local government, your response capability to industrial facilities must be clearly defined. Your local government may determine that for the economic good and safety of the community, your department will provide protection for the local industrial facilities. That level of response could be limited to providing

Industrial Fire Brigade Duties and Limitations

Duties	Duty Limitations Summary
Incipient Stage Fire Fighting	Not issued special protective clothing or SCBA. Trained to operate extinguishers and handlines up to 125 gpm (473 L/min)
Advanced Exterior Fire-Fighting	Offensive, exterior operations wearing PPE and SCBA. Handles hose streams up to 300 gpm (1140 L/min) and employs master streams
Interior Structural Fire-Fighting	Offensive, interior operations wearing PPE and SCBA. Handles hose streams up to 300 gpm (1140 L/min) and employs master streams
Advanced Exterior and Interior Structural Fire-Fighting	Offensive, interior operations wearing PPE and SCBA. Handles hose streams up to 300 gpm (1140 L/min) and employs master streams

Fig. 2–2. **Industrial fire brigade duties and limitations**

water supplies for the industrial firefighting forces working within the plant or actually having your personnel combat the industrial fires. As your response responsibilities increase, so does the level of risk for your firefighters and the potential liability for your department. As with other emergency response programs, personnel are required to be trained and qualified for the tasks they will be assigned. In addition, personnel are required to train on full-scale exercises and simulations.[2]

Fire departments must be prepared for many types of emergencies. As mentioned earlier, those missions must be specifically spelled out. As the fire chief or member of a fire department, you must carry out those missions to the best of your ability. Failure to safely and effectively manage any type of emergency that your department has been asked to prepare for could end in potential death or injury of firefighters and civilians, or monumental loss of property. This failure could have long-term negative effects on your department. Most important, you or your department may have caused or been party to causing harm when your main purpose for existing as an emergency response agency is to do good for the community. Furthermore, failure to properly prepare and effectively manage an emergency scene could result in civil or criminal charges. The defense of "we did the best we could with what we had" will not be very effective in court.

As a public servant, you probably will not be granted governmental immunity. Your defense will have to be based on approved governmental documents that define the assigned missions for your department. Your department's preparation to deal with assigned missions must be based on sources such as state and national laws, as well as industry standards such as the National Fire Protection Association (NFPA) standards. The days of the fire chief being right because of the prestige associated with his or her position, and his or her opinion being irrefutable, are long gone. Every department must base its policies and procedures on applicable laws and consensus and industry standards. If things go wrong, you and your departmental policies will be highly scrutinized in court. You must be able to cite page and paragraph on what basis the procedures were established. Basing your plans and procedures on applicable laws and standards is your best defense.

To aid you in preparing your department or assessing its current capability, the remaining portion of this chapter will give you an overview of some of the laws and publications that should be considered.

NFPA 1500

National Fire Protection Association (NFPA) 1500, *Standard on Fire Department Occupational Safety and Health Program,* is critical and central when you begin planning your department's industrial fire response capabilities. Even if your department has not adopted the NFPA consensus standards, these documents will provide great insight on the organizational requirements that are recognized within the fire protection profession. NFPA standards are based on national laws and industry-accepted practices. Applicable regulations will be discussed in more detail in the next paragraphs.

NFPA 1500 mandates that departments develop a Fire Department Organizational Statement that outlines the services that a department is authorized to perform. If your personnel are tasked to respond to industrial facilities, you should include this in your department's Organizational Statement. The Organizational Statement must be sanctioned by the governmental body over the department.[3]

The department must develop written plans and procedures for the roles that the department is expected to perform. In addition to the Organizational Statement, the department must develop a Risk Management Plan for the various jobs it is expected to fulfill. For example, if your department is required by the Organizational Statement to provide firefighting services to the local crude storage facility, those duties must be specifically defined and the risks analyzed. A risk analysis is required to be conducted for all tasks and duties.

It stands to reason that if your department is expected to extinguish industry specific fires at the facility, your personnel must be made aware of the associated risks and the proper techniques for managing those incidents. Many unaware firefighters have been killed because they did not know the associated hazards of an industrial facility. Let us say that your department is tasked to respond to and perform firefighting operations on the bulk storage tanks at the local airport and a local refinery. It might be assumed that fighting a fire in a large storage tank filled with refined hydrocarbons (e.g., jet fuel) should be the same as fighting a large tank filled with crude oil. In reality, however, the hazards are vastly different. The unfamiliar firefighter may even assume that jet fuel is more dangerous because it may have a lower listed flash point. He or she learned about flash point and volatile fuels in the basic firefighter course. What the municipal firefighter may not know, but the industrial firefighter at a crude facility knows very well, is that crude oil fires have the propensity to boil over. Boilovers are steam explosions that take place at the bottom of the storage tank when hot, heavy oil fractions contact water. These explosions can expel burning crude oil with little warning, up to 10 tank diameters after prolonged burning. That means that burning oil could land more than a half mile away from a 300-foot-diameter tank. This is just one example of a specific hazard associated with one type of industrial facility that firefighters have to be trained to combat. If your firefighters are expected to respond to these hazards, they must be aware of the hazard, be trained to deal with the hazards, and practice those skill sets at least on an annual basis.

As you can see, NFPA 1500 is an essential document to reference before you respond to an industrial fire incident, because it can help you to determine specific qualifications for your personnel. If the industrial response mission is not part of your approved Organizational Statement, and if you have not performed a risk assessment or properly trained and equipped your firefighters, then your department is not legally or responsibly prepared to respond to this special hazard emergency.

NFPA 600

National Fire Protection Association (NFPA) 600, *Standard on Industrial Fire Brigades*, covers the levels of training for on-site industrial firefighting personnel and the specific skills they must possess. This document applies specifically to fire brigade members who work directly for the private fire protection services for the industrial facility. This is not a standard for municipal firefighters, but knowledge of it will help you understand the qualifications of the industrial fire brigade members.

Industrial fire brigade personnel possess specific skills dependent on the nature of their plant. Before you respond to these facilities, you must know what their skill level is. These personnel may be trained only in exterior incipient skills or have no professional firefighting training at all.

In any case, these industrial brigade members will be vital to your response. They will know their plant thoroughly and know more about the associated hazards of the processes than your personnel because they work in this facility every day.

Industrial fire brigade members may not be full-time firefighters. Many times they work as plant operators and have been assigned or volunteered for fire brigade duties. You will have to find out what their skill sets and proficiencies are (fig. 2–2). Their actual capabilities could be exceptional, or they could possess specific skills that would be critical to a successful firefighting operation. By knowing the skill levels for the industrial fire brigade member covered in NFPA 600, you can gain some insight into their training (fig. 2–3) and qualifications and determine how to integrate them into the other emergency response forces.[4]

NFPA 1081

The requirements of National Fire Protection Association 1081, *Standard for Industrial Fire Brigade Member Professional Qualifications*, are linked to NFPA 600. NFPA 1081 outlines the specific skills that an industrial firefighter must master to be certified to the levels outlined in NFPA 600. Again, NFPA600 and NFPA 1081 are for industrial fire brigade members who perform firefighting duties at a specific plant of facility.

Fig. 2–3. Industrial fire brigade members training in exterior industrial firefighting *(Courtesy of Bloomington, IN F.D., Jeff Barlow, Fire Chief)*

NFPA 1081 is not a standard to train municipal firefighters. It should be consulted to determine what skills the firefighters at industrial facilities are likely to possess and how they fit into the overall response. Mutual aid agreements should spell out exactly what roles all of the responders are expected to play. Industrial fire brigade members will be part of the plant ERT for the facility. Some industrial facilities have existing mutual aid agreements with other surrounding industrial facilities that may respond with personnel and equipment (fig. 2.4). You will have to become familiar with the real capabilities of these industrial fire brigade members prior to any emergency. Do not underestimate these very important players since their information and assistance will be vital to your operations when handling a major incident.

Fig. 2–4. Industrial facilities may have mutual aid agreements with surrounding industrial or municipal fire departments. *(Courtesy of Williams Fire and Hazard Control)*

NFPA 1081 is the qualification standard for industrial firefighters, just as the other NFPA 1000 series qualification standards are intended for other levels of fire protection certification (fig. 2.5). Within this standard are the Job Performance Requirements from NFPA 600 and the Requisite Knowledge and Requisite Skills required to obtain certification.[5]

***Fig. 2–5.* Industrial fire brigade members training to NFPA 1081 qualifications** *(Courtesy of Bloomington, IN F.D., Jeff Barlow, Fire Chief)*

29 CFR OSHA 1910.156 FIRE BRIGADES

29 Code of Federal Regulations (CFR), Occupation Safety and Health Administration (OSHA), Subpart L, 1910.156 is the federal regulation covering fire brigade services in industrial facilities whether they are provided by the company or another private organization. The requirement to comply with this regulation is the same in any part of the United States. Individual states may elect to meet the requirement of federal OSHA regulations by complying with state OSHA regulations. If a state elects to adopt its own OSHA regulations, the regulations must be at least as stringent as federal OSHA. Compliance with minimum standards of federal OSHA is compulsory. If there is an accident, this will be minimum standard use to judge your department. States that use their own OSHA regulations must meet or exceed the federal OSHA regulations.

As just mentioned, 29 CFR 1910.156 applies specifically to private industrial brigades. The NFPA 600 and NFPA 1081 standards were developed to meet the minimum requirements of the federal OSHA regulations. 29 CFR 1910.156 only requires minimal training for industrial firefighters. This training is required on an annual basis for noninterior brigade members and quarterly for interior brigade members, and it must be commensurate with assigned duties. The regulation states that advanced training is required for team leaders and instructors. It also states that personnel assigned to interior firefighting duties must receive a physical examination. It goes into some detail on appropriate personal protective equipment (PPE), and most important, it requires the employer to follow 29 CFR 1910.134 for self-contained breathing apparatus (SCBA).

As you may have noted, 29 CFR 1910.156 only covers the very basics. The details are covered by the NFPA standards and in many cases by state OSHA requirements. You will have to determine facility by facility just how effective the fire brigades. If the facility complied only with the minimum requirements, it would not be very effective and capable of assisting with aggressive firefighting efforts.[6]

29 CFR OSHA 1910.120 HAZWOPER

As a professional or volunteer firefighter, you are probably aware of 29 Code of Federal Regulations (CFR), OSHA, Subpart L, 1910.120, or HAZWOPER, which stands for Hazardous Waste Operations and Emergency Response. With the exception of the definitions in paragraph (a), you can skip to paragraph (q) for emergency response requirements. As with the other CFRs, HAZWOPER is the basis for the NFPA hazardous materials response standards. To meet the minimum requirements of the federal regulation, the consensus standards had to satisfy the basic requirements of the federal regulation. Because this regulation is applicable to all firefighters, and thus you are probably already familiar with it, this CFR will not be covered in detail here.

It must be noted that personnel assigned to the plant or facility may follow paragraphs (b) through (p) for their routine hazardous materials operations. For example, if there was a diesel fuel spill of 2 gallons at the facility, then the facility would probably have a team that is specifically trained to deal with that hazard. There must be a remediation plan, and personnel must be properly trained and equipped. The plant personnel will handle most releases on a routine basis. Bigger or more complicated releases requiring a full response under paragraph (q) would also include the local brigade and hazardous materials team from an industrial facility. Again, as with the fire brigade qualification, you will need to know what the hazardous materials qualifications are for a specific plant or facility. Your department will follow either the state or federal regulation for hazardous materials response. The levels of response qualifications for hazardous material response personnel are defined in NFPA 471,

Recommended Practice for Responding to Hazardous Materials Incidents, and NFPA 472, *Standard for Professional Competence of Responders to Hazardous Materials Incidents*, as well as HAZWOPER. Your personnel must be trained to the Hazardous Materials Operations, Technician or Incident Commander level. Typically, the Hazardous Materials Specialist will be provided from inside industry. This specialist will be essential to determine the appropriate mitigation plan.

After an emergency response to a facility, you may turn the incident over to the specialized team for postincident remediation. Most fire departments do not deal with site remediation.

Additionally, these industrial response teams may have unique capabilities, especially the ability to deal with hazardous and toxic releases. These teams may be able to deal with the releases in an almost routine manner following industrial standards. Your department, which has to be prepared for almost any hazardous material incident, may take hours to set up and handle the same incident. You should be aware of these specialized teams, which are located at most industrial facilities. If possible, have the experienced teams quickly mitigate these situations.[7]

API 2021, MANAGEMENT OF ATMOSPHERIC STORAGE TANK FIRES

Another document that deserves careful examination as you prepare for a response to industrial facilities is American Petroleum Institute, Recommended Practice (API RP) 2021, *Management of Atmospheric Storage Tank Fires*. This document outlines the massive firefighting efforts necessary to extinguish a full surface fire on a storage tank. Most nonindustrial-oriented departments are not prepared to handle these resource intensive fires (fig. 2–6). API RP 2021 outlines the water volume, amount of foam concentrate and delivery densities required to extinguish these fires. It is a vital document when planning your response so that you do not waste your foam concentrate stores on a fire beyond your capability. It addresses the various types of fires that you might encounter on these tanks, including sunken and partially sunken roof, ground fires, and rim seal fires. It gives guidance on when to fight a fire and when it is time to withdraw. Although this document does not require specific qualification, it is essential for the preparation and training of firefighting forces expected to respond to large atmospheric storage tank fires. The potential for a catastrophic event is discussed in detail, and guidance is given on when to withdraw. API RP 2021 is an excellent document to discuss with your industrial counterparts during your preplanning. The industrial facility should be aware of, and may assist by, providing foam stores and large delivery devices required to control and extinguish these fires.[8]

Fig. 2–6. **Large petroleum storage tank fire. Is your fire department prepared?** *(Courtesy of Williams Fire and Hazard Control)*

CONCLUSION

As you have read in this chapter, firefighter qualifications are covered in numerous documents. The qualification standards could be in state, federal, or consensus documents. You must know about the qualification standards for industrial firefighters that you may operate with and also know what qualifications are needed for your municipal firefighters to mitigate industrial fires. You will encounter extremely proficient fire brigade personnel and others who may not be very proficient. In any case, plant personnel are a vital asset in assisting fire department personnel with combating industrial fires because of their knowledge of the plant processes, hazardous materials, and facilities. Get to know the qualifications of the personnel at any facility that you may have to respond to, and ensure that your personnel develop the skills to handle the types of industrial fires they may have to respond to in the course of their duties.

NOTES

[1]United States Department of Homeland Security (DHS). 2004. *National Incident Management System*. Washington, DC: DHS.

[2]National Fire Protection Association (NFPA). 2005. NFPA 1561, *Standard on Emergency Services Incident Management System*. Quincy, MA: NFPA. sect 5.1.

[3]NFPA. 2002. NFPA 1500, *Standard on Fire Department Occupational Safety and Health*, sect. 4.1.

[4]NFPA. 2005. NFPA 600, *Standard on Industrial Fire Brigades*, chaps. 5–8.

[5]NFPA. 2005. NFPA 1081, *Standard for Industrial Fire Brigade Member Professional Qualifications*, chaps. 5–8.

[6]United States Occupational Safety and Health Administration (OSHA). *Title 29: PART 1910—Labor-Occupational Safety and Health Standards, Subpart L—Fire Protection*. 1910.156 Fire Brigades.

[7]OSHA. Title 29: PART 1910—*Labor-Occupational Safety and Health Standards, Subpart H—Hazardous Materials. 1910.120 Hazardous waste operations and emergency response.*

[8]American Petroleum Institute (API). 2001. Recommended Practice (RP) 2021, *Management of Atmospheric Storage Tank Fires, 4th ed.* Washington, DC: API, 21.

3

Industrial Emergency Response Teams and Municipal Interface

INTRODUCTION

In many instances, members of the public will open their morning newspapers, or turn on the evening news, and see photos of a large industrial fire that had taken place or is still burning. What the public doesn't see is the interface between the municipal fire department and the industrial fire department or brigade. This interface takes place behind the scenes, out of the camera angle. Ideally, it is a seamless process. There are similarities and differences that can affect this interface either positively or negatively. This chapter will examine the similarities, differences, strengths, and weaknesses of this municipal–industrial interface.

ORGANIZATIONAL STRUCTURE AND PERSONNEL

Most municipal fire departments adhere to a semimilitary organizational structure. The department is usually led by a fire chief who reports to the mayor or city manager. Industrial fire departments or brigades may be full-time departments and also have a semimilitary organizational structure, with the fire chief reporting to senior plant management, usually within the health and safety group or organizational line. The industrial fire department or brigade may have a similar organizational structure and have the same operational duties as the municipal fire department, including inspectional functions. Management must outline not only what duties an industrial fire department or brigade can perform but also what duties they shall not perform.

Industrial fire brigades, where staffed, comprise employees of an industrial facility that have been trained to perform basic firefighting operations. These employees' full-time occupation may or may not be to provide fire suppression or related activities for their employer. In many cases, they will perform other duties such as plant maintenance or operations and fulfill their firefighting role during emergencies at the plant. They must also possess the knowledge and skills associated with these firefighting operations. The National Fire Protection Association's *Standard on Industrial Fire Brigades* (NFPA 600) contains the "minimum requirements for organizing, operating, training, and equipping industrial fire brigades."[1] It also outlines the minimum requirements for the occupational health and safety of the industrial brigade member while performing firefighting and related activities. In some cases, the industrial fire brigade may be referred to as the emergency response team (ERT), plant emergency response team (PERT), fire team, emergency brigade, or other suitable name.

NFPA 600 outlines the limitations of the industrial fire brigade during fire suppression activities. Industrial fire brigades may be limited to perform one of the following firefighting functions:[2]

- Incipient firefighting
- Advanced exterior firefighting
- Interior structural firefighting duties only
- Advanced exterior and interior structural firefighting duties

Chapter 2 highlighted the full requirements and individual qualifications for those functions, and as we reflect on their duties as outlined there, we can see that not all industrial fire departments or brigades are capable of performing all firefighting functions and duties. Municipal fire departments must be aware of the limitations of the fire departments or brigades that may respond with mutual aid, automatic aid, or normal response assignments. Knowing the brigade's qualifications and limitations will assist with the preincident response planning required to successfully mitigate an incident at the industrial facility. Information on the staffing levels and the support the industrial brigade can offer the municipal department should be included in the preincident response plans. In addition, when the municipal department responds to an incident at an industrial facility, the information regarding the attending brigade's qualifications will be useful in determining staffing assignments in the incident command system (ICS). Certain members of an industrial brigade that are not qualified to perform offensive actions can still be used to assist in support roles, freeing members of the municipal department for critical assignments.

NFPA 1081

The NFPA has also developed a *Standard for Industrial Fire Brigade Member Professional Qualifications* (NFPA 1081) (see chapter 2). NFPA 1081 "identifies the minimum job performance requirements necessary to perform the duties of an individual who is a member of an organized industrial fire brigade providing services at a specific facility or site."[3] Many industrial fire brigades are now requiring their members to attend courses given at training facilities accredited by the International Fire Service Accreditation Congress (IFSAC) or National Board on Fire Service Professional Qualifications (Pro Board) and be certified to NFPA standards by those facilities. During familiarization visits, municipal department members can ask about the number of brigade members that are certified to NFPA requirements. In addition, municipal departments with a variety of industrial facilities may want to consider training their members to NFPA 1081 qualifications so they could assist with handling a large incident at one of the facilities within their jurisdiction.

TRAINING

Training requirements for municipal and industrial firefighters may be similar yet vastly different (fig. 3–1). Basic firefighting tools and equipment; donning and doffing personal protective equipment (PPE); donning, doffing, and use of self-contained breathing apparatus; and hose training as well as hose-handling skills may be taught to both groups at a basic training or recruit academy (fig. 3–2). The duties that the industrial firefighter will eventually perform further define what additional training will be required. For example, an industrial firefighter may or may not be trained to fight interior structural fires. Conversely, a municipal firefighter may have an abundance of interior structural fire training but limited exposure to pressurized gas or petroleum liquid fires.

Municipal and industrial firefighters should partner in training opportunities. The industrial brigade should invite the municipal department to training sessions where industrial firefighting is taught, reviewed, and practiced. Municipal and industrial departments have a lot of information to share during these training sessions. Rapid intervention team (RIT) concepts, accountability systems, and search procedures are some of the subjects that can be shared by the municipal department (fig. 3–3). Confined space rescue, properties of pressurized gas fires, and large-scale flammable liquid fires are some of the subjects that can be shared by the industrial brigade.

Fig. 3–1. Industrial fire brigade training *(Courtesy of Bloomington, IN F.D., Jeff Barlow, Fire Chief)*

Fig. 3–2. **SCBA training may be taught to both municipal and industrial firefighters.** *(Courtesy of M. Barrett)*

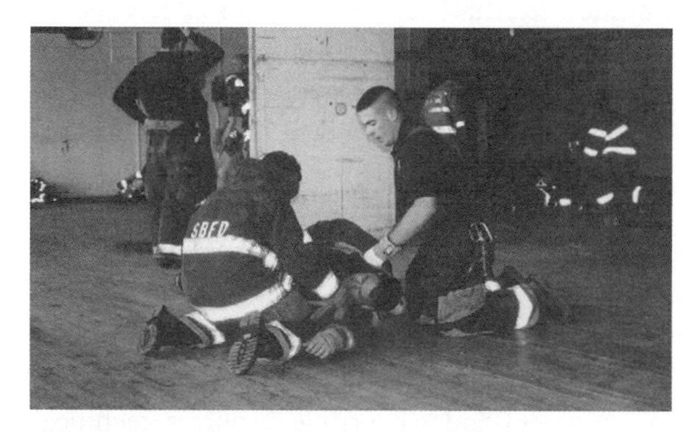

Fig. 3–3. **Rapid intervention team concepts can be shared by the municipal fire department with the industrial fire department.** *(Courtesy of Michael Barrett)*

Joint training sessions that exercise and reinforce mutual aid, automatic aid, or initial response duties for both the industrial brigade and municipal department should be conducted at frequent intervals. Municipal fire departments with industrial facilities within their response areas should take advantage of the training opportunities available at the industrial fire schools throughout the United States. Although many municipal departments frequently make use of their state, county, or regional fire schools, not many make similar use of an industrial school. The Louisiana State Fire Academy and the Texas Engineering Extension Services (TEEX) facility at Texas A & M University are two facilities that offer excellent industrial training. With liquefied natural gas (LNG) becoming a major concern for many departments, the Massachusetts Fire Academy as well as the TEEX facility offer excellent hands-on courses in LNG firefighting (Fig 3–4).

Fig. 3–4. **LNG training at TEEX, a member of the Texas A & M University System.** *(Courtesy of TEEX, Texas A & M System)*

Many of the industrial fire schools offer training to industrial brigades throughout the year. When partnerships are fostered between the municipal and industrial fire service, sharing classroom seats can be facilitated. Perhaps a few seats can be allotted when the industrial brigade attends an industrial school and vice versa when the municipal department attends one of their schools. Municipal departments should consider attending industrial academies on a scheduled rotational basis, with a concentration in the hazards they may encounter at local industrial facilities.

When municipal departments visit industrial facilities, contact should be made with on-site brigades or emergency response team leaders. Special firefighting considerations should be discussed with these personnel as well as other plant operations personnel. Additional discussions as to what their roles will be during an incident should take place.

During visits to industrial facilities, preincident response plans should be updated (fig. 3–5) (see chapter 5 for additional information). During these visits, a copy of the existing plan should be taken and reviewed with the plant personnel, and contact numbers, firefighting equipment, plant personnel responsibilities, and so forth should be updated. Access routes should be especially reviewed and verified. Has the plant altered access because of security concerns? Regarding security, this is a good point to mention that any preincident response plans should be included in the fire department's operational security (OPSEC) planning. Preincident response plans should not be accessible to the general public. They should be in a locked container or compartment on the fire apparatus. If they are included in on-board computer databases, these databases should be password protected, and the passwords changed frequently. After updating the plans, the old hard copies should be shredded. Remember, anyone can go through the trash outside the fire station.

Fig. 3–5. **During visits to industrial facilities, preincident response plans should be updated.**

Preincident response plans should contain an organizational chart. Pictures of the key players may also be valuable if they are attached to the chart. This way, when responding in to an emergency, municipal fire commanders can rapidly identify these key players.

SHARING OF RESOURCES

After we look at training, the next partnership that should be discussed is the resource partnership. In today's climate of fiscal restraint, many industrial organizations and municipal jurisdictions are looking at ways to save money. In many cases, one of the areas that is evaluated—and in many cases trimmed—is fire protection. In these situations, how can a department or brigade "add value" to its service delivery? One way is to share resources so that duplication of services can be eliminated and other services that may not already be present added. Some of these shared resources may include the following:

- Hazardous materials teams
- Special operations teams
- Incident management overhead teams
- Confined space and technical rescue teams
- Logistical and equipment support

In many instances, industrial facilities may have mutual aid agreements among participating facilities so that in the event of a large-scale emergency, additional resources can be called in to assist. This is particularly true in the petrochemical industry where large quantities of foam may be required in addition to the large-capacity appliances required to apply the foam. These resources should be identified by municipal departments, and agreements maintained whereby these resources could be used in the municipal arena where applicable and necessary. In addition to equipment and expendable resources, technical assistance can be included in these agreements.

In New Jersey, a partnership was formed between the New Jersey Fire Marshal's office, the state industrial mutual aid organization, and municipal entities to share resources in the event of a large-scale disaster or incident. The resources can be accessed through the County Fire Coordinator's resource database. The County Fire Coordinators can share this information electronically and can update the inventories daily if necessary. An example of the equipment and personnel that can be accessed and shared is as follows:

- 20 foam pumpers
- 20,000 gallons of foam concentrate
- 18 hazardous materials teams
- 10 technical rescue teams
- Numerous subject matter experts[4]

TACTICAL DIFFERENCES

At industrial incidents, there may be tactical differences between municipal firefighting and industrial firefighting (see chapter 7 for additional information). These differences may include a nonintervention mode. At industrial incidents, it may be necessary to evacuate the area and enter into a nonintervention mode where the risk versus benefit analysis indicates that the potential costs far outweigh any benefits. This type of situation may be indicated where a boiling liquid expanding vapor explosion (BLEVE) or tank boilover may occur (see chapter 31). We do not want to place our emergency or other support personnel at an unacceptable risk. It may be difficult for municipal firefighters to accept this strategy or tactic, but safety may dictate its use.

Municipal firefighters should also be aware that the cooling of exposures will be an essential part of operations (fig. 3–6). Initial operations may include this tactic while the fire is allowed to burn. During this period, plant operations personnel will be attempting to block in and shut down systems. At industrial incidents, these actions may have to be performed sequentially by the plant operators and coordinated with other areas of the plant or other plants so that these other areas or facilities are not affected. Alarming to the municipal firefighter may be the large amounts of combustible liquids and gases sent to *flare*. The sounds and visual effects may be alarming, but this is much safer than product spilling onto equipment or the ground. Cooling streams are essential to ensure that supporting structures do not collapse and create additional problems, such as leaking products that can add to the fire intensity or spread the fire.

***Fig. 3–6.* Cooling of exposures is an essential part of industrial fire operations.** *(Courtesy of Dave Cochran)*

During tactical operations, close coordination among the municipal department, the industrial brigade, and the plant operations personnel is essential. Without this close cooperation, the actions of one entity can have serious implications to the firefighting effort. It should also be noted that the organizational structure of the plant may place an operations supervisor as the incident commander. Preincident response plans should highlight this. After the arrival of the municipal department, who will be in command? Will there be a seamless transition to a unified command structure? Unified command is usually required when more than one agency will be sharing responsibility for the successful mitigation of the incident. This is especially true when the incident extends, or is threatening to extend, beyond a facility's property line.

Some of the key plant operations personnel that need to be located and consulted at an industrial incident include the following:

- Plant manager
- Hazardous materials specialists
- Process unit supervisor

These individuals will have the knowledge and experience to help the fire department isolate power and fuel sources to process and/or manufacturing operations and provide the necessary technical information to the incident commander.

The basic tactical priorities necessary for successful mitigation of municipal incidents are applicable at industrial incidents, and they include the following:

- Life hazard/rescue
- Extension prevention
- Confinement of the incident
- Extinguishment

For industrial incidents, the extension prevention may include underground systems such as sewer and stormwater systems.

Fig. 3–7. **Large-capacity industrial pumper-tanker**

Fig. 3–8. **Large-capacity portable pump.** *(Courtesy of Williams Fire and Hazard Control)*

EQUIPMENT

Municipal fire departments use equipment that has been designed for the hazards normally encountered within their districts. Conversely, industrial departments and brigades have equipment designed for their assigned duties and hazards encountered. By the nature of the hazards encountered, the industrial fire protection equipment may be a lot larger. For example, the typical municipal pumper may be 1,500 gallons per minute (gpm) capacity. An industrial pumper may be 5,000 gpm capacity or larger from a pressurized water source (fig. 3–7). In addition, these large-capacity industrial pumpers may carry large quantities of foam concentrate (up to 2,500 gallons). Other specialized equipment includes apparatus, such as foam tankers that have the ability to both transport and pump foam concentrate or solution.

In addition to large-capacity pumpers, large-capacity portable pumps and water or foam solution-delivery devices may be present at industrial sites (fig. 3–8). Monitor nozzles capable of delivering up to 14,000 gpm as portable devices or 4,000 gpm as an apparatus-mounted appliance can be seen in industrial applications. In the 1960s the New York City Fire Department (FDNY) had the Super Pumper in operation. This piece of equipment was able to pump 8,800 gpm. Today, the FDNY uses the Maxi-Water System. This system is a series of five 2,000 gpm pumpers with hose tenders, which can respond to large-scale incidents with the hose tender carrying large-diameter hose. Imagine what municipal departments could do with large industrial pumpers strategically placed within their jurisdiction for large-scale fires, including industrial responses. Municipal departments that respond to industrial facilities should be cross-trained on this large-capacity equipment and be ready to operate it when necessary.

To supply the large quantities of water required at industrial incidents (see chapter 7), large-capacity water systems may be present at industrial facilities. These water systems may supply more than 12,000 gpm and operate at higher pressures than a municipal department ordinarily encounters. A municipal fire water system may have static pressures as low as 25 pounds per square inch (psi) while an industrial facility may have static pressures as high as 170 psi. Is your pump operator/driver familiar with the operating procedures with these higher pressures? The pumper in this case may be used to reduce the pressures encountered.

Large-diameter hose will also be used in quantity at industrial facilities. Most municipal departments use 4- or 5-inch large-diameter hose (LDH). Industrial departments are currently using 6-inch, 7¼-inch, and, more recently, 10 and 12-inch hose. Does your department have the fittings and adapters needed to connect to these hoses and associated delivery appliances (fig. 3–9)? We are familiar with a large industrial fire where a mutual aid municipal department responded to provide water supply with a fireboat. The fireboat was capable of supplying 20,000 gpm; however, the largest supply hose carried was 3½-inch diameter. The industrial department stretched 5-inch supply hose for connection to the fireboat, but there were no fittings to adapt from the 5-inch hose to the smaller hose. In effect, the 20,000 gpm capacity was reduced to zero. Luckily, one of the mutual aid departments had the necessary fittings, and water supply from the municipal to the industrial department was accomplished. During training opportunities, identify the need for adapters and any other appliances that may be necessary.

In addition to large-capacity monitors and large-diameter hose, the industrial fire service has developed innovations that allow the source of foam concentrate to be placed remotely from the final delivery point for foam solution. Normally, foam concentrate must be placed within a certain distance from the nozzle, usually limited to 150 feet. Using a jet-ratio controller with a matched Hydro-Foam™ nozzle, the source of foam can now be placed as far away as 2,500 feet from the nozzle (fig. 3–10). Jet-ratio controllers are venturi-type devices that move the concentrate from a remote storage to the matched Hydro-Foam nozzle. Remember also that the foam supplies needed at industrial incidents are much more than the 5-gallon pails of concentrate carried on our municipal fire apparatus. Large quantities of foam can be in 55-gallon drums, but more likely 275-gallon totes, or large tanker trucks of 8,500 gallons. Municipal fire departments must be aware of the various delivery methods when using large volumes of foam concentrate (fig. 3–11). Municipal fire departments should be familiar with the aforementioned industrial firefighting equipment, including the monitors and all other associated equipment, so that in the event of an automatic or mutual aid response, they can be better prepared to assist the industrial department with its operation (fig. 3–12). The time to do this is before the incident and not during. Plant familiarization visits and multidepartment drills can accomplish this.

Fig. 3–9. **Large-diameter hose and high-volume monitors.** *(Courtesy of Williams Fire and Hazard Control)*

Fig. 3–10. **Jet ratio Controllers.**

Fig. 3–11. **Foam delivery using 275-gallon totes.**
(Courtesy of Williams Fire and Hazard Control)

Fig. 3–12. **Large-capacity nozzle showing foam inlet.**
(Courtesy of Williams Fire and Hazard Control)

CONCLUSION

To successfully mitigate a large industrial incident, cooperation must take place between industrial fire departments/brigades, municipal departments, private fire suppression experts, industry experts, and other entities. A municipal department must rely on the expertise of others and be able to blend into the suppression efforts, not be a hindrance. Industrial fire departments have the expertise in their field as well as the specialized training required to handle fires in the specific products produced, stored, and used at their facilities. In addition, they will have specific knowledge of the process equipment in the facilities serviced.

The resources that the industrial fire brigade or department can provide must always be considered by the municipal fire department. One of the buzzwords used lately in the fire service is *interoperability*. While it is at times associated with communications and interagency response, it is just as applicable to the municipal–industrial interface. We must be assured that all our equipment and operating procedures are compatible, and we must know this before the incident, not during.

When responding to an industrial incident, all available resources must be known and available. This includes specialized response teams, technical experts, plant personnel, and written documents including plant drawings and material safety data sheets (MSDS).

By working together and developing partnerships, the municipal–industrial interface will be a seamless boundary, with both entities working in harmony to achieve the end goal of protection of life and property.

NOTES

[1] National Fire Protection Association (NFPA). 2005. NFPA 600, *Standard on Industrial Fire Brigades*. Quincy, MA: NFPA, pp. 600–04.

[2] NFPA 600, pp. 600–05.

[3] NFPA. 2001. NFPA 1081, *Standard for Industrial Fire Brigade Member Professional Qualifications*, pp. 1081–85.

[4] Kanterman, R. 2005. MAC-SICS: Industrial Mutual Aid in New Jersey. *Fire Engineering*, November: 97–98.

4

Incident Management

INTRODUCTION

Municipal firefighters understand the need for an incident management system and should have such a system in place during operations. The National Incident Management System (NIMS) was expanded and restructured under Homeland Security Presidential Directive (HSPD) 5, *Management of Domestic Incidents.* The system provides a consistent nationwide template to enable federal, state, local, and tribal governments, as well as private sector and nongovernmental organizations, to work together to prepare for, prevent, respond to, and recover from domestic incidents.[1] One of the components of NIMS is Command and Management, which contains the organizational system of incident command (fig. 4–1).

Fig. 4–1. **The National Incident Management System contains the organizational system of incident command.**

A *system* is a unit of interrelated parts or functions designed to achieve a common goal. The incident command system (ICS) is a management system designed to enable effective and efficient domestic incident management by integrating a combination of facilities, equipment, personnel, procedures, and communications operating within a common organizational structure. It is a professional approach to managing emergency incidents. It can be said that incidents not managed properly or not at all become crisis events. At large-scale incidents, participants may bring to the table their own agendas (political or personal), organizational structures, and priorities. Without a system and structure in place, these competing elements may negatively affect the operations and outcomes. This system and structure is ICS.

NFPA 1500, *Standard on Fire Department Occupational Safety and Health Program,* requires that an incident management system (IMS) be utilized at all emergency incidents.[2] It further highlights that the incident commander (IC) shall be responsible for the overall management of the incident and the safety of all members involved at the scene. NFPA 1561, *Standard on Emergency Services Incident Management System,* defines and describes the essential elements of an incident management system that will meet the requirements in NFPA 1500 as noted earlier.[3]

In industry, some facilities have industrial fire brigades. An industrial fire brigade, as defined in NFPA 600, *Standard on Industrial Fire Brigades,* is "an organized group of employees within an industrial occupancy who are knowledgeable, trained, and skilled in at least basic fire-fighting operations, and whose full time occupation might or might not be the provision of fire suppression and related activities for their employer."[4] NFPA 600 further states that "an incident management system shall be established with written procedures applying to all members involved in emergency and training operations and shall be utilized to manage all emergency and training operations."[5] Fire departments with large industrial complexes should be aware of the command structure policies that may be in place at these facilities and adopt standard operating procedures (SOPs) that address these issues of command.

As we begin our discussion of incident command, we need to highlight some terminology that will be used throughout this book.

Preincident Response Planning

Preincident response planning is a "written document resulting from the gathering of general and detailed information/data to be used by public emergency response agencies and private industry for determining the response to reasonably anticipated emergency incidents at a specific facility (fig. 4–2)."[6] (See chapter 5.)

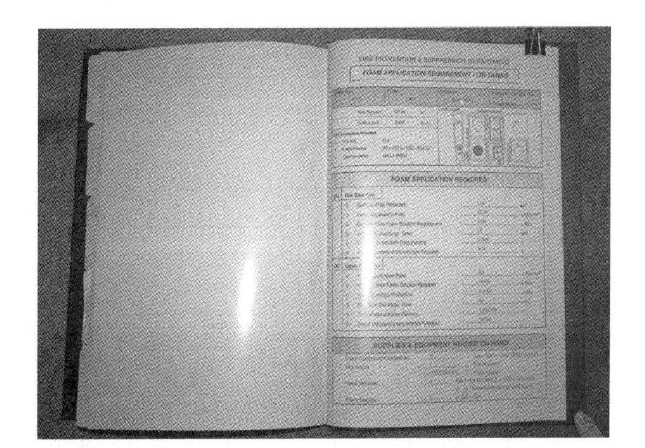

Fig. 4–2. **Preincident response plan for a storage tank facility. Other pages will detail other information as required such as additional resources including mutual aid assignments.**

Incident action plan

An incident action plan (IAP) lists the strategic goals, tactical objectives, and support requirements for an incident.[7] The IAP will contain specific actions for the next operational period during prolonged operations. When complete, the IAP may have a number of attachments such as a radio communications plan, a medical plan, incident organizational chart, and a safety message if not included on the initial IAP sheet.

Emergency response plan or emergency operations plan

The emergency response plan (ERP) or emergency operations plan (EOP) is the overview of how a jurisdiction, agency, or company will respond to and coordinate an emergency incident or large-scale disaster. As defined in NIMS, the EOP is the "'steady-state' plan maintained by various jurisdictional levels for responding to a wide variety of potential hazards."[8] Hazard-specific appendices are specific policies and procedures that may be followed by a jurisdiction or agency during a particular type of incident. In the case of an industrial facility, the parent company may have an ERP/EOP in place and also have site-specific appendixes for each of their facilities. Both a company's and a jurisdiction's ERPs must be known to the fire department and must be consulted when developing preincident response plans. The ERP/EOP identifies agencies, organizational structures, lines and methods communications, and resources that will be required at emergency incidents. One of the key elements of the ERP/EOP is how it will be activated.

As these plans are developed, jurisdictions should ensure that training and exercises are considered to test the plan. Training and exercises are essential to ensure all emergency response agencies work as a cohesive team on the day of the event. Familiarity with other response agencies, personnel, and equipment will ease the tension of all the agencies trying to work together.

INCIDENT COMMAND AT INDUSTRIAL INCIDENTS

How does the municipal fire department fit in to the industrial IMS? When there is no industrial fire brigade present, the municipal fire department may be the sole responder to an industrial incident (fig. 4–3). The municipal fire chief will usually be the IC on the scene; however, company policy may dictate that an employee of the affected plant or facility must remain in overall charge of incidents occurring on company property. This employee may be the plant manager or other ranking member of the facility's management team. In cases such as this, the municipal fire chief will work closely with the identified company employee, who will usually defer to the fire chief for the strategic and tactical decisions needed for the firefighting aspects of the incident but may retain control of the overall incident. The municipal fire chief or IC should always consult with industrial operations personnel regardless of who is in overall command (fig. 4–4).

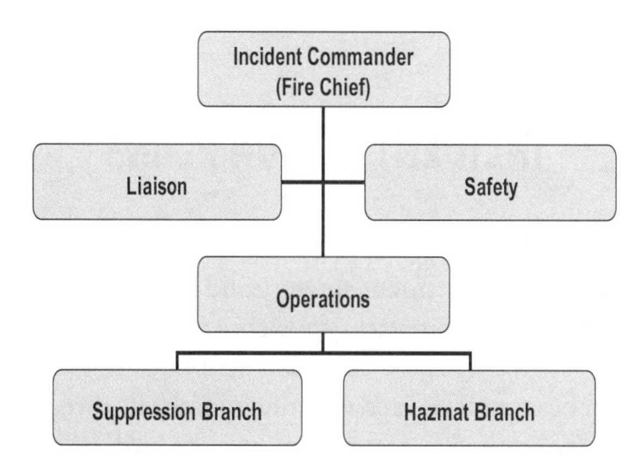

Fig. 4–3. **Incident command chart for a small incident where the municipal fire chief is the incident commander.**

Fig. 4–4. **Municipal fire chiefs or incident commanders should consult with industrial operations personnel. (Courtesy of Bloomington, IN F.D., Jeff Barlow, Fire Chief)**

In facilities that maintain an industrial fire brigade, the brigade's fire chief will most likely be the IC. But again, local company policy may dictate that another company employee act as the IC. There are large industrial complexes where the on-duty shift superintendent or senior ranking plant person is designated as the IC, with the fire department acting in a support role. In these instances, the fire chief will be acting as a technical advisor to the IC. Members of a municipal fire department responding to an industrial incident where a command structure is already in place will need to know their roles and responsibilities. For example, will the industrial fire brigade chief retain command, or will command be passed to the municipal fire chief? If command of the incident will be transferred, then the command transfer must be properly announced to all components to avoid confusion. In either situation, this should be discussed during the early stages of preincident response planning to avoid potential confusion during an incident. Similarly, if a portion of the command structure only is to be transferred, for example, fire operations, this too must be announced to all parties involved. This brings up another potential problem—communications. Is your communications system compatible with the industrial brigade's communications system? Have you included communications issues in your response SOPs?

Incident command may change several times as an incident escalates or declines. In the beginning of most incidents, one of the first responders will be in charge. For serious incidents, the first-responding IC would pass command to the appropriate level of officer. Appropriate procedures must be in place for passing assumption and command. No matter what stage the incident is in, everyone must know who is in command. Complete details on incident command systems and all components can be found in the Department of Homeland Security's National Incident Management System guidelines issued March 1, 2004 (www.dhs.com).

When responding to industrial incidents, chemical releases will be a strong possibility facing the responders. When fire departments respond to hazardous materials releases, 29 CFR 1910.120, *Hazardous Waste Operations and Emergency Response* (HAZWOPER) will apply. This regulation establishes requirements for public safety organizations that respond to hazardous materials incidents or hazardous waste emergencies (see chapter 2 for more information). The following are some of the areas that this regulation addresses:

- HazMat emergency response plan
- The establishment of an incident management system
- Training requirements for responders

It is important that those departments that will respond to industrial facilities where hazardous chemicals are used or stored are familiar with the above regulation.

STANDARD OPERATING PROCEDURES

It is impossible for any fire department to operate consistently and effectively without SOPs. This is particularly evident at large, complex, and infrequent events such as industrial incidents. SOPs allow an organization to develop a plan of action before an incident. SOPs support and implement the official policy of the department and act as the playbook during emergency operations. The development and use of SOPs greatly enhances the ability of the IC to make decisions during the incident. It is a fact that the more decisions made in an SOP prior to an incident, the fewer decisions that will have to be made during an incident.

There are two types of decisions that can be made at an incident: planned and innovative. It is much better to have planned decisions than innovative ones, and SOPs provide the mechanism for these planned decisions. SOPs drive the decision-making process in an organized and logical progression. Without SOPs the decision-making process is unorganized and key considerations may be missed.

Another benefit to having written SOPs regarding industrial incidents is that a well designed SOP can become the foundation for the IAP at a major incident. The IAP then becomes the written SOP for the incident. The fire department's personnel, as well as mutual aid resources and other participating agencies, can refer to the IAP for the operational objectives and strategies that are being used at the incident.

INCIDENT ACTION PLANS

An IAP is a plan that contains the objectives reflecting the incident strategy and specific control actions for the current and next operational periods. For small incidents, this plan will be simple and may be communicated verbally from the IC directly to the individuals carrying out the orders. For larger incidents, this plan will be written and will involve a planning section. This planning section will include technical information specialists that can assist with setting objectives and strategies based on their knowledge of the plant facility and process operation (fig. 4–5).

The IAP will ensure that all the response organizations as well as their supporting functions work together toward a common emergency response goal and that individual response agendas are coordinated and resources used efficiently. The IAP also allows us to plan ahead and not "fly by the seat of our pants" or "shoot from the hip."

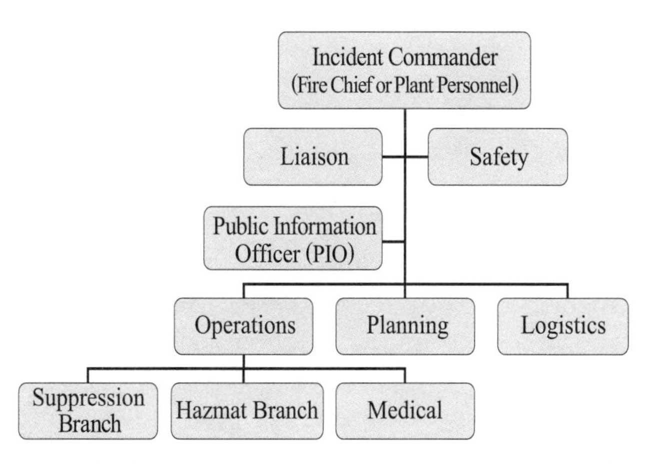

Fig. 4–5. **Incident command structure expanding with planning section activated**

UNIFIED COMMAND

At large-scale industrial incidents, a unified command system will likely be necessary (fig. 4–6). There may be multiple jurisdictions or agencies that have a role in managing a single incident due to the off-site impacts. An incident requiring a unified command may be one that extends (or threatens to extend) beyond a company's property line or involves multiple plants. How many incidents have we read about that required a large-scale community evacuation or environmental cleanup? In addition to the affected plant management, the Environmental Protection Agency, US Coast Guard, the Occupational Safety and Health Agency (OSHA), local emergency management, various state agencies, and other entities may be heavily involved. The events that trigger the need for a unified command would be when more than one agency is responsible for decision making within a single jurisdiction or where more than one jurisdiction is involved. All agencies with geographical, functional, or legal responsibility would participate in the management of the incident. As a whole, they will determine the overall incident objectives and strategies, then plan tactical activities jointly. This joint effort by the different agencies is accomplished without losing or abdicating agency authority, responsibility, or accountability.

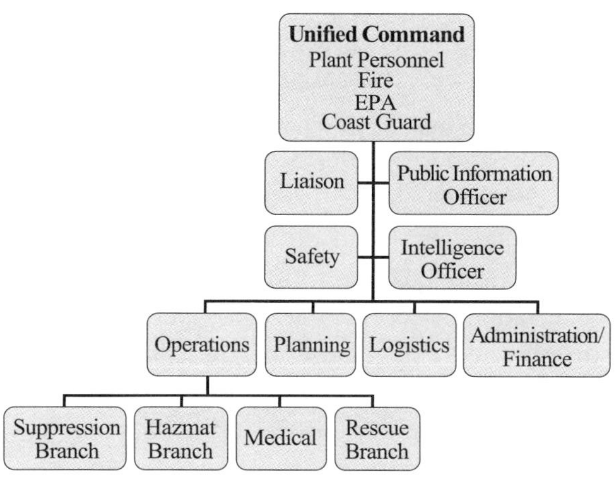

Fig. 4–6. **Unified command structure at a large incident**

EMERGENCY OPERATIONS CENTER

An emergency operations center (EOC) is a predesignated facility established by an agency, jurisdiction, or industrial complex to coordinate the overall agency, jurisdiction, or industrial complex response to support emergency, nonemergency, unusual, or large-scale operations. In some areas, the facility's EOC may be referred to as an emergency command center or a disaster command center. As the scope and magnitude of an incident gets larger, the facility or community EOC will most likely be activated. EOCs pull together people and resources to handle emergencies that exceed the scope of normal or expected operations (fig. 4–7). EOCs must have the resources and trained personnel to handle long-term operational requirements, and they must have the communications and creature comforts to accommodate sustained occupancy.

Fig. 4–7. **EOCs pull together people and resources to handle emergencies that exceed the scope of normal or expected operations.**

The location of the EOC may vary with the organization or community. Most EOCs will be predesignated and located in safe locations that have access to communications and other support equipment. Inside an industrial facility they may be located in the administration area or headquarters building. At a transportation incident, away from the industrial complex, an EOC may be set up at a hotel or other similar building close to the incident.

The overall command and logistical support as well as communications with the media, outside agencies, and corporate headquarters would be coordinated from this center. An EOC should be structured to offer support and provide policy guidance. At a facility's EOC, senior management personnel will be present in addition to engineering and plant operations staff. Process flow diagrams, instrumentation drawings, utilities drawings, and blueprints of the facility will also be located there. Additional documents that will be at an industrial EOC are emergency response guides, ERPs, material safety data sheets, telephone lists, mutual aid agreements, and resource (personnel and equipment) databases. Many industrial EOCs have computer or other instrumentation-monitoring capability to enable the support personnel to better gauge the condition of the facility's operating systems. In some cases, overall command of the incident will be transferred to this location. In this instance, it would be important that the fire chief or his/her designee report to the EOC and operate from there. If the fire chief reports to the EOC, then an operations chief would be designated to staff the incident command post (ICP) at the scene to manage on-scene response operations. If the fire chief elects to stay at the incident command post at the scene, then the designee assigned to the EOC must have the authority to make decisions at the EOC level. In all cases, however, there should always be a fire department representative in the EOC.

When the EOC is activated, the personnel assigned will focus on the strategic issues of the incident and coordinate logistical and resource support while the ICP will be responsible for tactical control issues pertaining to the on-scene response. Participants in EOC discussions and/or decisions may be located thousands of miles away and participate via teleconference or videoconference.

INCIDENT MANAGEMENT TEAMS

An incident management team (IMT) is a group of highly qualified individuals functioning as a preidentified team to respond to and assist with the management of incidents involving complicated environmental, life, property, tactical, or political influences. They provide strategic incident management support for incidents involving a large area, long duration, or technical complexity, or incidents that may have an extraordinary impact on the population at large. Some of the individuals assigned may be specialists in the various aspects of the ICS, such as command, safety, liaison, public information, operations, finance, planning, and logistics. Industrial complexes or industries geographically centralized may have such teams identified to respond and assist with, or assume the management of, a large-scale incident. These teams would be drawn from the resources of the various companies located in the geographical area or from an individual corporation's resources. Preincident response plans developed by municipal fire departments should include information on the existence of these teams. Training in large-scale operations should include the participation of such teams where they are in place.

STAGING

Staging is a location where unassigned incident personnel and equipment are gathered close to the incident, waiting for assignment to specific emergency operations. Level I staging is the initial location for emergency response units that respond to an industrial incident. From this location, the IC directs resources to a specific location or assignment to engage in emergency control operations. When an incident escalates past the capabilities of the initial responders, Level II staging will be set up. Level II staging is the location where the arriving units are initially sent when an incident escalates past the capability of the initial response. It is a tool reserved for large, lengthy, or complex incidents.[9] Level I or II staging is at a location close to, but still at a safe distance from, the immediate incident scene. At industrial incidents, staging may be located outside of the facility's fence; for example, the main parking lot of a plant may be used. However, one of the problems associated with using this location for Level II staging is that employee vehicles may be parked there already, which could hinder emergency operations and inconvenience employees. In addition, this parking lot may be designated as the safe area for employees to report to in the event of a major incident. Also, bear in mind that at industrial incidents, wind plays an important role in the designation of a staging area. The designated staging area should be upwind from the actual incident. For this reason, preincident response plans should identify at least two locations for staging areas based on the prevailing wind conditions, one upwind and one down- or crosswind.

A good staging area should have room for parking and maneuvering the fire apparatus and other support equipment that will respond. This would include groupings of task force or strike team apparatus. Ideally, it will have two entrances to facilitate vehicles entering and exiting. Security should be a consideration. Can the site be easily secured by law enforcement to protect the responders and keep unnecessary vehicles and onlookers out? Will there be shelter from the sun? Also, are sanitary facilities available, or can they be set up quickly? Access roads to and from staging and to and from the incident to staging must be sufficient to handle large numbers of vehicles. In the past, some facilities have erected signs designating these predetermined locations, but with homeland security and responder safety issues brought to the forefront, this practice is no longer recommended.

At some large industrial facilities, the ERP may call for the initial responders to report to a staging area away from the actual incident until the scene is investigated by plant personnel. This may be in a plant where a hazardous release may be the cause for the alarm. If responders are allowed to enter the site immediately, then exposure to the release may be possible by unsuspecting responders. In addition, if a flammable gas has been released, then the responding vehicles could ignite the gas, triggering an explosion. When staging areas are designated for initial responders, preincident plans must reflect this, and all responding departments must be made aware of this provision. Therefore, it is recommended that this response protocol be practiced during training events.

RESPONDING AS THE INCIDENT COMMANDER

As an IC responding from the municipal side, you must be ready to assume command upon arrival at the scene, unless otherwise outlined in preincident response plans, mutual aid agreements, and/or SOPs. You must be able to fully understand the situation. In general, the essential information needed to understand the situation can be found by asking the following questions:

- What has occurred?
- What progress has been made?
- What is the effectiveness of the current plan?
- What is the potential for escalation?
- What are the present and future resource and organizational needs?

To understand the incident, the IC must obtain information using a "global 360" perspective. As in structural fires, when we think of the six sides of a fire, we must think about all sides of an industrial incident. If we could place the incident and its surroundings inside of a clear globe and look at the entire incident from a 360-degree, global perspective, then we could gather as much visual information as possible. Incident commanders must gather this information from all sources possible. The use of tall structures such as higher process units, buildings, or even helicopters may facilitate this information gathering from above. When media coverage is present, the video footage or still photos taken from helicopters may be useful. Digital photography, video recording, and even real-time footage transmitted by video and thermal cameras must be used to their full potential. In addition to the immediate incident and surroundings, an even larger area will have to be considered where flammable and toxic materials may travel to other sections of a plant through common piping and/or process sewers.

Incident commanders must be able to acknowledge that the incident will not look the same for long periods of time. They need to ask themselves where the incident will be in the next 30 and 60 minutes. One of the best sources of information to assist with the strategic and tactical decisions at an industrial incident is the plant personnel. The information obtained from plant employees and plant control room technology, coupled with information obtained during preincident response planning, will be the foundation for the IAP.

ACCOUNTABILITY

One of the functions of the IMS is to track personnel and resources. NIMS has defined standardized mechanisms and established requirements for processes to describe, inventory, mobilize, track, and recover resources over the life cycle of an incident. This includes the on-scene accountability of responders as well as support personnel such as contractors, vendors, delivery truck drivers, and the like. It is imperative that fire departments responding to an industrial incident use their accountability system to the fullest (fig. 4–8). These incidents may be large in scale, and responders may be spread out over a large area, distant from the IC. In the event of a catastrophic failure of a process unit, the rapid identification and location of personnel must be established. If your department is assigned to respond to an industrial plant to assist the industrial fire brigade, ensure that you are familiar with its accountability system and that it is in use at all its incidents.

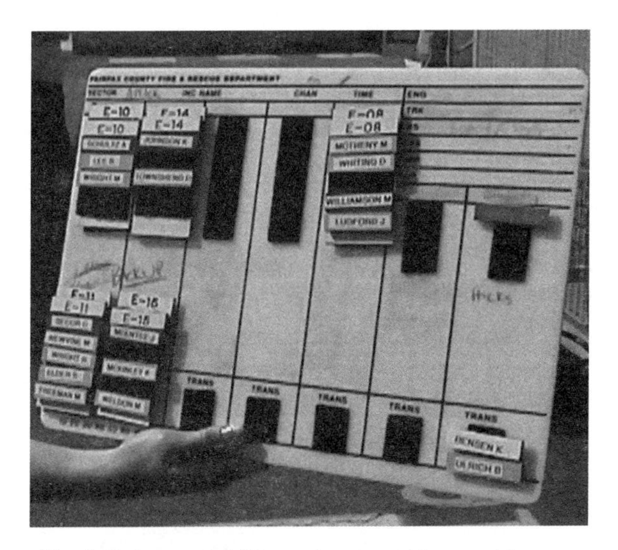

Fig. 4–8. **Accountability systems must be used for firefighting as well as nonfirefighting personnel operating at the scene.**

Electronic technologies are now available to assist with fire scene accountability. This technology enables ICs or accountability officers to monitor each firefighter's air supply, time on air, and other indicators when personnel are using self-contained breathing apparatus (SCBA). Emergency personal accountability safety systems (PASS) activations can also be monitored from a central location on the fireground. In addition, systems are being developed that will enable ICs or accountability officers to determine the exact location of downed firefighters at an incident scene.

RAPID INTERVENTION TEAMS

It may be reasoned by some that because many large industrial incidents use a defensive approach to firefighting that the rapid intervention team (RIT), also known as rapid intervention crew (RIC), concept can be ignored. This reasoning is far from the truth. An industrial incident can turn sour in an instant, and the defensive position could quickly become the center of the incident due to explosion or catastrophic failure of a pressure vessel. NFPA 1500, *Standard on Fire Department Occupational Safety and Health Program,* requires a fire department to provide personnel for the rescue of members operating at emergency incidents. The RIT should be located in a safe area, and fully equipped with SCBA and hoselines, and prepared to respond to the aid of firefighters in need. This area can be adjacent to the command post, so that the team leader can monitor the situation from the command post. In addition, other support or technical staff may be located at the command post. This support or technical staff will be able to provide valuable information for the RIT in the event the rescue of a downed firefighter becomes necessary.

The size of the RIT may vary, but the team should consist of at least four members. Downed firefighters may have to be carried over long distances and through damaged process units, so these probabilities should be planned for. Additional tools and equipment such as backboards, stokes, and Sked® stretchers should be part of the RIT complement.

Fire departments responding to industrial incidents where an industrial fire brigade is retaining command should set up their own RIT if none is being provided by the command structure in place.

TASK FORCE CONCEPTS

The IMS provides for the concept of task forces. A *task force* is a group of resources with common communications and a leader that may be preestablished and sent to an incident, or it may be formed at an incident. At a large-scale industrial incident, the task force concept may be used. Where many industries are in close geographic proximity, a system may be in place to form task forces using the resources from various industrial fire brigades to assist neighboring facilities. These task forces may be formed for a specific type of incident such as a large petrochemical storage tank fire where large amounts of foam and large-diameter hose are required. The task force may consist of large-capacity pumpers, foam tankers, and aerial ladder/platform delivery devices. Other equipment that may be included in a task force (or *strike team* if used) are vacuum trucks, skid-mounted large-capacity pumps, lighting units, and cranes. Mutual aid agreements and SOPs will address these task forces in most instances, but the municipal fire chief must be aware of their presence and the mechanism used to activate them.

COMMUNICATIONS

In today's fire service, there has been much written about interoperability of equipment among jurisdictions and their many departments, with particular emphasis on radio communications. Industrial personnel also need to examine their communications systems and how they communicate with emergency responders entering their facilities. Municipal fire departments that may be called on to respond to large industrial complexes should have provisions to communicate via radios to the industrial operations personnel and their fire brigade, if applicable. In a perfect world, the fire departments would have frequencies programmed into their radios to facilitate this, but some fire departments have multiple industrial facilities in their jurisdictions. It would be impractical to have all the frequencies available. However, one emergency radio frequency can be designated so that industry and emergency services personnel can have access to the channel during emergencies and drills. If this is not possible, then a department's preincident response plan should have a procedure in which a member of the industrial plant's operations department or emergency response team is radio-equipped and checks in with the fire department upon their arrival at the scene so that a bridge in communications can be established. This person should also be familiar with the technical language and jargon used by the particular industry.

CONCLUSION

The industrial incident command and management system is not much different from the one used in the municipal fire department. However, there are slight differences, and the scale of the system may be larger than the one that is commonly used at municipal incidents. An industrial incident will involve much more technical and support personnel than a standard municipal response. It would not be uncommon to have 10 to 15 technical persons participating in the IMS. Being familiar with the command structure of the industrial facilities in your response area, being familiar with NFPA 1500, NFPA 600, and NFPA 1561 as they relate to incident command, and developing SOPs to address the industrial response are essential when responding to industrial incidents and will ensure that operations are more efficient.

NOTES

1. United States Department of Homeland Security (DHS). 2004. *National Incident Management System.* Washington, DC: DHS.
2. National Fire Protection Association (NFPA). 2002. NFPA 1500, *Standard on Fire Department Occupational Safety and Health Program.* Quincy, MA: NFPA, sect. 8.1.3.
3. NFPA. 2005. NFPA 1561, *Standard on Emergency Services Incident Management System,* sect 1.2.
4. NFPA. 2005. NFPA 600, *Standard on Industrial Fire Brigades,* sect. 3.3.14.
5. NFPA. 2005. NFPA 600, sect. 4.2.1.
6. NFPA. 2003. *Fire Protection Handbook,* 19th ed., vol. I. Quincy, MA: NFPA, pp. 7–85.
7. Noll, G. G., Hildebrand, M. S., and Yvorra, J. 2005. *Hazardous Materials: Managing the Incident,* 3rd ed. Chester, MD: Red Hat, p. 588.
8. DHS, *National Incident Management System.*
9. Noll et al., *Hazardous Materials*, p. 151.

5

PreIncident Response Planning

INTRODUCTION

Emergency response planning (ERP) must be completed before incidents occur. The fire service is familiar with preincident response plans and standard operation procedures. Planning involves identifying credible scenarios that are most likely to occur within your jurisdiction or within a jurisdiction you may have to respond to. Preincident response planning must be performed by your department in consultation with outside organizations (fig. 5–1). These outside organizations could include other governmental agencies such as federal, state, county, city, or even tribal organizations. Additionally, planning must include private agencies, such as local industrial facilities, and not-for-profit organizations such as the American Red Cross. Only by proper planning can efforts be focused to achieve the strategic and tactical objectives required to mitigate complex emergency events.[1]

Fig. 5–1. **Emergency response planning must be performed in consultation with outside organizations.**

PREINCIDENT RESPONSE PLANNING

Preincident response planning is critical to ensure that your forces are properly prepared for a fire or emergency should one occur. A preincident response plan is "one of the most valuable tools available for aiding the fire department and on-site fire brigade if available, in effectively controlling a fire or other emergency incident."[2] This chapter specifically pertains to what the fire protection organization should do to prepare for an incident at an industrial facility. Fire service organizations must be prepared for many types of incidents, including fires, confined space, and medical or hazardous materials incidents. As the fire department personnel prepare for these responses, they must identify industrial processes, confined spaces, potential technical rescue locations, and hazardous materials they may have to deal with. A preincident response plan should include at least the following information:

- Life safety hazards
- Command structure to be used and key personnel to be consulted
- Structure (facility) size and complexity
- Building construction
- Occupancy
- Industry-specific common hazards
- Fire protection systems
- Access to facility and process areas
- Presence and location of hazardous materials

- Susceptibility to natural or man-made disasters
- Additional resources
- Staging areas and incident command post locations
- Communications plan
- Special firefighting tactics
- Review, training, and drill cycle
- Recovery of fire protection systems

A recommended first step in preparing the fire service preincident response plan is to review the plan that is in place. Also review other plans such as the local emergency planning committee (LEPC) emergency response plan for these facilities.

EMERGENCY OPERATIONS PLANS

Any emergency operations plan (EOP, see chapter 4) that exists for the industrial facility or surrounding area should be consulted. A review of the EOP by the fire department can provide valuable insight into the actions to be coordinated by both facility personnel and the fire department. Portions or all of the EOP should be incorporated into the fire department's preincident response plan, and the complete plan should be available at the dispatch center or with the responding units. Take some time to learn what actions will be expected by all participating response and support agencies, as well as the types of hazards that may be encountered. Next, make contact with plant management and operations personnel. Perhaps the site safety officer will be designated as the point of contact.

Many EOPs use credible scenarios as the basis for the plan. When reviewing the EOP, ensure that the expectations of the fire organization are achievable. If the expectations are achievable, and your department has been assigned responsibilities under the plan, then you must prepare your personnel for the response. However, if your department does not have the resources or training to perform the functions outlined in an EOP, then this information must be communicated to the proper authorities.

SITE VISITS

Site visits to industrial facilities are critical. You will never understand the response requirements from outside the facility fence. During your visit, you should take a look at industrial process equipment, hazardous materials involved, utility isolation locations, water supplies, confined spaces, fixed and semifixed systems, and access routes into and throughout the facility. Make notes on the existing plan, if one exists, so that it can be updated. If no plan exists, use your department's preincident response plan form to develop and document what you have learned.

The use of checklists will greatly reduce the chance that items invaluable for effective preincident response plans are missed and therefore not included in the plan. A well-thought-out checklist will ensure that all personnel gather the same information necessary for an effective preincident response plan. The size-up acronym COAL TWAS WEALTHS[3] can be used as a checklist when performing a site visit for preincident response planning (see chapter 7).

- Construction
- Occupancy
- Apparatus and staffing
- Life hazard
- Terrain
- Water supply
- Auxiliary appliances and aids
- Street conditions
- Weather
- Exposures
- Area
- Location and extent
- Time
- Height
- Special considerations

For complete descriptions of these terms, see chapter 7).

Most of the information represented by the acronym can be included in the plan, but some of the information will need to be modified for preplanning rather than size-up. For example, weather conditions cannot be determined during preincident response planning, but what-if scenarios can be included in the preincident response plan, such as, how will rain affect a spilled material?

When you visit these facilities, you should suggest to the facility management that the fire department should partner with the on-site personnel to conduct emergency response exercises. By conducting these exercises, both parties will gain more insight into the portions of the plan that may need to be addressed further.

In addition, locations and quantities of critical materials such as foam supplies, water pumps, portable monitors, and the like can be confirmed and documented. As part of this process, fire department personnel should ensure that the equipment is serviceable by, at the very least, a visual inspection and check of the maintenance or inspection records. During emergency response exercises, this equipment should be used so that serviceability can be assured.

DETAILS OF THE PLAN

The following information should be the minimum information contained in the preincident response plan:

Life safety hazards

Life safety considerations should be the first priority for preincident response planning.[4] The number of persons working within a facility may vary with the time of the day, day of the week, or the production schedule of the facility. This information should be included in the plan and updated as necessary when new information is obtained. In addition, the locations of the employees should be noted in the plan. During the site visit, note the system used by the facility for employee accountability during an emergency. How will missing employees be noted, and who will have the information for the emergency responders? The location of employees during evacuation should be noted, and fire departments should ensure that these locations do not interfere with emergency operations. We don't want to arrive at the scene only to find that 300 employees are gathered in the area where we need to set up or stage our apparatus.

Command structure to be used and key personnel to be consulted

Determining the command structure and who will be in charge during all phases of an emergency incident is essential. It will eliminate confusion for managers and overlapping of assignments. The failure to plan for and establish who is in charge and to effectively manage an incident can lead to ineffective fire and emergency operations, and result in loss of life and property. There are many examples of fires and emergencies that great confusion and consternation have resulted from the lack of sufficient foresight to determine who will be in charge at an emergency scene or event. It is not healthy, for example, for the fire chief and senior law enforcement commander argue at an emergency scene regarding who is in charge.

Not only should the incident command structure be predetermined, but also the individuals assigned a specific role must be carefully selected and the role defined. If the incident commander (IC) is not specifically trained to command firefighting forces and able to understand the advice from senior fire officials at the scene of an emergency, then things could go terribly wrong. The IC of a fire event does not have to come from the fire protection discipline. As noted in chapter 4, the IC may be a plant operations person. He or she still must be capable in the role of IC. The

IC and other supporting functions will also be identified in the incident action plan (IAP).

Emergency operations center (EOC) contact information and activation policies can be mentioned in the preincident response plan. If a member of the fire department will staff a position in the EOC upon activation, then this can also be included. See chapter 4 section on "Incident Command at Industrial Incidents" for further information regarding EOCs.

Structure (facility) size and complexity

The structure size and complexity can be contained in a line drawing of the facility or the actual drawings of the facility attached to the preincident response plan. Access points, entrances, exits, stairs, unusual layouts, fire protection equipment locations, and so forth can be highlighted in the drawings. Any anticipated difficulties that would impede emergency operations should be noted in writing in the plan. When evaluating access routes, ensure that any bridges within the plant are capable of supporting the weight of the responding fire apparatus. Fire department personnel should talk to the plant engineering personnel to assist with any question that may arise during the site visit, or any issues relating to engineering drawing interpretation.

Building construction

Construction features will affect how the building or process structure reacts to fire conditions. Many times, different areas of a facility have been built at different times, and different construction techniques and materials may have been used on each area. This should be noted. Also, note the locations of vertical and horizontal openings or hidden voids that may cause fire to spread. It will then be easier to "get ahead" of the fire and prevent or reduce fire extension. When determining the areas required for apparatus placement, consider building construction and the collapse potential of such construction. If our primary area to set up for fire operations is within the anticipated collapse zone, we should rethink this position during preincident response planning, not during the incident.

Occupancy

The type of occupancy will have a distinct effect on how the incident objectives and tactical operations are developed and implemented. We have already noted that the life hazard is our first concern. Occupancy type will dictate the number of employees present at a given time. A printing establishment that prints a newspaper for early morning publication will have more workers during the evening and night shifts. An industrial facility that is completely automated will have few employees present during operational hours.

The type of occupancy will also give information as to the type of hazards or hazardous materials present. Certain industrial processes use a wide variety of chemicals as noted in subsequent chapters. The occupancy will dictate what chemicals and processes are used.

Industry-specific hazards

The industry involved will dictate the type of hazards common to the industry. As noted in subsequent chapters, some hazards may be normal hazards, meaning they are present normally by virtue of the process involved, for example, furnaces and molten steel in steel plants. These hazards may be present in the facility yet not be a part of the incident. On the other hand, the hazard may be the reason for the incident.

Fire protection systems

Knowledge of the presence of fixed and semifixed fire protection systems is important to the IC in developing the IAP. Also, be sure to note the requirements for fire department supply of these systems. The New York City Fire Department (FDNY), recognizing this, clearly spells out the requirements for standpipe and sprinkler system supply when responding to incidents. Preincident response plans should denote which engine company (or companies) will supply the systems.

The locations of fire department connections should also be noted. Many times, these connections are not visible from the street or blocked by equipment or vehicles.

When considering fire protection systems, hydrant locations and other water supplies must be noted. Not only should we note the presence of water supplies and their location, but we should also note whether the available water is sufficient for firefighting operations, including any fixed systems that may be operating as well as hose streams that will be used for firefighting and exposure protection.

Other protection systems such as dry-chemical, CO_2, or foam extinguishing systems should be noted. The presence and location of the systems should be included in the preincident response plan. Any special operating instructions for these systems should also be highlighted.

Access to facility and process areas

Access considerations to not only the facility but also the internal process areas must be included in the preincident response plan. The actual access points should be highlighted, as should any other conditions that may hinder access such as weight limitations, height limitations due to pipe racks, narrow streets, or a road configuration that is smaller than the turning radius of the apparatus (figs. 5–2 and 5–3). During the preincident response plan process, apparatus access should be tested to all areas of the plant or facility where applicable. Remember, construction or repairs may change the original information that we received when the plan was initially developed. Also, access may change because of the incident itself or the wind direction. As noted in chapter 10, access for emergency responders may change to accommodate facility security plans. Maintain a liaison with facilities that you may respond to so that the fire department is aware of any changes to access.

Fig. 5–2. **Access routes in plants may be narrow. Once apparatus is positioned, streets may become blocked.**

Fig. 5–3. **Overhead clearances must be taken into account during emergency response planning.**

Presence and location of hazardous materials

The presence and location of hazardous materials is a must for preincident response plans. During site visits, the material safety data sheets (MSDS) must be consulted, and chemical information must be included in the plan. Because some industrial facilities have many hazardous materials present, it may not be feasible to attach the MSDS to the preincident response plan, or list the chemicals present in the plan. In this case, the location of the MSDS and related information must be indicated in the plan. Many facilities use a locked box system where the plans are contained, and this box is accessible to the fire department upon arrival at the facility. The box is usually locked with the fire department maintaining the key.

Susceptibility to natural or man-made disasters

Certain facilities may be more susceptible to natural or man-made disasters. This should be highlighted in preincident response plans. For example, during heavy rains, certain areas of a facility may be prone to flooding. If this flooding affects response, it should be highlighted. Certain facilities may be high-value targets for terrorist activities. This should be noted in the plan so that the responding units are aware of this fact and can verify the cause of the incident before committing personnel to event mitigation. (See chapter 10 for further information.) During hurricanes or other severe weather, certain processes may be affected at a plant. This too should be noted in the preincident response plan. Other plans held by a fire department should coincide with an individual facility plan as well. For example, during severe weather, if fire department units are assigned to perform *windshield surveys*, certain facilities may need to be given priority for inspection during the survey. Preincident response plans should be consulted during this survey, and any temporary changes should be made to the plan.

Additional resources

Many industrial facilities prestage resources for handling major incidents. These resources may include foam, hose, vehicles, high-capacity monitors, and portable pumps. These resources may be intended for use by more than one facility. Cooperative mutual aid agreements greatly reduce the cost for a single facility or response agency. In addition, private emergency response agencies and foam manufacturers may have stores in close proximity to industrial complexes. A logistical plan must be developed to move these critical materials from their storage locations to staging areas or equipment dumps.

Additional resources, including both material and human resources, that may be required to mitigate an industrial incident must be noted in the preincident response plan. Private companies that supply expertise and staffing in the event an incident is beyond the capabilities of the local fire department should be listed in the plan. Complete contact information should be included, and verification of its correctness should take place annually.

Additional equipment resources and their suppliers should be listed, as well as any procedures that may have to take place to procure this equipment. Complete contact information, as well as a verification policy as noted earlier, must be included.

Staging areas and incident command post locations

The plan should identify staging areas for personnel and equipment. Staging areas must be located close to the emergency scene and be of an adequate size to accommodate the emergency equipment expected at an incident. The amount of equipment and personnel that may be in a staging area could be significant. Some experienced ICs recommend that you request twice as many resources as you think you need during your initial assessment. Planning for a staging area to accommodate these resources is essential. Remember, the area that you select today may be filled with automobiles on the day of the incident. Anticipate this during the planning process, and specify alternate staging areas.

The plan should also identify incident command post (ICP) locations where practical. The ICP must be in a safe location and easily identifiable. Depending on the type of incident, it may be necessary to identify two potential ICP locations. Upwind and uphill is always the preferred location for the ICP. Because winds vary seasonally, two locations must be identified. One of the worst things that can happen on at the scene of a major emergency is to set up your ICP then determine it must be relocated because of changing conditions or worse yet that you located it down wind. Relocating the ICP could create chaos just as you are trying to establish order at the emergency scene.

Communications plan

Communication at an emergency scene is always a challenge. Responders from the same agency typically have communications challenges, and multiple agency operations compound the problem. A single agency may not have enough command, operations, and support channels. Heavy radio traffic makes it difficult for the IC to monitor all the radio transmissions. There are cases where the IC missed emergency traffic from the firefighters, and those firefighters perished. Add the complexity of additional jurisdictions and agencies responding, and effective communications can become even more difficult. Plan to have assistance for the IC to monitor the communications traffic.

A communications plan can be included in the preincident response plan that will outline which frequencies that responders and support agencies will use to communicate. Ideally, this will prevent departments or agencies from operating on channels that cannot be received by others. In the event that agencies must operate on channels that cannot communicate with other agencies, then the plan should note how the communications will take place.

Special firefighting tactics

Because of the unique nature of many industrial facilities and processes, specialized firefighting tactics are required in many cases (highlighted in subsequent chapters). These may consist of the need for specialist response companies, other extinguishing agents such as dry-powder, or specific application techniques. These specific application techniques may be subsurface foam injection on tank fires or foam solution supply to semifixed extinguishing systems.

Review, training, and drill cycles

Although this information is not used during an actual incident, the preincident response plan should include a section regarding the time frame for plan review and the frequency for drills and exercises to test the plan.

Recovery of fire protection systems

As mentioned previously, the preincident response plan assists the IC in making strategic and tactical decisions. Part of the strategy should be planning for the recovery of any fire-extinguishing systems. The preincident response plan should include information on who will perform this service and contact numbers for these individuals or service companies. The need for a fire watch to be established until the systems are back in service should be noted in the plan.

Other considerations

Firefighting vapor dispersions and cooling streams can result in millions of gallons of runoff for an emergency scene. This effluent must be managed. Most process areas have berms, dikes, drainage, or sloped surfaces that are designed to channel runoff to appropriate processing systems. This outflow could also go to holding tanks or other vessels. During extreme malfunctions or breakage of the piping, these flows can overwhelm the collection and treatment systems and result in excessive flows that must be otherwise contained.

Some large plants have holding ponds for just this purpose. In other cases, expedient dikes and temporary ponds will have to be set up. The contamination from outfalls and runoffs can be extensive and expensive to remediate. It may be possible to dig a holding pond and line it with nonporous material to handle the runoff and prevent mass contamination of the area around the emergency site. Smaller amounts of contaminate may be absorbed or captured by HazMat booms or pigs. Preincident site drainage maps and runoff area should be developed in the planning stage and contained in the preincident response plan to assist the IC.

COORDINATING THE PLAN

After you and your crew have gone to the trouble to develop a comprehensive preincident response plan, you should share and coordinate the plan with others in the department. Note the development or updates of the plan in *pass-on log* books, and brief other members of the department during shift change roll calls and in-station training sessions. If possible, other shifts or units that will respond should visit the facility after reviewing the plan, so that they can gain firsthand knowledge. In reality, however, you may have so many facilities in your primary and mutual aid response area that everyone may not have a chance to visit each facility. It is also important to share these plans with other fire departments that may respond on mutual or automatic aid assignments.

OPERATIONAL SECURITY

It is important to note that the operational security (OPSEC) of these plans must be maintained. During the development of the plan, any draft copies or notes used in the development must be destroyed before they are discarded. We have all read where sensitive records have been discarded, and someone has picked up the discarded documents, causing deep concern among agencies or persons whose confidential information was contained in the documents. In addition, these plans, when carried as hard copy on the responding apparatus, must be secured in a locked box (fig. 5–4). Would it be easy for someone to remove them from your apparatus and use the information in a way that it is not intended? If your department uses computers to display preincident response plans, then the computers should be password-protected. How many fire departments keep a set of preincident response or emergency operations plans on a shelf in the department, company, or unit office? Are these areas secured when the fire station is not occupied? Some departments respond to an incident and leave the fire station unattended and unlocked. The rural department that feels it is located in a safe community may not realize that the plans for the local power plant in a binder on the shelf in the station office is a potential security breach. In today's climate of international and domestic terrorism, the use of OPSEC principals and practices cannot be overstressed. These practices include our firefighting personnel using caution to not inadvertently disclose sensitive information obtained during the preincident response planning process. Remember, "loose lips sink ships."

***Fig. 5–4.* Use OPSEC principals and store preincident response plans in a locked container on the apparatus, where they will not be open to public access.** *(Courtesy of M. Barrett)*

CONCLUSION

Preincident response planning is essential to professionally prepare for both large and small response scenarios. Identification of response resources such as agencies, personnel, expertise, and physical assets will greatly ease the stress on the day of the big event. The best plans are the ones that are never used. The more information the responding units have, the better prepared they will be to handle the incident. During the development process, we need to remember the importance of working side by side with the plant operations people as well as the on-site fire brigade or department. Use their experience to gain knowledge into the processes involved and the specialized techniques required to mitigate an incident. It is during the preincident response planning process that the gap between the municipal fire department and industry can be bridged.

When a large-scale incident occurs, the public and industry will usually look to the fire protection organization to bring order to the chaos. It is much easier to pull out your book of plans and follow predetermined actions, as opposed to trying to devise a safe and effective action plan from memory at 2 o'clock in the morning. Remember, the preincident response plan will be the foundation for the IAP. Plan well, and you will sleep better.

NOTES

[1] United States Environmental Protection Agency (EPA). 1986. *Title 42, The Public Health and Welfare*, chap. 116, *Emergency Planning and Community Right to Know.*

[2] National Fire Protection Association (NFPA). 2003. *Fire Protection Handbook,* 19th ed., vol. I. Quincy, MA: NFPA, pp. 7–85.

[3] Terpak, M. A. 2002 (August). Size-Up: Updating an Old Acronym. *Fire Engineering,* pp. 69–80.

[4] NFPA, *Fire Protection Handbook,* pp. 7–87.

6

Drills and Exercises

INTRODUCTION

Drills and exercises are essential for any emergency response organization to prepare to effectively handle their role in mitigating industrial fires or emergencies. Depending on your organization's assigned role in an industrial incident, you could be either in a participatory function or the overall incident commander (IC). Whether in the participatory or IC role, you will have to practice your assigned tasks. Furthermore, you will have to practice your assigned role in conjunction with the other emergency response organizations. Remember that failure of any one part of the emergency plan could result in failure of the entire mission.

RECOMMENDED PRACTICES

From the authors' own practical experience, combined with regulatory and recommended guidance, this chapter will cover methods of preparing your department for challenging industrial emergency incidents. Depending on your department's structure and anticipated responsibility in an emergency operation, you will have to determine what methodologies to use to practice and ensure that your department is prepared. Some of the types of exercises include individual tasks, team or company tasks, and overall incident management. Full-scale drills and exercises must test every aspect of your response plan, but this doesn't mean that individual or unit tasks should not be practiced as part of your proficiency training program. Individual members and team units must seamlessly fit into the overall plan. These tasks are the tools in the planning toolbox that the IC can use when needed. Choosing which tool to use is a critical part of the IC's job: The wrong tool or a dull tool will result in failure.

Comprehensive plans are vital but alone do not ensure success. You could have a comprehensive plan that details every element of the required response capability, but real capability is developed only by thorough training and testing of the plan. It is relatively easy for a fire chief or emergency planner to develop a picture-perfect plan with all of the i's dotted and the t's crossed, but it is much more difficult to develop real operational capability that will function on the day or night of the emergency. Many organizations are falsely confident that their paper plan will somehow work when needed, even though they have not devoted the necessary resources for training or to test that plan. You have to be able to "walk the talk."

FUNDING AND COST CONTROL RESTRICTIONS

Funding and cost restrictions depend on the cost of the drills and exercises conducted and your department's budget for such events. As the complexity of the drills and exercises increases, so will the costs. Station or company drills are relatively inexpensive and easy to conduct. Little coordination is required, so few costs are incurred. As additional agencies and resources are called upon, the coordination requirements and execution costs go up. Just scheduling the time to practice with multiple agencies can be expensive. Adding to these expenses are the costs associated with keeping three or more shifts of emergency responders proficient in their expected duties. Not only will there be associated costs with bringing all the necessary personnel to the drill or exercise, but also bringing the equipment, vehicles, and expendables (e.g., foam concentrate) will carry additional costs. Furthermore, departments should test their ability to actually flow large-volume cooling and/or foam streams. Actual production of these streams may show deficiencies in fire-water supplies, foam induction methods, logistics of moving hardware, foam stores, and staffing to conduct such an operation. These large-flow streams should be adequately tested long enough to demonstrate the capability to sustain the operation for the planned duration. Water is much cheaper than foam concentrate, so these streams could be maintained after foam capabilities have been exhausted. Environmental restrictions may prevent you from actually producing foam. If this is the case, you should find an alternate, environmentally friendly method of practicing this capability.

In some cases, even large-flow water streams will not be permitted to be discharged onto the exercise hazard (such as refinery vessels), so departments will need to practice in an appropriate location. Where possible, it is highly recommended that you operate in the actual industrial facility. Testing of the fire-water systems and working with high-capacity, high-pressure systems will be a challenge for departments that are used to operating from domestic fire-water systems of approximately 1,000 gallons per minute (gpm) at 80 psi.

In addition, there may be costs associated to properly handle foam or oily water produced during the exercise. Because of the associated costs of these exercises, careful planning will have to be done to provide for expected expenses. You should budget for these drills and exercises in your 5-year plan. Overtime and shift recall will need to be considered. Your local government would not appreciate a large cost being incurred unexpectedly. Furthermore, there may be state or federal money available to support your efforts. Don't forget about funding from industry for these exercises; most companies have a budget to support emergency planning. It is in their interest that real capability be developed.

We recommend that these exercises be conducted as often as possible. Adequate long-term planning is necessary to overcome the numerous challenges. Even though we have listed some of the challenges to conducting these exercises, you should not be discouraged. The lessons learned will be well worth the costs and difficulties you will encounter. Experience shows that individuals and organizations may be slow to support your efforts but will be overwhelmingly appreciative after an exercise is accomplished.

TABLETOP EXERCISES

Tabletop exercises are an excellent means to practice emergency response plans (fig. 6–1). Response agencies can come face to face and play out their expected roles. Many times, the people sitting at the table are not the ones who wrote the plans. Shortcomings of the plans are sometimes brought out as the frontline emergency responders examine the reality of the plan. Additionally, these individuals may also uncover anything the plan writers might have overlooked, and thus be able to proactively enhance the plan before an emergency event.

Fig. 6–1. **Tabletop exercises are an excellent means of practicing emergency response plans.** *(Courtesy of Bloomington IN F.D., Jeff Barlow Fire Chief)*

Participants in tabletop exercises should use the current plans, standard operating procedures, checklists, and maps. The revision numbers on any distributed documents should be checked to ensure that they are the latest. Also, standard emergency response maps should be checked. All responders should use the same revision and scale. Standardized grid maps are essential when establishing entry control points, casualty collection points, and staging areas. The lack of standardized maps is a common cause of confusion among emergency responders.

To further enhance these tabletop exercises, some agencies use topographical scale models, sometimes including figures and vehicles. Physically visualizing the setup locations of vehicles and crews will help everyone to get a realistic understanding of how these resources will be deployed at an emergency scene. Terrain and weather conditions such as wind direction can easily be changed during a tabletop exercise to allow response personnel to determine how they would handle changing conditions.

Tabletop exercises are relatively cheap to conduct and can flesh out many potential shortcomings of a written plan. In addition, personnel will have a chance to become acquainted with their counterparts from various response agencies. These exercises greatly enhance communications and understanding of expected missions. Although tabletop exercises are extremely valuable, practical exercises must be conducted to further evaluate the plan.

PRACTICAL EXERCISES

Practical exercises test capabilities and hone skills for the real event. All the planning and talk do little to measure how emergency forces will actually perform on an emergency event. These exercises should be as real as possible. In some cases, actual fires can be set. Many training facilities can simulate aircraft and industrial fire scenarios, which adds a sense of realism and excitement. Adding some stage effects such as smoke and *moulaged* victims adds additional realism.

Exercises should be allowed to run through their full course of events. When exercises are conducted in industrial and aircraft facilities, they should be initiated by actual personnel who work in the facilities and not be terminated until casualties are treated at medical facilities. All aspects of the expected emergency response should be practiced. Failure to practice any aspect or any emergency response will seriously limit the ability of the exercise to test the real response capabilities of the associated agencies. As will be discussed later, you must plan for the postincident critic as you develop your drill scenario.

Saying that you must practice every aspect of an emergency response is not saying that you must do this every time or that there is no need to perform practical exercises of certain parts of the emergency response. For example, security personnel would conduct routine training within their agency to secure an emergency scene. Personnel from all agencies should hone their skills during individual exercises in order to do well during major exercises.

MUTUAL AID

Most departments depend on mutual aid to one extent or another. Industrial and aircraft fires are complicated, and they normally require response and resources from many agencies located in the geographical area. Mutual aid exercises will acquaint response personnel with personnel from other emergency response agencies whom they expect to work with at large fires. Fire departments and other emergency response organizations are typically organized very differently from organization to organization. In addition to expecting that multiple agencies would respond to an industrial fire incident, these departments could also be expected to work with the plant emergency response team (PERT), fire brigade, or fire department. These plant personnel are great resources because of their knowledge of the facilities and processes. They could have specialized equipment or foam stores for your department to use. Without joint practical training and exercises, it is doubtful that plant personnel and your personnel will work well together on the day of the event. In addition, the specialized equipment would be of little use if no one knows that it exists or how to use it.

Response capabilities, training, and equipment varies widely from agency to agency. Some fire departments may be equipped with exceptional equipment that personnel do not know how to fully use or with specialized equipment that they train with infrequently. This is true of both volunteer and paid organizations. As mentioned, if the agencies do not perform continuous practical training, it is doubtful that they will function well on the day of the emergency.

Some of the lessons to be learned during mutual aid training are how the command structure works for the other organizations. In addition, accountability systems, breathing apparatus procedures, resupply operations, and rehabilitation should be reviewed. Details have to be checked, such as interoperability of breathing apparatus cylinders and the necessity of using adapters to connect hoses. Large-diameter hoses may be 4, 4.5, or 5 inches, or larger. Without proper adapters, water cannot be delivered to an intake manifold of a different size. Not only is it a good practice to drill with mutual aid agencies, but it is also a regulatory requirement.

REGULATORY REQUIREMENTS

The National Incident Management System (NIMS) requires that "incident management organizations and personnel must participate in realistic exercises—including multidisciplinary, multijurisdictional, and *multisecular* interaction—to improve integration and interoperability and optimize resource utilization during incident operations."[1] The requirement for realistic drills and exercises is spelled out very well in this document. It requires that all entities that may have to come together to handle an emergency prepare together. The optimization of resources in this paragraph cannot be overemphasized. If departments and agencies can obtain specialized equipment and share that equipment through effective planning and training, then this will enhance the overall readiness of the entire geographical region. An example may be a mobile breathing apparatus refilling or foam trailer. Agencies should look at the big picture and obtain resources that will enhance response capabilities and not duplicate little-used equipment that could be shared. Effective sharing of resources needs to be practiced. In any case, Homeland Security Presidential Directive (HSPD) 5 directs that all organizations work together. In this day and age when vast national resources are being spent to ready our emergency organizations to combat natural and man-made scenarios, we owe it to the taxpayers to share those resources.

The requirements to conduct exercises and evaluations comes from various other sources. The National Fire Protection Association (NFPA) requires organizations to realistically evaluate all aspects of the plan. This includes individual skills and also requires corrective action to be taken for deficiencies noted during the drills.[2] The

NFPA repeatedly requires that all agencies that will be required to respond to a facility under a plan must practice during exercises. NFPA 1620, *Recommended Practice for Pre-Incident Planning*, also requires that participants from various organizations wear armbands or other visible identification that indicates what their role is. It also states that there should be a unique signal for the eventuality should a real emergency occur during the exercise. As many other documents state, a postdrill critique is essential and extremely valuable to capture the lessons learned.[3]

For airports, NFPA 403, *Standard for Aircraft Rescue and Fire-Fighting Services at Airports* requires that tabletop exercises be conducted annually and full-scale exercises be conducted every two years.[4] Large-scale aircraft disasters should be practiced on and off the airfield. With an aircraft accident, there is a potential for more than 500 casualties. It is physically impossible for any airport fire department to handle a large-scale emergency on its own. Large numbers of ambulances and firefighting forces will need rapid access to the airport premises to save lives. This is not an easy feat in a post–September 11, 2001, environment. These plans must be practiced to ensure that your forces could actually get in to assist.

SOURCES OF TRAINING

There are many sources of specialized industrial fire protection training available to departments around the country. Some state agencies provide hazardous materials training. Many fires in industrial facilities involve hazardous materials. Departments should also take advantage of training opportunities that could be available at local industrial facilities. These facilities will have process and materials experts that will normally be glad to provide this training to personnel in your department. Additionally, a wealth of information and training is available from organizations that handle materials such as natural gas, liquid petroleum gas, ammonia, and chlorine.

Manufacturers' Workshops

Other specialized training at schools around the country is conducted by manufacturers of industrial firefighting equipment and agents. Organizations such as Task Force Tips conduct training on their products. Firefighting agent and hardware manufacturers such as Williams Fire and Hazard Control, National Foam, and Ansul have foam schools. The courses are conducted by experienced industry experts who work hand in hand training firefighters in the specialized techniques required in industrial process and storage facilities. If your department is asked to respond to industrial incidents such as hydrocarbon storage tank fires, it pays to attend these training sessions and gain as much knowledge and support as possible. The practical experience available from these experts will pay tremendous dividends to any department.

Other Sources of Industrial Training

Universities such as Texas A&M and Louisiana State University conduct special industrial-fire-training classes (fig. 6–2). These classes are hands-on firefighter-oriented classes that deal with subjects such as process unit fires, tank fires, hazardous materials, fire brigade, aircraft firefighting and rescue, and incident command. Like the manufacturers' training, these classes are conducted by some of the most experienced experts in the field. The importance of gaining experience from these experts cannot be overemphasized. Aircraft and industrial accidents are few and far between. So training and practice is the only way for the vast majority of firefighters to develop the necessary skills to handle complex industrial and aviation accidents.

Fig. 6–2. **TEEX, Texas A & M System Industrial Firefighting training** *(Courtesy of TEEX, Texas A&M System)*

Fig. 6–3. **A field critique can be conducted at the scene of the drill to include those persons who may not be able to attend a formal critique procedure.**

DRILL CRITIQUES

Critiques are essential to capture lessons learned during drills and exercises. Critiques are an objective evaluation of the effectiveness of all aspects of the practice event (fig. 6–3). It goes without saying that you should also critique real life events and take advantage of those lessons as well. It is important to have as many of the responders as possible feed their observations back into a critique. This goes for all, from firefighters to fire chiefs and similar players from all response organizations. Valid observations and recommendations can be obtained from all levels of an organization. For practice events, evaluators are normally assigned for each participatory organization. These evaluators each have the responsibility to observe and evaluate their agency's performance.

In addition, they should evaluate how other agencies performed in relation to meeting the overall objectives. Extreme diplomacy must be exercised when critiquing another organization. You may not have the technical expertise to critique another organization, or that other organization may not appreciate your criticism even if you are right. In cases such as this, you would have to gain credibility by expressing your concerns to qualified evaluators and asking for their assistance.

One thing that you may discover during drills and exercise is that your department or mutual aid departments may not be able to realistically handle an expected event. In this case, it is wise to inform the local managers and government. When you do this, you should offer options on what resources are needed to accomplish the mission. In some cases, you will find other mutual aid agencies to assist. Other options are to approach industry leaders to see if they can supply needed resources such as foam, apparatus, or additional personnel. Most of the major foam manufacturers offer emergency assistance (for a price) to source and deliver large quantities of foam to an emergency site. Sometimes you may need more than foam or materials; you may have to hire an emergency response company to assist. An emergency response company responds with needed personnel, hardware, and extinguishing agent to handle specialized fires or scenarios. Most important, because their personnel fight many industrial fires, they have the experience and confidence to make a plan work. The downside is that you pay heavily for their expertise, personnel, and hardware. It is a good idea to put into the plan the possibility that an emergency response company may have to be called. Once a potential contingency agreement is established, the specialist company may come and survey and provide some valuable advice. Having a contingency agreement in place and knowing how to initiate it in case it is needed will make the decision process easier when it seems as if everything is going wrong. As mentioned previously, you should take the opportunity to work with the specialist response companies in advance. If requested, they may even provide free training and advice to establish a relationship in the hopes that they will be the company of choice in the case of a major event. Inclusion of these experienced emergency response personnel in your local drills and exercises could greatly enhance your emergency response capabilities and help to identify shortcomings in advance.

DRILLS WITH LAW ENFORCEMENT

Drills with law enforcement organizations can be very challenging. Problems can arise from the nature of both organizations. Both law enforcement and fire protection organizations are tasked with responding to events that are out of control. This could be a crime scene for law enforcement or a major fire for fire protection organizations. In both cases, the responders must rapidly assess what has happened and gain control. This is a difficult job, and many times aggressive or offensive actions must be taken to gain control.

As law enforcement and fire protection forces converge on the scene and take actions to gain control, there could be—and have been—many heated exchanges as both agencies try to perform their duties. With the best intentions, these forces sometimes clash. Add multijurisdiction agencies such as private, city, county, state, and federal, and misunderstandings will occur. All organizations have the best interest of the public in mind but are likely to focus on their primary objective. For example, many fire scenes are also crime scenes, and law enforcement may suspect that a crime has been committed. The overall goal is to save as many lives as possible and to ensure those responsible are held accountable for their actions.

Because of the propensity for these misunderstandings and overlapping responsibilities, it is essential to work out these relationships ahead of time. A full understanding of the roles and how to pass command is required. Drill and exercise with as many organizations as possible. It may not be possible to get the Federal Bureau of Investigation (FBI) to play a role in a practice event, but someone could play the role of an FBI agent. It is much better to work out the rough edges of the plan ahead of time than it would be to have a dramatic encounter during an emergency event. Your drills and exercises may even be able to define the roles of law enforcement and prevent misunderstandings during a real event. Remember that you will have your hands full with the fire and rescue operations. Any distraction from your primary duties will be detrimental to your operations. Being proactive by conducting drills and exercises is in the best interest of all.

CONCLUSION

As described in this chapter, there will be many challenges to conducting realistic drills and exercises. Proper planning, budgeting, and coordination are required. Inclusion of all the expected players is crucial. During your journey, you will encounter the skeptics and those afraid of the challenge of the event. Organizations or individuals who are not prepared and have been hiding behind a paper plan for years may not be enthusiastic about having their shortcomings exposed. Preparation for these drills will require organizations to work hard and devote time and resources. No one wants to be embarrassed during these drills. This preparation probably would not occur without the scheduling of major drills and exercises. The business of conducting drills and exercises is challenging but well worth the efforts that will result in team building and better preparation of all involved.

NOTES

[1] United States Department of Homeland Security (DHS). 2004. *National Incident Management System*. Washington, DC: DHS.

[2] National Fire Protection Association (NFPA). 2004. NFPA 1600, *Standard on Disaster/Emergency Management and Business Continuity Programs.* Quincy, MA: NFPA, sect. 5.13.

[3] NFPA. 2003. NFPA 1620, *Recommended Practice for Pre-Incident Planning.* Quincy, MA: NFPA, chap. 19.

[4] NFPA. 2003. NFPA 403, *Standard for Aircraft Rescue and Fire-Fighting Services at Airports.* Quincy, MA: NFPA, chap. 4.

7

General Firefighting Tactics and Response Concerns

INTRODUCTION

Later in this book, we review specific industries, processes, and fire hazards, along with the associated strategies and tactics used for firefighting. In this chapter, some general firefighting strategies and tactics that may be applicable at most industrial incidents will be presented. Some may say that "firefighting is firefighting," but handling an industrial incident while using the same general guidelines has some differences, concerns, and key points that the municipal firefighter should be aware of. We feel that the best way to approach this chapter is to look at a standard size-up acronym and use it to highlight the differences in firefighting tactics and response concerns. The acronym we will use is COAL TWAS WEALTHS.[1] Each letter represents a particular area of size-up that will highlight these tactics and response considerations:

- Construction
- Occupancy
- Apparatus and staffing
- Life hazard
- Terrain
- Water supply
- Auxiliary appliances and aids
- Street conditions
- Weather
- Exposures
- Area
- Location and extent
- Time
- Height
- Special considerations

In addition, this same acronym can be used in the preincident response planning process.

CONSTRUCTION

At all incidents, construction features affect the response to fires and the tactical objectives. With industrial occupancies, responders will find for the most part larger structures that contain large open areas of manufacturing equipment, such as is commonly found in automotive manufacturing (chapter 34) or steel production (chapter 26). There, you might encounter buildings that are more than 80 acres under a single roof (fig. 7–1). In process plants, expect to find large exterior areas containing process units, pipe racks, supporting equipment, process vessels, and storage tanks. These large areas and process equipment may be protected by automatic suppression systems, but if not, or if the system is out of service, this large open area may cause rapid fire spread requiring large-caliber streams immediately upon arrival. Do not be afraid to place these streams in operation early for exposure protection or fire attack. Departments with large commercial occupancies within their response districts should have a preconnected large-caliber stream device available on the apparatus, such as a Blitzfire™ 500 gpm nozzle. These devices enable the rapid application of large volumes of water on an incident and thus rapid control of the fire while additional lines and appliances are being placed into operation. Remember: "big fire, big water."

Fig. 7–1. **Large open areas at a steel manufacturing plant**

With large, open areas, we must consider that truss construction will most likely be present. This construction presents its own unique concerns, and the presence of this construction should be indicated in preincident response plans. It is advisable to have the buildings marked with this information also, even if such marking is not required by code. Large-area buildings will also predominantly use steel in the construction, so this factor must also be taken into consideration.

Many outdoor processing plants such as refineries use considerable amounts of steel to support process units as well as piping and utility racks throughout the facility (fig. 7–2). This steel, while protected with fireproofing in some areas, should be considered vulnerable to the effects of heat or explosions. Protection of exposed steel with copious amounts of water should be a general tactic. Minimum application should be at least 350–500 gpm. If the steel fails, it may injure or kill firefighters, create a larger hazard due to the failure of piping with additional fuel or hazardous materials leaking, or it may block access or emergency egress from the area.

Fig. 7–2. **Overhead piping and utility racks**

Other construction features that require consideration during firefighting operations are the framing systems, building materials, and interior and exterior finishes of the structures.[2] Construction techniques may create voids where unexpected fire travel can occur. As mentioned earlier, large, open areas may be present, but in some industrial processes large openings may have been created for passage of materials or equipment.

The construction of exposures must be considered in preincident response planning as well as firefighting size-up. Although the construction materials of the fire building may be of one type, the materials used in the construction of exposed buildings and structures may be of a different type, requiring much different firefighting tactics.

OCCUPANCY

When we consider industrial buildings or sites, we need to evaluate what is the occupancy use and contents. This will provide information on the processes that take place, the hazardous chemicals that may be on-site, and the materials and quantities being stored. Some processes and chemicals may be water reactive. This must be taken into consideration. Information regarding the chemicals should be contained in preincident response plans and material safety data sheets (MSDS) should be part of the these plans also. Additional information regarding chemicals and specific hazards present that may affect the use of standard firefighting procedures and strategies can and should be obtained from plant personnel. If hazardous materials are involved, a standard hazardous materials incident action plan (IAP) should be implemented, and assistance from hazardous materials experts must also be obtained. The effects of the smoke must be considered, and precautions and/or evacuations of surrounding areas should be considered.

APPARATUS AND STAFFING

When looking at apparatus and staffing requirements for industrial fires, we need to look at the processes and chemicals or materials affected. We also need to evaluate the size of the facility. A typical process unit fire may require at a minimum 6,000 gpm of fire-water flow. Does your apparatus have the capability to deliver such large volumes? If not, have your preincident response plans identified this shortfall and compensated by the use of mutual or automatic aid?

Where large volumes of petrochemicals are involved, large amounts of foam delivery will be required in case of fire or large spills. Again, we must ask ourselves if our apparatus has the capability to deliver foam solutions at the required percentages and flows. The typical 200-gpm line eductor will fall very short of these required foam percentages and flows. Mutual aid agreements must be considered if your fire department does not have the resources available to fight a large industrial fire. It may be best to classify your mutual aid resources so that only equipment essential for fighting this type of fire is sent to the scene. Normal mutual aid agreements usually call for a certain number of vehicles and a certain amount of staffing to respond when requested. Classifying resources by type would allow the correct type of apparatus to respond with correct staffing levels. For example, an industrial mutual aid response may call for only pumpers capable of pumping 1,500 gpm or more responding with a minimum of four or five firefighters per apparatus. It may also only request pumpers with Class B foam delivery capabilities.

Other apparatus that may be required at large industrial fires would be command vehicles, mobile breathing air compressors, and rehabilitation units. These incidents will be of "campaign length" duration, and even though they may be exterior, defensive operations in many cases, due to the chemicals that may be involved, large numbers of self-contained breathing apparatus (SCBA) will be required for operations personnel.

Large industrial fires require large staffing assignments, not only for suppression operations but also for logistical support. Long lays of large-diameter hose will be required in most instances (fig. 7–3). When using large quantities of foam, support personnel will be required to ensure the

Fig. 7–3. **Long lays of large-diameter hose will be required.** *(Courtesy of Williams Fire and Hazard Control)*

foam is placed at the correct point for operational effectiveness. Totes (275 gallons typically) will require forklifts and forklift operators to position them (fig. 7–4). If 55-gallon drums are used to supply foam, additional staffing will be required for logistics and opening of the drums to allow the foam to be delivered from the containers. Do not forget to consider where you will put the empty drums. Foam tankers may also be available.

Fig. 7–4. **Positioning totes with forklift** *(Courtesy of Williams Fire and Hazard Control)*

Apparatus positioning must be considered at industrial fires. Respond from uphill and upwind where possible. Only in unusual circumstances where the terrain or highway configuration may prevent responding from uphill should you approach from any other direction than upwind and uphill. If you must approach from downwind, use distance and atmospheric monitoring to increase your safety.

Ensure that you have the most up-to-date information on the chemicals or processes involved, and use SCBA if needed. When arriving at the scene, it may be wise not to commit all of the responding apparatus into the plant or facility until the area where they will be positioned is surveyed and deemed safe. In addition, apparatus should not be placed in a hazard zone and should be positioned facing evacuation/escape routes. Evacuation routes should be identified during preincident response planning, verified early in the incident, made part of the IAP, and kept open during the incident.

The initial staffing levels of the fire brigade or fire department must be considered. Based on these initial staffing levels, it may only be possible to provide exposure protection or rescue. When municipal departments arrive at the scene in an industrial facility, information concerning the number of firefighting forces already deployed and their actions must be obtained and used in the ongoing size-up process.

LIFE HAZARD

As with any emergency response, life hazards should be our first priority. Firefighter's lives are included, and the response and initial apparatus positioning should ensure this. Upon arrival, responders need to account for plant personnel who may be trapped in the area and begin rescue operations after conducting a risk assessment and only when it is safe to do so. Contact should be made with plant personnel to determine if any of them are missing. There may also be contractors who were working in the area. *Work permits* can be used to determine who was working in the area at the time in addition to normal plant operations personnel. Risk versus benefits should be employed when attempting rescues or searching for missing plant or contractor personnel. Full protective equipment should be used, including SCBA. If hazardous materials are present, the proper protective equipment for the hazardous material should be used.

Rescues may have to be performed at elevated locations, in congested process areas, or in areas with high noise levels. Rescued persons, as well as the rescue personnel, may have to be decontaminated due to the chemicals involved in the fire or accident.

Accountability systems should be in place whenever fire department personnel are operating in or have the potential to operate in *immediately dangerous to life or health* (IDLH) atmospheres. A scene accountability system must be used at all operations, even those with a strictly defensive posture. Accountability systems in use by industrial fire brigades should be known by municipal fire departments that may respond to their facility (fig. 7–5). This will aid in determining who is operating on the scene and who may be missing.

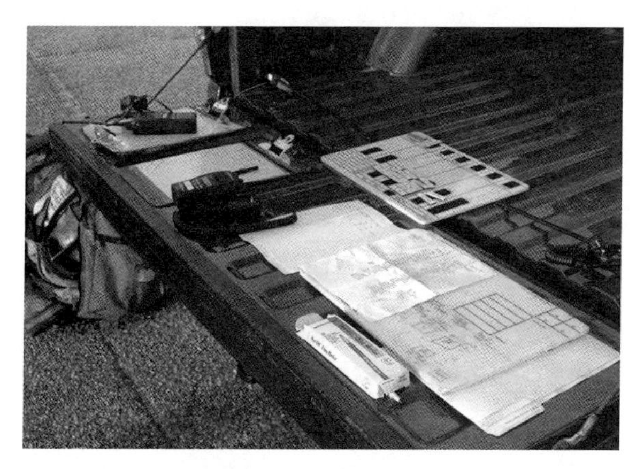

Fig. 7–5. **Accountability systems used by industrial brigades should be familiar to municipal fire departments that respond to the industrial facility.**

Rapid intervention crews (RICs), teams (RITs), or companies should be established at industrial operations. Due to the size and complexity of the area of operations, additional RICs should be established to ensure coverage of all operating personnel in case of an emergency. These RICs should be equipped with SCBA and have rescue tools available in case of structural collapse. Air-monitoring equipment would be a required tool for the RICs to have to determine safe working atmospheres in areas they may be called on to operate in.

To protect our responding firefighters, site monitoring should be performed at industrial fires and emergencies. This monitoring should be performed using at a minimum a multigas meter that measures flammability, hydrogen sulfide (H_2S), oxygen (O_2), and carbon monoxide (CO). In addition, monitoring capability must be set up for any specific industrial gases that were identified during preincident response planning. Assistance and advice in this effort may be obtained from the plant personnel. Monitoring will assist with hazard zone identification, selection of personal protective equipment, and evacuation or shelter-in-place decisions. The plant personnel may be able to assist with site monitoring and may have on-site monitoring teams available. Monitoring of the immediate area, as well as the facility fence line and downwind communities that may be affected, should be performed and records kept.

TERRAIN

Terrain issues may be of concern when responding to industrial incidents and may affect apparatus placement and tactical actions. Containment and control of water runoff or other by-products of the chemical process being used will be affected by terrain. At many industrial processing facilities, there are containment areas to assist with containing hazardous materials, but during firefighting operations, the large volumes of water used may overflow containment areas. If the topography of the area allows this overflow to travel, responders must know where it is going and be prepared to contain it further using sand or other methods that may be appropriate. At smaller industrial buildings, the topography may affect apparatus placement, for example, a street grade that is too steep will prevent the placement of aerial apparatus to be used as an elevated master stream.

WATER SUPPLY

Water is critical at most fires. Many industrial fires will require large volumes of water and large-caliber stream operations for extinguishment, exposure cooling, and/or vapor suppression (fig. 7–6). It is not improbable to require volumes of 2,000–12,000 gpm, with much higher volumes for large petrochemical storage tank fires. Although most industrial complexes will have a dedicated fire-water system, it may be possible that in the case of a catastrophic fire, this system may have been damaged or not designed for the large-volume flows necessary. How will your department supplement the water supply? For large petrochemical fires, large-volume flows may be required to overcome the thermal updrafts created.

***Fig. 7–6.* Large-volume manifolds may provide water supply at industrial facilities.** *(Courtesy of Williams Fire and Hazard Control)*

If a municipal department responds to an incident requiring a large volume of water delivery, what method of delivery is provided? Is your department's large-diameter hose (LDH) compatible with the LDH used in the industrial complex? If not, what adapters will be required? Industrial fire brigades and some municipal departments are now using 12-inch hose for large-volume water delivery. The authors are familiar with a large storage tank fire that many mutual aid departments responded to. One large department provided the services of a fireboat capable of pumping 10,000 gpm. This should have been a great help to the industrial fire brigade, supplying the additional water supplies necessary to help extinguish the fire. However, the industrial brigade was using 5-inch LDH and was not able to connect to the fireboat. Fortunately, another department on the scene had the necessary adapters to connect the hose to the fireboat. When your department has an industrial fire brigade or department working within its jurisdiction or mutual aid response district, inquire about what size hose they are using and whether any additional adapters be required for compatibility and interoperability.

Industrial facilities and their support network may have large-capacity portable fire pumps available for use. Once again, if this is a possibility, does your department have the necessary fittings and adapters to connect to these pumps if they are used? It would be wise to drill with industry on the operation of these pumps so that any interoperability issues can be resolved prior to the incident, and not during the incident.

Fire flow requirements and available water supplies should be noted in the preincident response plans developed by municipal departments. These fire flow requirements should also take into account exposure cooling requirements. Industrial water supplies are also high-pressure systems compared to municipal systems.

AUXILIARY APPLIANCES AND AIDS

Auxiliary appliances and aids are the systems and equipment that may be located on the premises or complex of an industrial facility. These include but are not limited to fire-suppression equipment such as sprinkler, foam, CO_2, and standpipe systems. Departments should know the location of these systems and how to manually activate and supply them with additional extinguishing media when required. Can the fire department augment a fixed foam extinguishing system if required? Does the department have the resources available to provide this additional foam supply? What type of foam is in use (fig. 7–7)?

Auxiliary appliances also include fire detection equipment (see chapter 9). When performing inspections and preincident response planning, fire service personnel should take note of the

***Fig. 7–7.* High-expansion foam system used in an aircraft hangar** *(Courtesy of Ansul)*

locations and types of detection equipment. This information may give clues to the type of fires that may be expected of the product being monitored.

Fire departments should be aware of the location of control valves for fire extinguishing systems. There may be times when the system must be isolated; or if the system is found not to be working, the valves may have to be checked to make sure they are open. Knowing the type of system is also invaluable. Is the system a semifixed type that requires fire department supply?

In this section, we also address aids and assistants. When arriving at an industrial facility, plant personnel who can supply information or assistance should be considered valuable resources and potential aids to the incident commander. These individuals possess valuable information regarding the chemicals or process equipment present, and their participation can mean the success or failure of the operation. Persons that should be consulted might include the plant safety manager, industrial fire brigade chief, process foreman, facility manager, and hazardous materials response specialist. It is important that these individuals are brought into the picture immediately upon your arrival on the scene. Sometimes an action that you may consider may not be the correct action to take. Consult the experts.

STREET CONDITIONS

General accessibility must be considered under street conditions. Many large industrial complexes may be constructed with narrow streets and low hanging pipe or wire crossings. Depending on the size of your fire apparatus, this may or may not affect response and placement. While preincident response planning, ensure that the turning radius, height and weight restrictions, and width of your vehicles can be accommodated within the plant grounds or surrounding streets. Identify any and all obstructions or response restrictions in preincident response plans. Just as with a municipal response, construction may affect the response within an industrial facility complex. Departments should be placed on the notification list when street modifications or temporary closures are performed by industrial facilities within the plant grounds or on adjacent streets. When identifying response routes and staging areas, street conditions that will affect the response of your department as well as mutual aid departments must also be considered. During an evacuation of the facility, how will the response of emergency vehicles be affected? Does your response plan use the same roads as the facility's evacuation plan? During preincident response planning, these issues must be addressed. Two ways in and out should be identified if possible. Remember not to block escape routes with LDH, emergency response or other vehicles, support materials, foam supplies, and the like.

WEATHER

Weather-related elements can affect normal fire department operations and will affect an industrial response even more so, especially if hazardous materials are present. Strong winds can carry hazardous vapors long distances, or the lack of strong winds can keep hazardous vapors in the immediate area, making mitigation efforts that much more difficult. Wind can also spread fire and create severe exposure problems.

Rain and/or high humidity may also affect fire department operations at industrial incidents (fig. 7–8). Many chemicals are water reactive, so if they are exposed to rain or high humidity, they may become that much more of an exposure problem to firefighters, support personnel, and the civilian population at large. Humidity and heat will also have an effect on firefighting personnel. Officers should be concerned and arrange for early and frequent relief of firefighting personnel.

***Fig. 7–8.* Rain can affect firefighting operations by creating flooding conditions that may limit access.** *(Courtesy of Williams Fire and Hazard Control)*

Extreme cold has debilitating effects on firefighters just as extreme heat does. Firefighters exposed to extreme cold conditions may be subject to frostbite or other cold weather medical conditions. Firefighters dressed for the extreme cold will find it difficult to move with the heavy layers of clothes necessary. Movement may be slower, and additional personnel may have to be called to compensate. In many years of response experience, the authors have noted that firefighters tend to underdress rather than overdress. What we mean is that the normal fire department response is of short duration, so firefighters dress for this short duration. Remember that many industrial incidents are of long duration, and if the firefighters are not dressed to be exposed to the elements for a long-duration incident, they may have to be relieved early so that they can return to the station and dress appropriately. Rehabilitation is just as important during cold weather incidents as it is during hot weather incidents. Extreme cold may also affect water supplies. Fire-water systems that are not maintained properly can be subject to freezing, seriously hampering firefighting operations. Fire apparatus may also be affected. There have been cases where pumps and water piping on aerial apparatus have frozen, even though the apparatus had responded from heated fire stations. Cold weather decontamination may have to be performed before apparatus can leave the scene. This will involve the use of steam equipment to thaw out the equipment. Many municipal departments located in areas where extreme cold conditions are expected have this equipment in their department. Industrial plants may have steam throughout the plant that can be tapped for this purpose. Consult with plant personnel if this is required.

EXPOSURES

Immediately after addressing life hazard issues, exposure protection must be undertaken. With industrial incidents, this must be a major tactical objective. Critical equipment must be cooled sufficiently to prevent the incident from escalating. Supporting columns and beams must also be cooled to prevent failure (fig. 7–9). If adjacent supports fail, additional products may be added to the fire equation.

***Fig. 7–9.* Cooling of supporting steel and exposures will be required at industrial incidents.** *(Courtesy of Dave Cochran)*

Protection of exposures will help contain the fire and potential damage to one area. The immediate area of the fire is probably seriously damaged already. Protection of exposures at industrial fires is many times the appropriate action while machinery is deenergized, and flammable gas and liquid supplies are isolated or shut down, allowing them to burn off.

At petrochemical incidents, a rule of thumb can be applied that will also work at other industrial facilities. This rule of thumb is the

5–10–15 minute rule. Instrumentation and electrical systems such as cable trays and major distribution systems can fail in approximately 5 minutes if proper cooling protection is not initiated by either the plant personnel or fire brigade. Flame impingement on pressure vessels can initiate the initial stages of a boiling liquid expanding vapor explosion (BLEVE) if cooling streams are not applied within 10 minutes. In addition, structural steel components can begin to weaken when exposed to flame impingement in as little as 15 minutes. These numbers are generalizations, but other factors and conditions may alter the times, either up or down.

When pressure vessels are exposed to direct flame impingement, the general rule of thumb is to place 500 gpm on each point of flame impingement. When multiple points of flame impingement are present, additional waterlines must be used to effectively protect and cool the points of flame impingement. Water supplies of over 1500 gpm may be necessary. It would be wise to use multiple water sources so that if one source becomes inoperative, the backup will already be in place. It is also wise not to supply all critical hose or monitor lines from one pumper.

AREA

When we think of the fire area, we usually equate this to building area (length × width). This is correct for structures, but many industrial sites include processing units as well as buildings. The area should be defined as the square footage involved, including any area that may threatened by the fire. Water flows and resource deployment should be calculated on these dimensions. The best time to perform these calculations is during preincident response planning. During initial size-up, the total area must be determined. During the *global 360,* information on the square footage involved and threatened must be given to the incident commander.

LOCATION AND EXTENT

When fire departments arrive on the scene of an industrial incident, the location and extent of the fire must be determined. This might not be easy to determine because obvious fire may not be the only fire or hazard to be found (fig. 7–10). Many times, when fires occur in chemical- or petrochemical-processing units of plants, problems may arise in other sections of the plant due to the shutting in of the process unit on fire. Process units are usually set up to shut down operations sequentially. During an emergency shutdown, however, this may not take place, causing more problems in other areas. It is not uncommon to have additional fires in other sections of the plant. This must be factored in during the size-up, and other areas must be continually monitored by plant personnel. Flammable liquids and gases may be sent to flare. This procedure may look like the end of the world to a municipal firefighter but would be normal to plant personnel. Fire departments should not let their guard down, even when it appears the fire is being contained or under control.

Fig. 7–10. **Determining the extent of the fire may be difficult at times.** *(Courtesy of Dave Cochran)*

TIME

As with any fire or emergency, the time of day or year can affect industrial fires. On the one hand, industrial buildings may not be occupied at night, with the exception of night watch personnel. There may be a limited production staff on duty during night operations. In a large plant that uses personnel to staff a fire brigade, how will this affect their response? On the other hand, an industrial plant may be fully staffed during night hours, contributing to a larger life hazard. Preincident response plans should indicate varied staffing levels where applicable.

The time of day will have an impact on the evacuation of surrounding communities if this process is required. During the night hours in a residential community, most of the residents will be at home. During the day, bedroom communities may have fewer people to evacuate. Does your community have a reverse 911 or a similar system that can be used to alert residents that an evacuation is necessary? Will additional staffing be required during evening hours to ensure adequate evacuation guidelines?

During evening hours, industrial storage facilities may not have any personnel on duty. This condition may affect the discovery of a fire. How will it affect the structure itself? How will it affect your initial firefighting tactics? The fire may be greater in size because of the delayed discovery, causing greater structural damage.

HEIGHT

A building or industrial process structure's height can have an effect on firefighting operations. Fires occurring on the upper floors of a building, in particular those that occur above the reach of aerial devices, must be fought from the interior. Exterior ventilation will be restricted, and the time from arrival on scene to fire attack will be delayed. This time delay will allow the fire to grow in intensity. Smoke also behaves differently in tall buildings due to the *stack effect*.

Height can also be a factor at industrial plants. Process equipment may extend several hundred feet into the air (fig. 7–11). A fire located at these heights will be inaccessible even to those departments with aerial devices. Rescue operations may also have to be carried out on tall process equipment. High angle rescue gear must take these heights into account. Once again, information regarding the height of process equipment must be included in preincident response plans.

Fig. 7–11. **Process equipment may extend several hundred feet into the air.** *(Courtesy of Dave Cochran)*

SPECIAL CONSIDERATIONS

In this section, we look at other considerations that the fire department should take into account when responding to industrial incidents.

Hazardous materials

Most industrial facilities use, store, or manufacture hazardous materials (see chapter 11). Hazardous materials should be indicated in preincident response planning, and the MSDS should be part of the plan. Decontamination of fire department personnel as well as exposed plant employees should be considered and resources allocated. The presence of hazardous materials can cause problems such as health hazards and accelerated fire spread. The presence of hazardous materials may trigger shelter-in-place orders or the evacuation of surrounding communities. It may also halt firefighting operations in some cases. Early identification of hazardous materials, that is, during the preincident response planning phase, is crucial to an effective and safe firefighting operation. Product containment and control must also be considered by the incident commander. Plugging and patching of piping as well as product diking and damming may be required. Ensure the response of a qualified hazardous materials team when hazardous materials are present.

Process control

At chemical and petrochemical facilities, process operations must be controlled during firefighting. Most of these facilities contain interconnecting process units. This interconnection consists of piping, valves, process equipment, instrumentation, and related equipment. The safe shutdown of these interconnecting units must be performed by qualified on-scene plant operations personnel who are familiar with the process operations and equipment. Consultation and coordination between fire department command personnel and the plant operations personnel is a must. Limiting the exposure of fire department, plant, and civilians to hazardous situations must be a priority of this shutdown. At times, products will be transferred by plant operations personnel to reduce the fuels contributing to the fire or to limit the fuel that may be contained in an exposed process unit or storage tank. Plant operations may also introduce nitrogen or stem where applicable to assist with fire control and extinguishment.

Specialized extinguishing techniques

At industrial facilities, in addition to the structure and contents fire, fire departments may encounter a variety of fire situations. These may include a pressure-fed fire, flowing fuel fire, and in-depth fuel fire. For these fire conditions, a primary strategic objective would be to reduce the available fuel by isolating the fuel source where possible. In the case of the pressure fire, isolating the fuel source is a must. This isolation may be done remotely by process control operators, or it may have to be performed by firefighting personnel closing a manual valve under the protection of hose streams (fig. 7–12). At times, hose streams must be used to push the fire away from the valve that is to be isolated. These techniques must be practiced prior to the fire. Municipal departments with industrial facilities should send personnel to industrial fire schools to learn and practice these techniques. For an in-depth fire, fuel may have to be pumped off by plant personnel.

Fig. 7–12. **Firefighting personnel isolating a control valve under the protection of hose streams** *(Courtesy of TEEX, Texas A & M System)*

Fuel spills and in-depth fuel fires will require the use of foam application. At times, high-volume foam streams will be required. Has this been identified in preincident response planning, and have adequate resources been allocated? Exposure cooling will also be required. When large volumes of foam are required, the use of 5-gallon containers or 55-gallon drums for the supply of foam concentrate will not be adequate. Larger delivery methods such as *foam totes* or *foam tenders* will be required. The time to practice the setup and use of high-volume foam delivery systems is before the fire, not during.

Flowing fuel fires (three-dimensional fires that include pressure and jet fires) require that the fuel be stopped or that an extinguishing medium such as dry-chemical be used. Foam can be used to blanket the fuel that is already on the ground, but the flowing fuel must be extinguished at its source. Dry-chemical technology in conjunction with foam and cooling will extinguish these types of fires. Hydro-chem technology whereby a single nozzle can be used to deliver water or foam, and dry-chemical is excellent for this type of fire.

At times, the best control technique may be a controlled burn. Extinguishing a gas or flammable liquid fire may create more of a hazard if the gas or liquid cannot be stopped immediately. Shutting off fuel supplies and letting the fire burn itself out may be the best course of action. Exposure cooling and environmental concerns must be considered.

When flammable or explosive vapors are present, fog streams can be used for vapor dispersion. The use of unstaffed monitors is recommended. Fire departments should practice the rapid deployment of these monitors to limit the exposure time of personnel during incidents.

LCES

(LOOKOUTS, COMMUNICATIONS, ESCAPE ROUTES, AND SAFETY ZONES)

The wildfire acronym LCES should be adopted for industrial firefighting. Lookouts must be experienced and must be able to see the fire and the firefighters, and they must be able to recognize the risk to firefighters. They should be posted in strategic areas to ensure that the fire will not spread behind involved firefighters through drainage pits, pipe ways, sewage piping, or other means. These lookouts should also be vigilant in checking that overhead structures are not weakening and possibly beginning to collapse.

Communications must be maintained with all personnel operating on the scene as well as plant operations personnel. A plant operations person with a radio should be at the command post to ensure that the IC has radio communications with the plant's control room, and that all actions by the fire department and the plant operators are coordinated to prevent mishaps. Radios must be intrinsically safe for use in certain industrial settings. Personnel operating in remote locations must be kept informed of any change to operational tactics. Lookouts must maintain communications with operating personnel, also.

As discussed previously, escape routes should be established, and all personnel informed of these routes during safety briefings. Apparatus should be positioned facing these escape routes. Two escape routes should be identified and must lead to safety zones.

Safety zones must be established. In case of emergency, a safe area of refuge should be designated that personnel can report to and accountability established. This should be upwind and uphill of the fire area. A clear evacuation route should be designated from this safe area so that personnel can be further evacuated from the safety zone if conditions deteriorate to a level that makes this area unsafe.

CONCLUSION

Industrial fires present additional problems and considerations that municipal firefighters are not normally accustomed to dealing with. Incident command must be established and size-up performed, but with the additional considerations outlined in this chapter. The use of on-scene plant operations personnel for advice is a must. They should be consulted early and often to ensure that the fire department's actions do not create a larger problem. In addition, the standard tactical objectives such as rescue, exposure protection, confinement, extinguishment, and overhaul must be performed, with particular attention to the additional hazards involved.

The key to any firefighting operation is preincident response planning. Identify the process and hazards involved, specialized fire tactics that may be required, and resources available before the incident happens. Remember the acronym COAL TWAS WEALTHS. It will help with your preincident response planning prior to any incident as well as assisting with your size-up in the event of an incident.

NOTES

[1] Terpak, M. A. 2002(August). Size-Up: Updating an Old Acronym. *Fire Engineering*, pp. 69–80.

[2] National Fire Protection Association (NFPA). 2003. *Fire Protection Handbook,* 19th ed., vol. I. Quincy, MA: NFPA, pp. 7–87.

8

Fixed and Semifixed Suppression Systems and Water Supplies

INTRODUCTION

Fixed and semifixed fire protection systems are used extensively in industry to protect critical equipment and processes and to mitigate damage from fires should they occur. These systems can be designed to protect tanks and vessels, conveyor belts, structural members, transformers, gas turbine generators, pipes and pipe racks, air-fin coolers, aircraft hangers, truck and rail loading racks, and combustible solids, to name a few. In addition, systems can be used for dust and vapor reduction or exposure protection. Gaseous suppression systems are also fixed systems that operate by replacing the flammable vapors with inert or nonflammable gases, cooling the vapors, and/or interrupting the chemical chain reaction. Dry-chemical fixed suppression systems operate in the same manner as gaseous systems and work by interrupting the chemical chain reaction. Fixed systems may be designed for local application discharge (for the envelope around a specific hazard such as a pump) or as a total flood system. There are too many specific applications for individual hazards for fixed and semifixed systems to cover in this chapter, but most types will be discussed sufficiently for you to understand how they protect a hazard in your response area. In addition, water supplies will be discussed because they are critical to support water based fixed systems.

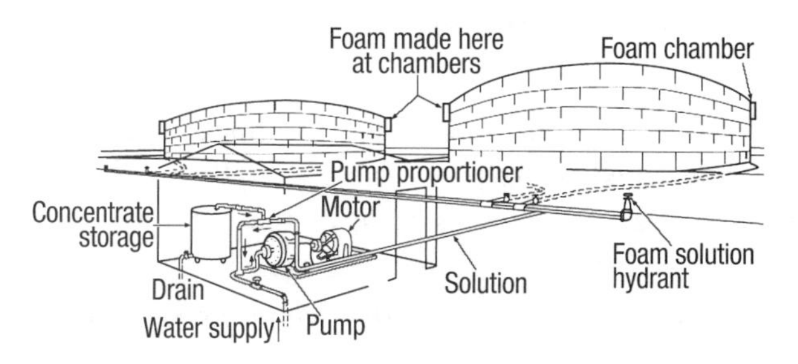

***Fig. 8–1.* Fixed system overview** *(Reprinted with permission from the* Fire Protection Handbook, *19th Edition, Copyright © 2003, National Fire Protection Association, Quincy, MA 02169)*

WATER- AND FOAM-BASED SYSTEMS

This chapter describes some of the common types of fixed and semifixed water and foam-based suppression systems that you may encounter in an industrial facility. Typically, these systems protect against hazards such as storage tanks, pumps, critical valves and pipes, and vertical and horizontal pressure vessels. Water- and foam-based fixed systems may be designed to be either manually operated or automatically initiated through devices such as smoke, heat, or flame detectors. The difference between a water-based fixed system and a water-based semifixed fire suppression system is that a fixed system has all the components to produce water or foam discharges without the aid of a fire vehicle (fig. 8–1). A semifixed system is required to be supported by a fire vehicle (fig. 8–2). These systems could be water or foam systems.

Fig. 8–2. **Semifixed foam connection** *(Courtesy of National Foam)*

Manual operation or actuation could mean several things. There could be a detection system that would sound an alarm, but plant personnel would be required to manually start the system. Manual actuation could be in a remote location (with plant operators' panel in a control room or security office), operated through a manual pull station or directly on the valve itself. Further manual activation could mean that an operator has to physically open a valve to start the water or foam flow.

CLOSED HEAD SPRINKLER SYSTEMS

One type of fixed fire suppression system that you are probably already familiar with is a closed head sprinkler system. Closed head sprinkler systems installed in industrial applications are very similar in operation to a typical sprinkler system (fig. 8–3). The difference between a typical sprinkler system that you would find in a municipal structure and one protecting an industrial hazard could be in the heat at which the sprinkler head is designed to operate, a protective covering on the head, or higher flow rates; and possibly, it could be a dry pipe system to prevent freezing. High-temperature sprinkler heads are needed in some installations, such as paint ovens and industrial furnaces, because the processes that they are protecting produce extreme temperatures. The flow rates on many types of sprinkler heads protecting industrial hazards exceed the flow rates that you may encounter in common structures. The heads may also be coated with a wax type substance designed to protect the head from corrosive environments, and this wax substance will melt off at a specific temperature. These specialized closed head nozzles operate under same principle by the melting of a fusible link or bursting of a bulb. Water, and sometimes foam, flow from the affected nozzles to extinguish or control a fire.

Fig. 8–3. Examples of closed head sprinklers
(Reprinted with permission from the Fire Protection Handbook, *19th Edition, Copyright © 2003, National Fire Protection Association, Quincy, MA 02169)*

Finally, some closed head sprinkler systems are designed as dry pipe systems (fig. 8–4). These systems are dry downstream of the alarm check valve, which is held closed by air or nitrogen. The valve may be opened by activation of the sprinkler heads and subsequent release of the air or nitrogen holding the alarm check valve closed. Water replaces the air or nitrogen in the piping and flows from the nozzles.

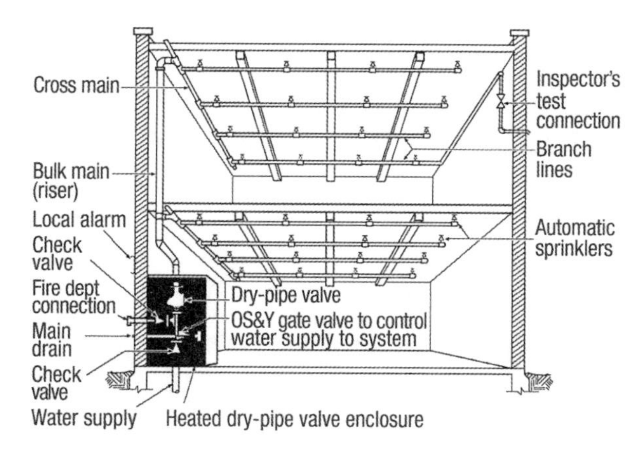

Fig. 8–4. Example of dry pipe sprinkler system
(Reprinted with permission from the Fire Protection Handbook, *19th Edition, Copyright © 2003, National Fire Protection Association, Quincy, MA 02169)*

Similar to the dry pipe system is the preaction system (fig. 8–5). This is a dry pipe system with a detection system that operates the alarm check valve. Once the detection system operates, it allows water to flow into the sprinkler system piping. The sprinkler heads will still have to operate to release water onto the fire. As the air or nitrogen is released from the piping, water fills the pipes and discharges out of any of the activated sprinkler heads. Some systems also have an interlock system that requires the associated detection system to open the alarm check valve to release the system. All types of dry systems take significantly longer to actuate than wet systems because air must be exhausted from the piping.

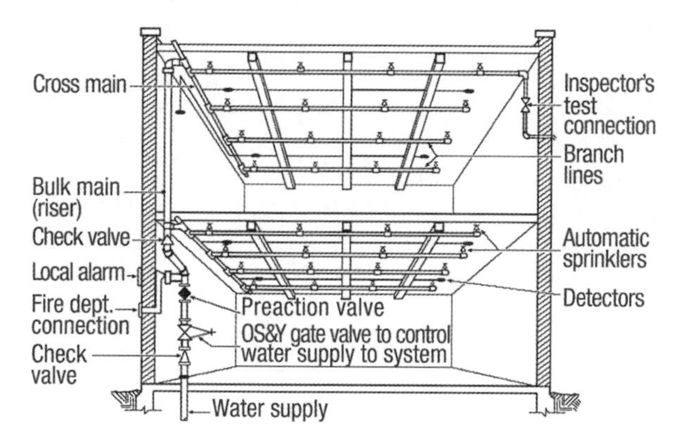

***Fig. 8–5.* Example of preaction sprinkler system**
(Reprinted with permission from the Fire Protection Handbook, *19th Edition, Copyright © 2003, National Fire Protection Association, Quincy, MA 02169)*

DELUGE SYSTEMS

A deluge system (water or foam) is another type of fixed or semifixed system. Deluge systems operate differently than the typical closed head sprinkler systems (fig. 8–6). In a deluge system, all heads are open (fig. 8–7). For the automatic fixed systems, flow is controlled by a deluge valve that normally activates by an automatic detection system. Automatic deluge systems also have manual releases on the deluge valve and in many cases are activated electrically through a pull station. Many aircraft hangars are protected by foam deluge systems. These systems are used extensively in other industrial applications (fig. 8–8).

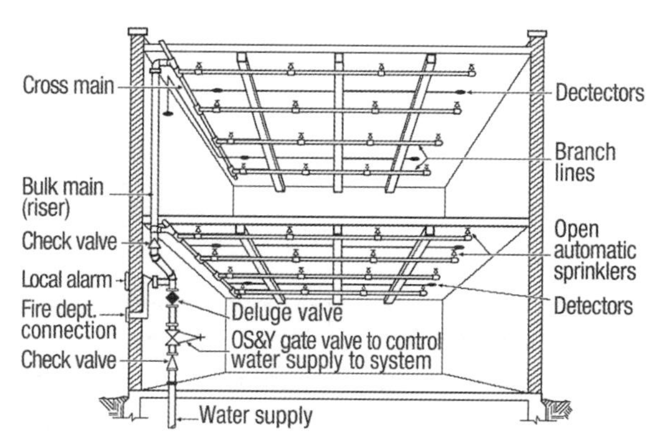

***Fig. 8–6.* Example of deluge sprinkler system** *(Reprinted with permission from the* Fire Protection Handbook, *19th Edition, Copyright © 2003, National Fire Protection Association, Quincy, MA 02169)*

***Fig. 8–7.* Examples of water spray heads**

Fig. 8–8. **Foam deluge system installed on a loading rack** *(Courtesy of National Foam)*

By definition, a deluge system should have a deluge valve; however, some deluge systems require manual activation by an outside screw and yoke (OS&Y) valve or by supply from a fire department connection. The open head deluge system is closely related to the water spray system.

WATER-BASED SPRAY SYSTEMS

Water-based spray systems are used to protect exposures, control the burning, or extinguish fires on ignitable liquids with a high flash point. They can also be used to protect high-voltage electrical equipment such as transformers. Spray systems are equipped with special spray nozzles that control discharge variables such as application rate, particle size, and spray angle. These systems are commonly installed to protect flammable gas storage vessels where you do not want to extinguish the fire. Extinguishment of a gas fire by the system would not be desirable before the flow of gas can be shut off. The water spray system is used to control the situation until the flammable gas can be shut down, and the fire is extinguished from lack of fuel. Other spray systems such as foam systems are designed to extinguish fire or control ground fire. Water-based spray systems are designed to provide a specific application density for the surface area of a given hazard, such as a tank or vessel. Densities could range from 0.10 gpm/ft^2 for metal decking, to 0.25 gpm/ft^2 for pressure vessels, and 0.50 gpm/ft^2 for the envelope around a compressor. In some cases, a water spray system could be installed to extinguish fire in flammable solids, high-flash-point flammable liquids, or water-soluble flammable liquids. As with deluge systems, these systems normally have a deluge valve but may be a semifixed design that requires support from a fire vehicle.[1]

STORAGE TANK PROTECTION

An atmospheric storage tank is one that contains either a refined product such as gasoline or an unrefined product such as crude oil. The reference to atmospheric just means that it is not under pressure for storage. The main pressure in an atmospheric storage tank comes from the head pressure created from the elevation of the liquid being stored and somewhat from the liquid's vapor pressure. The most common type of system encountered on an atmospheric storage tank is rim seal protection installed on either an open top or covered floating roof tank. A rim seal system is designed to protect the most vulnerable area of an open top floating roof tank—the rim seal (fig. 8–9).

Fig. 8–9. **Rim seal fire** *(Courtesy of Williams Fire and Hazard Control)*

The rim seal will have a 12- to 24-inch foam dam that is designed to hold the foam in the rim seal area and extinguish a fire along the seal. These fires are most likely caused by lightning strikes.

Rim seal protection systems can be either fixed or semifixed. The fixed systems could be automatically actuated but will most likely be manually actuated. The manual actuation procedures could include starting the foam pump and either opening an OS&Y, a post indicating valve (PIV), or a deluge valve. As mentioned earlier, a semifixed system requires a fire vehicle to connect to a lateral terminal connection and supply the foam chambers with foam solution at a specific pressure. Common operating pressures for foam chambers are 30 to 100 psi at the inlet to the foam chamber. This means that the fire apparatus will have to operate at sufficient pressure to supply the foam chambers with the required inlet pressure, and it will have to overcome the pipe friction loss and elevation loss (fig. 8–10). Foam chamber manifold inlet pressures should be established during preincident response planning. Depending on the product most of low-expansion Class B firefighting foam is effective on a rim seal fire when discharged through foam chambers. As stated, foam chambers are most commonly installed for rim seal protection but could be installed to extinguish full surface fires in both floating roof and covered storage tanks (fig. 8–11).

Fig. 8–10. **Floating roof foam maker discharge** *(Courtesy of Ansul)*

Fig. 8–11. **Air-foam chamber installed on a floating roof tank** *(Courtesy of National Foam)*

Another type of water-based system that may be encountered on a covered roof atmospheric storage tank is subsurface injection (fig. 8–12). Subsurface injection cannot be used on floating roof tanks because the floating roof will keep the foam from reaching the surface of the fuel, even if the roof should partially sink. A specific type of foam maker, a high back pressure foam maker, is required for these systems. The high back pressure foam maker will be installed on a dedicated pipe downstream of the fire department connection. It is designed to aspirate the foam as it passes the foam maker. As indicated in the system description, the foam will be discharged as finished foam into the fuel. The foam must have a high tolerance to the fuel so that it will rise to the surface of the fuel. Some aqueous film-forming foams (AFFF) have low fuel tolerance and will not effectively rise to the surface. Subsurface injection systems could be either fixed or semifixed in design. In some cases, a portable high back pressure foam maker may be installed on a product line for the injection of foam. Remember that foam must be injected above the water line in tanks. Foam injected below the water line will dissolve into the water and not rise to the surface. Water is common, especially in crude oil tanks, so be sure to ask the plant personnel what the water level is before you select a product line to use for subsurface injection or use the semifixed system.

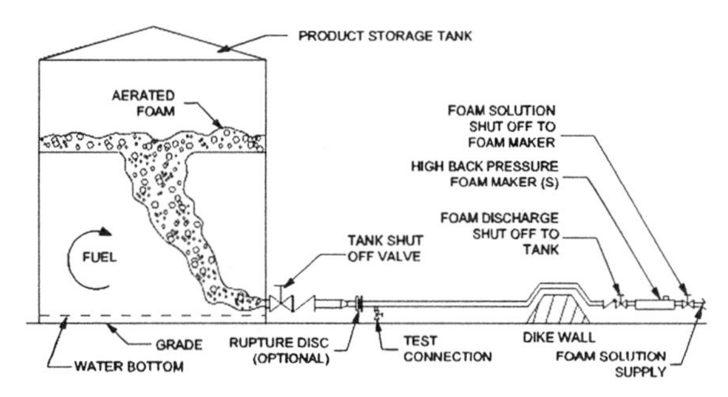

Fig. 8–12. **Subsurface foam system overview** *(Courtesy of National Foam)*

A simpler type of fixed or semifixed system commonly installed on atmospheric storage tanks is a tank cooling system. These cooling rings are designed to discharge water onto the exterior of a tank and protect it from the radiant heat being generated from a fire in close proximity. Many times these cooling systems will have to be manually actuated.

Fig. 8–13. **Water spray system protection for LPG sphere**

GAS STORAGE SPHERE AND SPHEROID PROTECTION

A variation of the deluge system is used to protect spheres, spheroids, and some vertical vessels. These systems are designed to discharge water onto the top of the vessel at a specified flow rate (fig. 8–13). Water runs down all sides of the vessel and cools the tank surfaces. Some systems are designed with an overflow *weir*, or crown, which allows water to run over the entire surface of the sphere (fig. 8–14). This type of system typically has a fire department connection and may also be controlled by a deluge valve or a manual OS&Y valve. These systems could also be fixed or semifixed in design.

Fig. 8–14. **Water spray system on LPG sphere** *(Courtesy of Williams Fire and Hazard Control)*

Fixed monitor nozzles may also be present to protect various types of hazards (fig. 8–15). They are typically used for exposure protection in industrial areas to protect tanks, as well as process and storage areas. These monitors may be manually or remotely operated. They could also be connected to a foam skid. Depending on the design, there may be a fire department connection. These monitors could also be connected to a deluge valve (such as inside of an aircraft hangar) and automatically oscillate in a predetermined pattern.

***Fig. 8–15.* Monitor nozzle to protect LPG sphere**
(Courtesy of Williams Fire and Hazard Control)

Water injection can also be used on liquefied petroleum gas (LPG) vessels to extinguish fires by displacing the LPG, because LPG is less dense than water. Thus, water can be pumped into an LPG tank (either above or below the liquid product level) until it exceeds the level of the LPG leak. Typically, there will be a fire department connection, check valve, and a block valve such as an OS&Y valve. To start the process, you need to know the pressure of the LPG inside the tank. The pressure supplied to the tank (after friction and elevation losses for piping and hose) must exceed the pressure of the LPG. Once the water pressure exceeds the pressure of the LPG, the control valve can be opened. Water should be pumped in at a steady rate until it discharges out of the leak. Once water is flowing out of the leak, you should balance the amount of water being put into the tank with the amount of water being discharged from the leak.

LOW-EXPANSION FOAM SYSTEMS

There are several types of common foam-generation systems installed with fixed systems. These systems include a balanced pressure (pump type) (fig. 8–16), in-line balanced pressure (pump type) (fig. 8–17), balanced pressure bladder tank (fig. 8–18), and pressure tank proportioning system. All of these systems operate on a balanced pressure theory where foam concentrate is supplied to the *ratio controller* (RC) at the same pressure as the water pressure. Foam is educted into the RC at the proper proportion through a fixed orifice by the slight negative pressure created by the *venturi* effect of the water as it passes through the RC. As indicated by the name, the pump type of foam-proportioning systems have foam pump(s) mounted on a skid. This pump supplies foam concentrate at a pressure greater than the water pressure. Through a balancing valve, the foam concentrate pressure is lowered to equal the water pressure and supplied to the RC (sometimes called the proportioner). This balancing can take place on the deluge valve risers (in-line balanced proportioning system) or at one injection point (balanced pressure pump type).

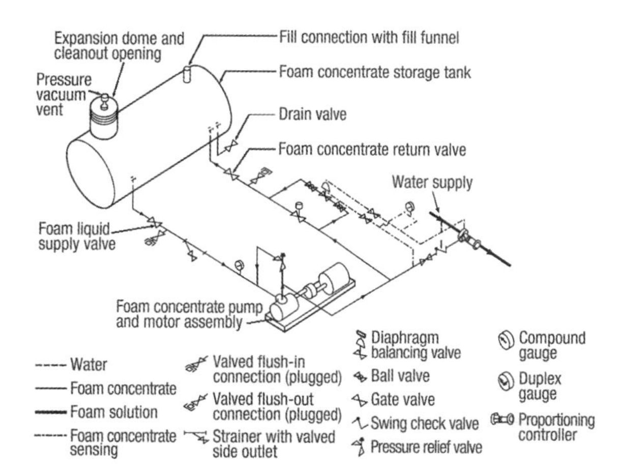

***Fig. 8–16.* Balanced pressure (pump type)** *(Reprinted with permission from the* Fire Protection Handbook, *19th Edition, Copyright © 2003, National Fire Protection Association, Quincy, MA 02169)*

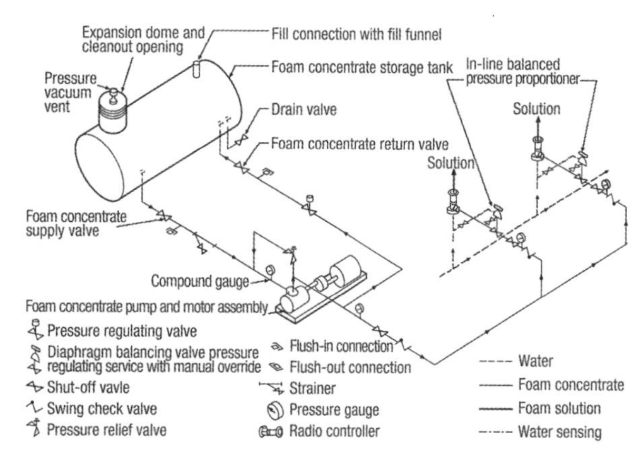

***Fig. 8–17.* In-line balanced pressure (pump type)**
(Reprinted with permission from the Fire Protection Handbook, *19th Edition, Copyright © 2003, National Fire Protection Association, Quincy, MA 02169)*

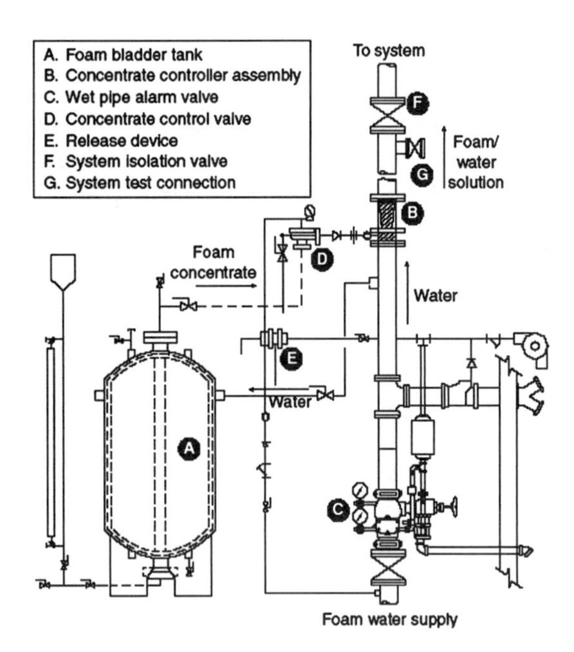

Fig. 8–18. **Balanced pressure bladder tank** *(Reprinted with permission from the* Fire Protection Handbook, *19th Edition, Copyright © 2003, National Fire Protection Association, Quincy, MA 02169)*

In the pressure tank and bladder tank system, the water from the fire-water system is used to expel the foam concentrate from the storage tank. The foam concentrate is expelled at the same pressure as the water, and foam is again supplied to the RC at the same pressure as the water. These systems are fairly simple and reliable if the valves are left in the ready position before the emergency.

HIGH-EXPANSION FOAM SYSTEMS

High-expansion foam systems are used to extinguish Class A and B type fires. Many storage occupancies and aircraft hangers can be protected by high-expansion foam systems. These systems are typically supplied by a foam skid system with a pump or a bladder tank system, and they are mechanically similar to the low-expansion foam systems. The difference is the high expansion foam makers are most commonly run by water motors operating off the foam solution line. High-expansion foam systems are most commonly connected to a detection system for automatic actuation. You will recognize these systems by the high-expansion foam generators mounted in the hazard area (fig. 8–19). These generators require large amounts of air and operate like large fans (fig. 8–20). Foam solution is sprayed against a mesh screen, and air from the fans expands the foam (fig. 8–21). Efficiently expanded finished foam may reach a 1,000-to-1 expansion ratio. Support for these systems may only require manual activation of the system.

Fig. 8–19. **High-expansion foam generator** *(Courtesy of Ansul)*

Fig. 8–20. **High-expansion foam generator** *(Courtesy of Ansul)*

Fig. 8–21. **High-expansion foam discharge in aircraft hangar** *(Courtesy of National Foam)*

SUPPORTING WATER-BASED SYSTEMS

For fixed systems to operate properly, they must have minimum pressure at the base of the riser or inlet manifold. These systems were designed to deliver a specific amount of water or foam to the protected hazard. Most spray and foam nozzles must have 30 psig at all nozzle inlets to discharge the required flow. Failure to maintain minimum pressure on the system will reduce the flow below the design density and possibly cause the system to fail. Many of these systems have fire department connections, but some do not. Some deluge systems, for example, operate at such a high capacity that a pumping apparatus could not supply it.

There are several other types of systems that are difficult for municipal pumping apparatus to support. Early-suppression fast-response (ESFR) closed head sprinkler and in-rack sprinkler systems may require flows as high as 3,000 gpm. These high-flow systems are designed to control fires in high rack storage and warehouses. The best way to support these systems is to ensure that all required valves are open, and any dedicated fire water pumps are operating.

PREINCIDENT RESPONSE PLANNING

It is essential that preincident response planning takes into account the pressure required on the system. A design plate is required to be mounted on the system to indicate minimum pressure required to support the system. If the design plate is missing, have the plant engineer access the design drawings and technical specifications to determine design pressures and flows. If the fixed system is activated, be sure that you do not draw the water pressure down below the required system support pressure. One of the first things to check when responding to a water-based fixed system is the water supply to the system. Ensure that the PIV and OS&Y valves are fully open. Closure of valves is sometimes accidental but could have been done maliciously to prevent the system from operating. It may be as easy as opening a PIV to allow the system to operate correctly.

SUPPORTING FOAM SYSTEMS

Support of foam-suppression systems is more complicated and may not be possible with standard municipal equipment. Most municipal pumping apparatus carry only a small amount of Class B foam for common nonindustrial hazards. The type of foam is also critical. For example, if conventional AFFF is used on a water-miscible fuel (e.g., alcohol), the foam will simply dissolve into the fuel. The type of foam must be appropriate to the fuel.

What about mixing various types of foams during application? Several factors must be considered. NFPA 11, *Standard for Low-, Medium- and High-Expansion Foam,* and other references state that foam concentrates from different manufacturers and of different types cannot be mixed. You would never mix a fluoroprotein (FP) and an AFFF concentrate in a tank before proportioning into a system. You could, however, support foam chambers with AFFF solution if they were protected with a fixed FP foam system that had exhausted its foam or if the foam system failed.[2]

There are several factors to consider that are unique to foam systems. Most foam chambers, fixed nozzles, and monitors are compatible with AFFF. However; if a discharge device was designed for AFFF, it may not be able to successfully aspirate foams requiring special aspirating nozzles. Points to remember when supporting foam systems are that the type of foam must be compatible with the hazard, that you have enough foam to sustain a discharge, and that the foam discharge devices are appropriate to the type of foam on your vehicle. Sometimes firefighters may have to manually start foam pumps, open foam valves, or balance foam and water pressures on the balancing valves. Foam may also be available on-site for emergency use. The limitations of municipal pumping apparatus will be discussed next.

INDUSTRIAL WATER SUPPLIES

Industrial water supplies are often much different from municipal water supplies. Typically, industrial water supplies are designed to supply large quantities of water at high pressures. A properly designed system will supply the fixed or semifixed systems, as well as utility and manual firefighting requirements. The municipal department will have to be prepared to take advantage of these large water supplies and work with them safely.

Many industrial fire-water systems are independent of city or county water systems. Industrial facilities can have millions of gallons of water storage, complete with pumps and piping to supply their fire-water system. Water supplies can also come from elevated and ground storage tanks, ponds used to collect reclaimed water, natural reservoirs (ponds and lakes), rivers, or saltwater sources. There also may be an emergency crossover from the municipal water system.

INDUSTRIAL FIRE-WATER SYSTEM DESIGN

In most industrial fire-water systems, pumps will be actuated by a pressure drop in the system. As the fire-water demand increases, additional pumps will come online to support the operation. It must be noted that even though fire pumps should start automatically, they may fail to start or be positioned in the manual setting. Plant personnel sometimes place the pumps in the manual setting because of a malfunction of the automatic controls or to prevent inadvertent pump starts. Fire officers should ensure that plant personnel operate and monitor the operation of these pumps. In some modern plants, the pumps can be monitored in the control rooms. Sophisticated systems can start and stop pumps, monitor system pressures and flows, and even control the discharge of the pressure relief valves. It is recommended that plant personnel respond to the fire-water pump house to ensure that the pumps are operating properly in support of systems and apparatus. Plant personnel are usually very familiar with the fire-water pumps, associated instrumentation, and power supplies. Adequate water supply is critical to control or extinguish any industrial fire, and its failure can quickly cause the situation to deteriorate. Lack of water could lead to a boiling liquid expanding vapor explosion (BLEVE), rapid spread of a surface fire, or failure of lines deployed to protect firefighters.

Industrial water supplies will support large water spray or foam systems intended to protect vessels, tanks, piping, and key equipment. Many of these suppression and exposure systems can deliver more than 3,000 gpm. The water supply may operate at greater than 150 psi. If your department uses low-pressure supply hose, you will have to ensure that either a hydrant or an intake relief valve is used to safely lower the pressures to usable pressures. The apparatus operators will have to be trained on how to properly manage the high pressures. Dangerous downstream pressures can easily be created if pressure management systems are not properly used. Without pressure relief systems, pressures downstream of the apparatus could easily exceed 250 psi while the apparatus is at idle. In addition to the theoretical training, all operators must receive training in high-pressure industrial fire-water systems.

Fire hydrants and manifolds

Industrial fire hydrants are designed with 6- and 8-inch risers. They have multiple outlets and possibly more than one large-diameter connection. Most plants use dry barrel hydrants that are not as susceptible (as wet barrel hydrants) to freezing, damage by vehicles, or explosions. Water supply could also come from wall and yard hydrants, manifolds, or headers designed to be used by the fire department (figs. 8–22 and 8–23). The connections could be national standard thread (NST) or national standard (NS), Storz, or, in some cases, a bolted flange. Ensure that your department vehicles are compatible with the industrial systems that you must respond to.

***Fig. 8–22.* Manifold** *(Reprinted with permission from the* Fire Protection Handbook, *19th Edition, Copyright © 2003, National Fire Protection Association, Quincy, MA 02169)*

***Fig. 8–23.* Fire hydrant**

As mentioned earlier, the capacity of the fire-water system should be designed to deliver sufficient water to supply the fixed and semifixed systems, utility water, and manual firefighting operations. The piping could be 20 inches in diameter or greater and will provide large quantities of water at very low pressure losses in the piping and hydrants. Most systems are designed so that if a catastrophic event occurs, such as an explosion, the affected areas can be isolated and the remaining system will be usable. Underground valves will have to be located and closed to restore pressure to the rest of the system. Because of the complexity of these incidents, you will have to work closely with plant personnel to effectively bring the firewater system back online. Hydrants should be color coded in accordance with NFPA 291, *Recommended Practice for Fire Flow Testing and Marking of Hydrants,* to indicate the water available from each hydrant. Remember that a Class AA, light blue industrial hydrant will probably supply much greater than 1,500 gpm. It is critical that industrial fire-water systems be tested regularly to determine what flows are available from the hydrants. Industrial system tests are different from municipal water tests because multiple hydrants will have to be opened to drop the pressure low enough to operate all pumps and get true results. Flow capabilities should be indicated on preincident response plans for all areas.[3]

Fire water pumps

Fire pumps should be installed in accordance with NFPA 20, *Standard for Installation of Stationary Pumps for Fire Protection.* As discussed earlier, fire-water pumps should be arranged to actuate automatically on pressure drop or fixed-system actuation. When pumps are started by pressure drop, each pump is required to operate even in the event of failure of the others. Fire pumps will be rated at their 100% and 150% capacity. For example, a fire pump rated at 3,000 gpm at 150 psig will actually provide 4,500 gpm at approximately 65% of its 100% rated pressure. The pump in the example will actually provide 4,500 gpm at 97.5 psig. [4]

Normally, multiple pumps are provided. Using the pump in this example, if a system was provided with three 3,000 gpm pumps, then the system would be capable of supplying 13,500 gpm at approximately 97.5 psig. In fact, these pumps will provide additional water down to 20 psig. As explained later, it may be hazardous to drop the fire-water system to low pressures if the system is supporting a fixed system.

The pumps could have electric motor or diesel engine drivers. In many cases, the fire-water system will have emergency generators or a secondary power source to run the electric-motor-driven fire pumps. In other cases, diesel engine fire pumps will provide water pressure in the event of a power outage.

GASEOUS, WATER MIST, AND DRY-CHEMICAL FIXED SYSTEMS

Other types of fixed systems include halon, carbon dioxide, FM-200®, and various other clean agents such as inerting agents. Examples of systems you may encounter in industrial facilities include FM-200 in computers and telecommunication facilities, marine applications, petrochemical, and other locations that may be normally staffed. High-pressure water mist systems may be found in applications such as gas turbines and machinery spaces. Carbon dioxide and local application systems are used extensively in areas and spaces that are not normally staffed. These applications include power generation, metal production, and processing, for example, automotive and marine systems.

Clean agent systems

Clean agent systems such as FM-200 (HFC-227) are covered in NFPA 2001, *Standard on Clean Agent Extinguishing Systems.* In addition to numerous HFCs (halocarbons), such as FM-200 and Sapphire™, there are inerting agents such as nitrogen and argon (fig. 8–24). These systems are typically connected to a detection and automatic activation system (fig. 8–25). The firefighter must be aware of these systems and that they are designed to protect specific hazards. Manual activation of these systems may be necessary through pull stations, remotely from a control room, or by mechanically depressing a lever on the cylinders. In some cases, operators have intentionally placed these systems in a manual operation mode. In other cases, the automatic mode may have failed or been accidentally left in the manual mode. These systems will commonly have a discharge delay and alarm to ensure that personnel can evacuate the area prior to gas discharge. In other cases, where agent discharge delay would allow the fire to rapidly grow, the discharge delay will not be used. Some clean agent and other gaseous systems designs allow for an abort switch. Abort switches require constant pressure to keep them activated. Releasing the abort switch will continue the discharge timer. The other gaseous systems operate in a similar manner to the clean agent systems.[5]

Fig. 8–24. **Sapphire clean agent *(Courtesy of Ansul)***

Fig. 8–25. **Sapphire clean agent and control panel**
(Courtesy of Ansul)

Carbon dioxide systems

Carbon dioxide (CO_2) systems are common in many industrial occupancies. NFPA 12, *Standard on Carbon Dioxide Extinguishing Systems,* covers these systems. Flooding CO_2 systems may only be used in normally nonoccupied spaces such as gas turbines. Carbon dioxide is a colorless and odorless gas that is 1.5 times heavier than air. It acts by displacing the oxygen in the space and somewhat by cooling. Design concentrations are above 34%, depending on the hazard. Since it is heavier than air, it collects in low areas. Carbon dioxide may be stored in high-pressure cylinders (fig. 8–26) or in low-pressure storage tanks. Low-pressure system components consist of a large insulated storage tank and refrigeration unit to keep the CO_2 cool (fig. 8–27). In addition to the flooding system, there may also be CO_2 hose reels and nozzles that protect specific hazards such as pumps. System activation may be through automatic, manual electric release or direct manual release from the pilot cylinder means.[6]

Fig. 8–26. **High-pressure CO_2 system** *(Courtesy of Ansul)*

Fig. 8–27. **Low-pressure CO_2 tank** *(Courtesy of Ansul)*

Fig. 8–28. **Low-pressure CO_2 system at steel plant**

Fig. 8–29. **Dry-chemical skid** *(Courtesy of Ansul)*

Dry-chemical systems

Dry-chemical fixed systems are also commonly installed to protect industrial hazards and are covered by NFPA 17, *Standard for Dry Chemical Extinguishing Systems.* These systems are similar to gaseous systems, and are designed as local application and total flood systems. Dry-chemical agents are effective on both B, C and A, B, C type fires. Agents that are effective on A, B, C fires are based on ammonium phosphate powders. One of the most effective B, C–rated dry-chemical agents is potassium bicarbonate (PKP). In addition to being able to extinguish two-dimensional liquid fuel fires, they are also effective on three-dimensional and flowing fuel fires (fig. 8–29).

Dry-chemical agents are considered nontoxic, but they do present a respiratory irritation and will obstruct vision when applied in nonvented areas. They are nonconductive but will corrode and contaminate electrical, electronic, and other vulnerable components. They can be discharged automatically, manually through an activation station, or mechanically at the fixed system. Large dry-chemical extinguishing systems use expellant gas (normally nitrogen) to push the powder out of the storage cylinder. Although dry-chemical is considered nontoxic, the gases given off by the burning material may be deadly.[7]

Fig. 8–30. **Dry-chemical system on heavy equipment** *(Courtesy of Ansul)*

Fig. 8–31. **Dry-chemical discharge on heavy equipment** *(Courtesy of Ansul)*

Water mist systems

Another type of system that may be encountered is water mist. These systems have become more popular for a number of reasons. Water mist systems are covered in this section because they have been installed in hazard areas that were protected by Halon 1301 in the past. With the phase out of the halon, water mist technology was developed. It is safe for occupied spaces. Water mist is environmentally friendly and can be used on a variety of hazards. These hazards include oil and gas applications, turbines, power generation, manufacturing, marine, and other industrial hazards. Water mist systems work to extinguish and control fire in a number of ways. The ultrasmall droplets of water (100–120 microns) are discharged at approximately 350 psi and are extremely effective in absorbing heat of combustion. The droplets also interfere with the transfer of heat from the flames. Last, the steam and water vapor produced by converting the liquid to gas act to exclude oxygen and smother the fire. Water mist systems are covered in NFPA 750, *Standard on Water Mist Fire Protection Systems.*[8]

Halon systems

Finally, Halon 1301 and 1211 systems have, for the most part, been phased out but generally operate in the same manner as clean agent systems. The halons are, however, more toxic than FM-200 and other HFCs. Various fire suppression agents are under development and seeking listing to replace the halon. FM-200 is only one of many fire clean-agent suppression gases that have come onto the market in an effort to replace highly effective halon agents. While there is no direct replacement, several agents show great promise from their ability to suppress fire and are even less toxic to firefighters.

FIREFIGHTER SUPPORT OF GASEOUS, DRY-CHEMICAL, AND WATER MIST SYSTEMS

As with other types of fixed systems, properly designed and maintained gaseous systems normally work very well to do their job in the event of a fire. If the doors to the room are closed and the discharge control panel has not been put into the bypass position, the system will normally initiate as designed. In addition to closing doors that may have been blocked open or activating the system if the detection system has been bypassed, gaseous systems sometimes have features such as a sustained discharge and may have a second reserve bank of agent. If a fire is still burning after the initial system discharge, sometimes it is possible and beneficial to discharge the second bank of agent. After the discharge, all doors and windows should remain closed while the gasses, dry-chemical, or mist do their work. This will help to extinguish deep-seated fires and allow hot components to cool below their ignition temperatures.

While the agents are doing their work, it is prudent to deploy hoselines or other types of extinguishing means in case the fire is not extinguished by the fixed system. Again, firefighters must wear full personal protective equipment (PPE) and self-contained breathing apparatus (SCBA) when entering these spaces. Extreme caution must be used before allowing firefighters to enter industrial spaces. In some cases, they should never enter the spaces, and in other cases, they must wait while equipment cools and shuts down. In some installations, such as an interior or a building or a nonventable space, exhaust fans may be provided for use after the discharge to clear the space. Adequate ventilation time should be allowed to ensure that personnel are not exposed to hazardous gases.

A note of caution concerning gaseous systems is appropriate. Gaseous systems that are initiated by smoke detection are very sensitive to inadvertent activation. Smoke detectors will initiate because of water vapors and dust particles. Even the act of opening an external door on a humid day and allowing outside air to enter the protected space could and has in many cases activated the gaseous suppression system. In the case of dust activation, rarely entered spaces such as pipe tunnels are subject to inadvertent activation. Firefighters in the investigative mode have inadvertently activated these systems. Most of these systems require two detectors to activate to initiate a discharge. Firefighters may be investigating an alarm activation and accidentally activate a second detector by disturbing dust in the protected space. This is why firefighters must always wear breathing apparatus in the investigation mode when entering spaces protected by fixed systems. Lockout/tagout systems should be used where appropriate.

CONCLUSION

Here are some final notes on fixed and semifixed suppression systems. Over time the occupancy may have changed. A recent fire in a scrap magnesium storage facility proves this point. The facility was designed to store a material that could be extinguished by water sprinklers. After the occupancy change, this was not taken into consideration. When the magnesium caught fire, the sprinkler system activated. The water intensified the fire, and it burned out of control. In addition to occupancy changes, systems could have been poorly designed and installed in the past. Municipalities should carefully scrutinize systems installed in their jurisdiction, create an accurate preincident response plan, and practice that plan. Fixed and semifixed systems in industrial facilities may be somewhat more complicated than the systems protecting municipal properties, but they are still the best method of controlling fires. Operational and properly supported systems are the best tools to keep firefighters out of dangerous situations.

NOTES

[1] National Fire Protection Association (NFPA). 2007. NFPA 15, *Standard for Water Spray Fixed Systems for Fire Protection*. Quincy, MA: NFPA, sect. 6.4.3.

[2] NFPA. 2005. NFPA 11, *Standard Low-, Medium-, and High Expansion Foam,* sect. 4.4.

[3] NFPA. 2002. NFPA 291, *Recommended Practice for Fire Flow Testing and Marking of Hydrants,* chap. 5.

[4] NFPA. 2003. NFPA 20, *Standard for Installation of Stationary Pumps for Fire Protection,* chaps. 6–7.

[5] NFPA. 2004. NFPA 2001, *Standard on Clean Agent Extinguishing Systems,* sect. 4.3.1.

[6] NFPA. 2005. NFPA 12, *Standard on Carbon Dioxide Extinguishing Systems,* sect. 4.5.

[7] NFPA. 2002. NFPA 17, *Standard for Dry Chemical Extinguishing Systems,* sect. A5.6.

[8] NFPA]. 2003. NFPA 750, *Standard on Water Mist Fire Protection Systems,* chap. 4.

9

Fire and Gas Detection Systems

INTRODUCTION

A fire alarm or detection system is a key component in a building's or facility's overall fire protection and life safety features. A properly designed, installed, commissioned, operated, and maintained fire detection system can help reduce fire losses by allowing early notification to occupants and fire departments. *Fire detection systems* are designed to allow for building and facility occupants to be notified prior to a building becoming untenable due to smoke, heat, and gases. This initial notification alerts occupants about what they need to do to evacuate the building or relocate to a safe area. At the same time, depending on the design of the fire detection system, local first responders and fire departments are alerted to a potential fire or other emergency and can respond. Fire detection systems can also be designed to activate or initiate water-based fire protection systems such as deluge systems or clean agent systems such as FM-200 or Intergen. Fire detection systems can also shut down building systems, recall elevators, open security doors and locks, and activate closed circuit television (CCTV) cameras.

Gas detection systems, when activated by the presence of a fugitive gas, can shut down process equipment, activate fans, initiate fixed systems, and notify first responders (fig. 9–1). A gas detection system, used in conjunction with a fire detection system, can provide helpful information about what

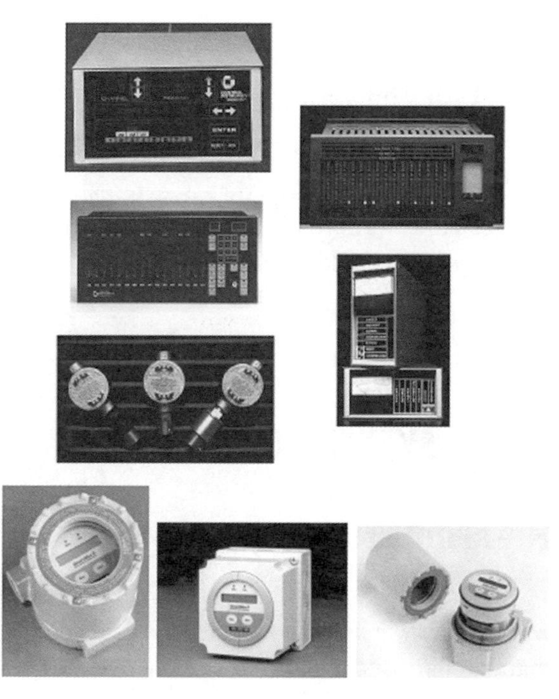

Fig. 9–1. **Gas detection system components** *(Reprinted with permission from the* Fire Protection Handbook, *19th Edition, Copyright © 2003, National Fire Protection Association, Quincy, MA 02169)*

is going on inside an industrial occupancy. Typical gases that might be monitored include hydrogen sulfide (H_2S), carbon monoxide (CO), ammonia (NH_3), chlorine (CL), and a base hydrocarbon such as pentane or hexane (C_5H_{12} or C_6H_{14}).

Fire and gas detection systems are readily available to provide the early detection and warning of a fire, or gas or vapor release. One of the main objectives and purpose of the fire detection system in an industrial occupancy is to warn of possible impending events that threaten life safety, property, and continuity of business. A fire and gas detection system can supplement the process information systems in a plant. In general, process controls located in the control rooms only provide feedback for normal conditions within the process system. The feedback that is usually provided includes the following:

- Process and product temperatures and pressures
- Position of valves
- Product levels
- Process alarms and indicators
- Utility information

Fire and gas detection systems have the ability to feed back information that is external to the process and warn operators and occupants of conditions that could be harmful.

Fig. 9–2. **Fire alarm control panel** *(Reprinted with permission from the* Fire Protection Handbook, *19th Edition, Copyright © 2003, National Fire Protection Association, Quincy, MA 02169)*

FIRE DETECTION SYSTEMS OVERVIEW

Fire alarm circuits

A fire detection system consists of one or more *alarm-initiating circuits* connected to fire detectors, manual alarm boxes or pull stations, water flow devices, and other alarm-initiating devices, for example, a gas detection system. These circuits carry an alarm indication to a fire alarm control panel (FACP), which is powered by a main (primary) power supply and a standby (secondary) power supply (fig. 9–2). Additional components of fire detection systems also include notification appliances such as horns, strobes, bells, evacuation and voice systems, and smoke control or other building management functions (fig. 9–3).

Fig. 9–3. **Fire alarm system components** (*Courtesy of Ansul)*

Fire detection systems have three basic types of circuits: *initiating device circuits* (IDCs), *signaling line circuits* (SLCs), and *notification appliance circuits* (NACs). All or some of these circuits are connected to the FACP and provide detection, notification, and control of equipment in a defined building or facility.

Initiating device circuits. IDCs, also called conventional or nonaddressable circuits, connect alarm and supervisory initiating devices to the FACP. An IDC is a nonintelligent circuit that monitors electrical activity through the circuit by using an end-of-line resistor. When the resistance changes, this signifies an alarm to the FACP. An IDC is defined by the 2002 edition of NFPA 72, *National Fire Alarm Code,* as "a circuit to which automatic or manual initiating devices are connected and where the signal received does not identify the individual device operated."[1] IDCs are connected in groups of initiating devices, called a zone configuration (fig. 9–4). For an example, an IDC might be used in a multistory building to monitor smoke detectors. Each individual floor would be wired into a zone, which would identify only the floor on which the smoke (or heat) detector activated, not the exact device or location. This lack of detector location information could cause a delay in response to the fire's origin if valuable time was spent trying to locate the fire.

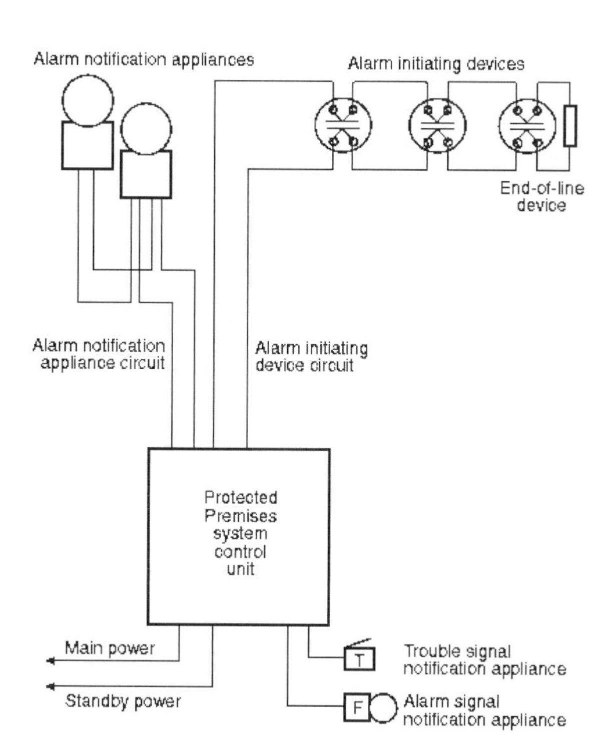

***Fig. 9–4.* IDC overview** *(Reprinted with permission from the* Fire Protection Handbook, *19th Edition, Copyright © 2003, National Fire Protection Association, Quincy, MA 02169)*

Signaling line circuits. The next type of circuit is the SLC, which also connects alarm- and supervisory-initiating devices to the FACP. This type of circuit is called an addressable circuit. A SLC is an intelligent circuit that monitors data being transmitted through the circuit and looks for information feedback from the devices. The devices are also addressable and provide a wealth of information back to the FACP. Many times, SLCs are called smart devices, because they communicate information back to the FACP. Information can include the status of the detector, for example, alarm or normal, and whether the detector is dirty or needs maintenance or replacement. NFPA 72 defines the SLC as a circuit or path between any combination of circuit interfaces, control units, or transmitters over which multiple system input signal or output signal or both are carried (fig. 9–5).[2] In the same example for the IDC, let's look

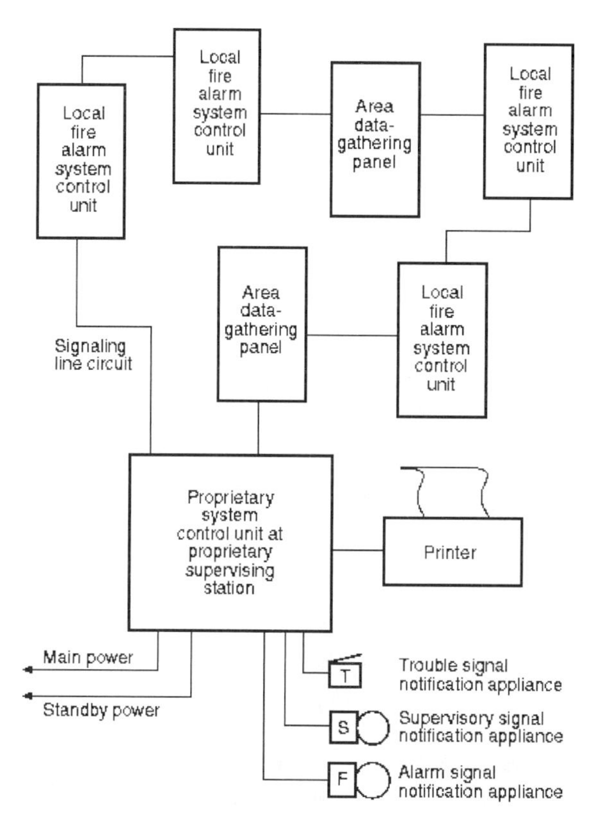

***Fig. 9–5.* SLC overview** *(Reprinted with permission from the* Fire Protection Handbook, *19th Edition, Copyright © 2003, National Fire Protection Association, Quincy, MA 02169)*

at a multistory building using an SLC circuit to monitor smoke detectors. When the detector goes into alarm, the specific address and location are displayed on the FACP, indicating the precise location of the smoke detector (or any other initiating device). This allows first responders and the fire department to respond directly to the area in question and begin looking for the cause of the alarm. Unlike in the previous IDC example, addressable systems could save several valuable minutes that would otherwise be spent looking for the fire's location.

Notification appliance circuits. The third type of circuit in a fire detection system is the NAC. The NAC circuit is a group of appliances whose intended purpose is to alert building and facility occupants of a potential problem in the building. Notification equipment can include audible appliances (such as bells, horns, speakers, or chimes), visible appliances (such as strobes, lamps, or printer), and vibrating appliances or appliances that combine some or all of these different types (fig. 9–6).[3] These requirements are based on *public-mode notification,* in which the occupants of the building receive evacuation signals, or *private-mode notification,* in which the signal is transmitted to a person responsible for taking action or who might notify occupants when to evacuate, such as in a hospital or assisted-care facility.

Fig. 9–6. **NAC equipment** *(Reprinted with permission from the* Fire Protection Handbook, *19th Edition, Copyright © 2003, National Fire Protection Association, Quincy, MA 02169)*

You could also see this in an industrial facility where the access is limited to nonemployees, and evacuation is not the primary objective. Typically, a site siren or warning system is used in these types of occupancies, where evacuation is not an immediate priority. The main feature of this type of system is the ability to notify the occupants of the entire plant or large sections of the plant. Even if personnel cannot escape from an area involved in incident, they are at least notified and aware that an incident has occurred. Consideration should be given to ensure adequate means of escape from all buildings, process areas, elevated structures, and offshore installations. Alarms should be heard or seen in all areas of the plant, irrespective of distance or conditions. The way to determine the number and location of both audible and visual devices is to distribute them strategically throughout the plant versus placing them in a large central area only. These alarms should be initiated by the local or main control room for the location involved in the incident. There should be a different sound, or variation in the audible and visual alarm, to differentiate between a gas or vapor leak, fire, and other predetermined incident type. Every person, either regular employees or visitors, should be aware of the different types of alarms to avoid conclusion during a real incident.

INITIATING DEVICES

Fire detection devices

The most common type of initiating device used on an IDC or SLC is a fire detector. These devices initiate several processes related to detecting fires and monitoring functions. They can activate sprinkler systems, monitor room conditions, shut down or recall elevators, or turn off electrical equipment. Fire detectors are designed to detect and identify fire signatures. These characteristics consist of the by-products of the combustion process, smoke, gases, light, and heat. The most common fire detectors used are ones that detect heat or smoke. Other detectors designed to identify other fire signatures include gas, cold temperatures, pressure, radiant heat, and carbon monoxide. The most common signatures are smoke, thermal release, and energy release.

Selection of the detector is based on the detector's capability to search for and identify the anticipated or expected fire signature in the compartment in which it is installed. These detector types include the following (figs. 9–7 and 9–8):

- Smoke detectors—smoke signatures, particle sampling, and air aspiration
- Heat detectors—fire (thermal) signatures including fixed-temperature and rate of rise- and line-type detectors
- Ultraviolet/infrared flame detectors (UV/IR)—flame (radiant energy) signatures
- Pressure sensors/detectors—detect and measure pressure variations or buildup in tight enclosures
- Gas-sensing detectors—relative changes in the level of gas concentration in a compartment
- Carbon monoxide detectors—measure and alert to unsafe levels of carbon monoxide in a compartment or area
- Cold sensors—detect extremely cold temperatures indicative of a liquefied natural gas (LNG) leak

Smoke detectors are commonly used in offices, schools, homes, and other business, education, and mercantile type occupancies. Although they are certainly the most common form of fire detector, their use in process or other industrial type settings is somewhat limited due to the nature of environment. Generally, in industrial occupancies, heat detectors, flame detectors (UV, UV/IR, IR/IR), pressure sensors/detectors, and gas-sensing detectors are the most common.

Heat (thermal) detectors respond to energy emissions from the fire in the form of heat. Heat detectors are typically activated via the convectional form of heat transfer, similar to sprinkler heads. Due to the means of activation, the time it takes to achieve an alarm stage is slower compared

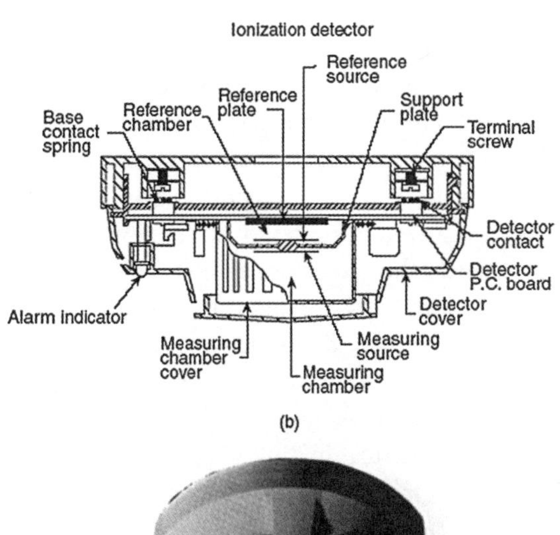

Fig. 9–7. **Ionization smoke detector** *((Reprinted with permission from the* Fire Protection Handbook, *19th Edition, Copyright © 2003, National Fire Protection Association, Quincy, MA 02169)*

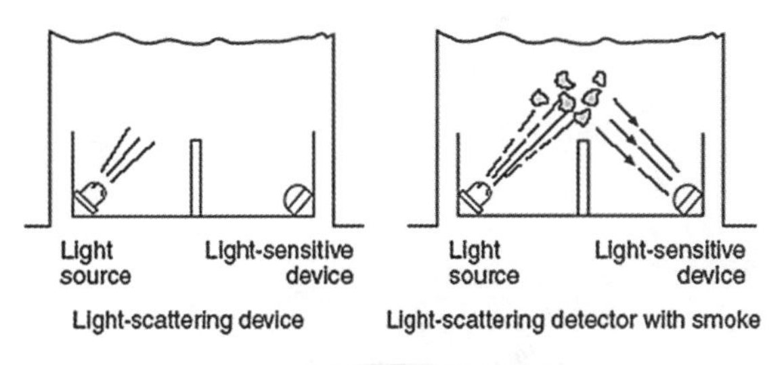

Fig. 9–8. **Photoelectric smoke detector** *(Reprinted with permission from the* Fire Protection Handbook, *19th Edition, Copyright © 2003, National Fire Protection Association, Quincy, MA 02169)*

to the average smoke detector. There are two common types: fixed-temperature and rate-of-rise dectectors. Fixed-temperature detectors signal an alarm when the element located inside the detector is heated to a predetermined temperature point (fig. 9–9). Rate-of-rise detectors signal an alarm when the surrounding temperature rises at a rate that exceeds the preset rise rate for that detector. Combination fixed-temperature and rate-of-rise heat detectors provide an advantage over single-activation detectors because they will respond to a rapid rise in temperature but will tolerate a slow rise in the surrounding temperatures without signaling a false alarm. Also, heat detectors are more reliable than other types of fire detectors, which means fewer false alarms.

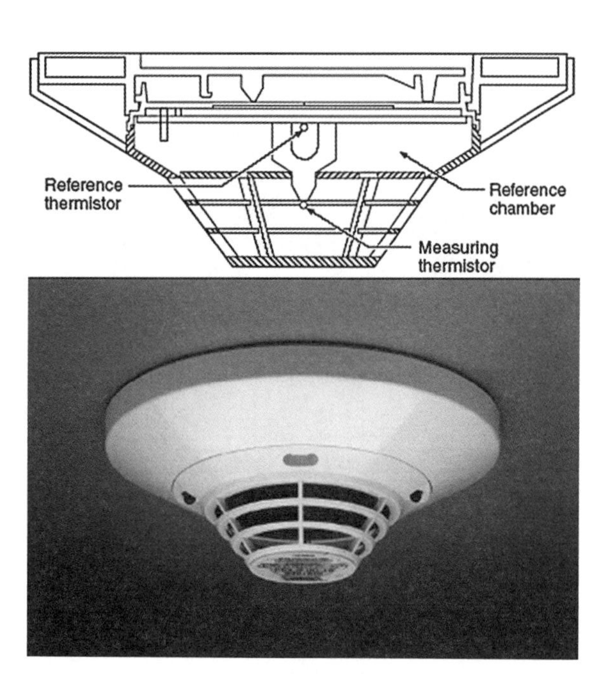

Fig. 9–9. **Heat detector** *(Reprinted with permission from the* Fire Protection Handbook, *19th Edition, Copyright © 2003, National Fire Protection Association, Quincy, MA 02169)*

Heat detectors are commonly used to activate foam or water deluge or water spray systems, protecting vertical process vessels such as distillation columns, fractionation towers, or other similar types of vessels where there is a presence and large amount of a flammable liquid or vaporizing flammable liquid. They are also used in horizontal process and storage vessels such as heat exchangers, LPG bullets, and similar vessels. Additional applications include turbine packages, process units, and product pump stations.

Flame (radiant-energy) detectors detect the presence of light from flames in the ultraviolet or infrared range and signal an alarm back to the FACP. These types of detectors are designed to recognize the typical light flicker of a flame or a radiant energy signature from a fire. Because of transient light sources such as reflections from the sun, these detectors are usually equipped with a time delay to eliminate false alarms. The most common types of flame detectors used in industrial settings are the following (fig. 9–10):

- Ultraviolet (UV)
- Single-frequency infrared (IR)
- Multifrequency (dual or triple) infrared (IR/IR or IR^3)
- Ultraviolet/infrared multiple-band detectors
- Ultraviolet/infrared–simple voting (UV/IR)
- Ultraviolet/infrared–ratio measurement (UV/IR)

Flame detectors are used when the activation time of a fixed fire protection system is critical, and the need for rapid intervention is immediate. The use of these types of flame detection devices can be frequently found in flammable liquid loading and unloading facilities such as gasoline tankers at a bulk plant, rail loading, or in onshore and offshore process facilities such as drilling and production rigs and platforms (fig. 9–11). They are also used quite regularly in other process areas where escaping flammable liquids could readily ignite, and rapid intervention from a fixed fire protection system is required to prevent further spread and escalation of the fire.

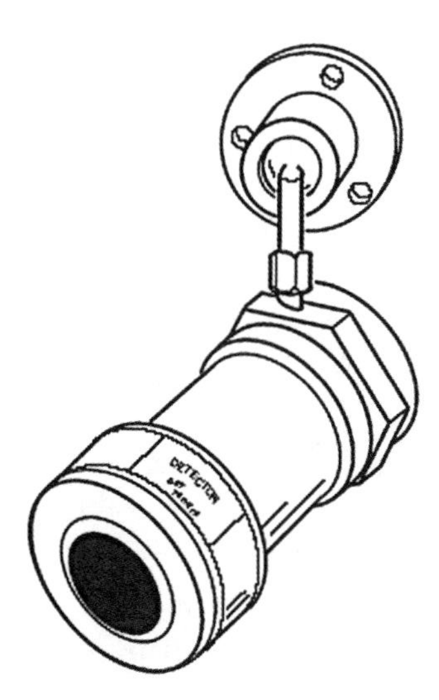

***Fig. 9–10.* Flame detector** *(Reprinted with permission from the* Fire Protection Handbook, *19th Edition, Copyright © 2003, National Fire Protection Association, Quincy, MA 02169)*

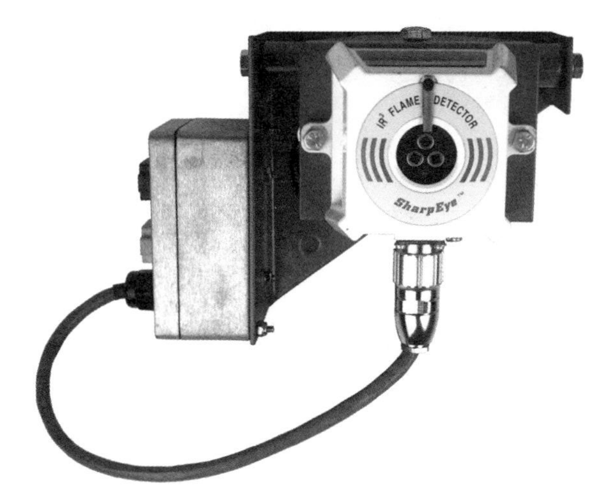

***Fig. 9–11.* IR Flame detector** *(Courtesy of Ansul)*

Pressure and gas detectors

In industrial occupancies that do not have a flammable liquid hazard but instead a dust hazard, pressure sensors or detectors are used to detect and measure pressure variations or buildup in tight enclosures such as grinding or milling equipment (see chapter 15 for equipment details). These detectors sense the buildup of pressure from an explosion and can be designed to activate an explosion protection system.

Gas-sensing detectors measure the relative changes in the level of gas concentration in a compartment (fig. 9–12). They are typically used in the petrochemical and oil/gas industry to warn operators of a possible leak. The intent of this early detection is to prevent the formation of a combustible gas or vapor mixture that could cause an explosion, generating an overpressure blast. The most common gas detector is the catalytic detector. More recently in industry, IR beam detectors have been used for special line-of-sight applications such as monitoring fence, perimeter, or property boundaries. A gas detection system is installed in areas that are most likely to be sources of accidental releases. The gas detection senses the leak and activates alarms or fixed fire protection systems to prevent ignition of a gas in an attempt to mitigate the effects of a flash fire or explosion.

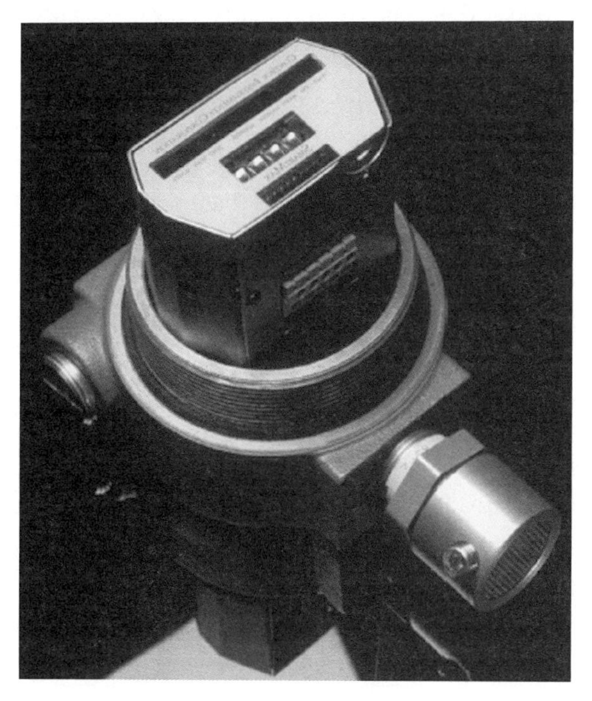

***Fig. 9–12.* Gas detectors** *(Reprinted with permission from the* Fire Protection Handbook, *19th Edition, Copyright © 2003, National Fire Protection Association, Quincy, MA 02169)*

Emergency shutdown systems

In addition to the systems already discussed, a third system exists that controls the shutdown capability of a processing facility. *Emergency shutdown devices* (ESDs) control the plant's capability to shut down in the event of any type of emergency or process abnormality. ESDs are installed in petroleum, petrochemical, paper-processing, and similar process facilities. ESD systems can be automatic, controlled manually, or operated remotely. An ESD system is a method or protocol for rapidly isolating parts of a process or an entire unit or plant. It is intended to stop various operations of a process and isolate them from the overall process to limit or reduce the likelihood of an uncontrollable event escalating. The goal or objective of an ESD system is to protect plant personnel, property, and the environment from process events that could have a severe and negative impact on the community.

An ESD system is different from fire or gas detection systems in that an ESD system responds to a hazardous situation by isolating or eliminating a fuel source to a process unit that otherwise could destroy an entire facility. It may also shunt power to specific pieces of equipment within the area. Imagine if you could not shut down process pumps to an area involved in a fire, hydrocarbons and other fuels could supply the fire for days and days.

Most ESD systems are designed to initiate a facility shutdown in several ways such as manual, automatic, or both, for example:

- Manual activation from the main control room
- Manual activation from strategic locations within the plant
- Automatic shutdown from a fire or gas detection system
- Automatic shutdown from process operating parameters. (i.e., if a process temperature falls outside a specific set point, the ESD will shut down the system)

ESDs are classified by their activation action. It is common to find four or five levels within a single plant or facility. These levels activate certain emergency measures for an increasing area of the facility. For example, a Level 5 ESD might only isolate control alarms and nonprocess equipment. A Level 3 ESD could shut down a pump, piece of equipment, or process, and is considered to be a major shutdown. A Level 1 ESD typically shuts down the entire facility. This would be considered catastrophic and is rarely used. Responding firefighters should never operate ESD systems unless ordered or directed to by plant management. Prematurely activating an ESD could have harsh consequences and cause further downstream complications, much worse than by not activating the ESD system.

CONCLUSION

It is important for responding firefighters who arrive at the scene of an industrial fire to be aware of the types of systems present that can provide essential information needed to make an informed decision about how best to proceed. Before entering any facility, especially an industrial facility, check with the control or operations personnel of the area affected. The information that can be obtained from the fire and gas detection systems can aid in the development of a strategy for the incident. Because the fires that can occur in industrial occupancies are considerably more severe than ordinary combustible fires, the objective of a fire and gas detection system is to rapidly detect a fire where personnel and high value or critical equipment may be involved, so that rapid intervention can be achieved.

NOTES

[1] National Fire Protection Association (NFPA). 2002. NFPA 72, *National Fire Alarm Code*. Quincy, MA: National Fire Protection Association, p. 19.

[2] NFPA 72, p. 22.

[3] NFPA 72, p. 20

10

Terrorism and Security Issues

INTRODUCTION: TERRORISM DEFINED

Terrorism has been defined in many different ways. For this textbook, we define terrorism as the use of force or threats by a person or organized group to harm, demoralize, intimidate, or coerce individuals, societies, or governments, often for ideological or political reasons. Terrorism is a technique used to inflict casualties, disrupt lives, upset social norms, and create panic or fear. It may be domestic, meaning that it is carried out in one country and does not involve other countries or citizens of another country, or it may be international. International terrorism means that the act is planned or carried out by parties of another country or sponsored by another country.

METHODS OF TERRORISM

The methods that terrorists may use can be varied. They may include suicide bombings, flying hijacked planes into buildings, or bombing crowded venues such as sports stadiums, nuclear power plants, or industrial complexes. Other methods may include derailing trains carrying hazardous chemicals, using snipers to cause panic and death, and attacking industrial facilities on the ground. There are an almost infinite number of ways to commit this crime, limited only by one's imagination.

A major tactic of terrorists is to create fear. In many cases, especially in the oil and gas industry, explosions are the weapon of choice.[1] The size of the explosive charges will most likely be large enough to cause maximum damage to equipment, so industrial facilities must consider this possibility and take appropriate precautions.

INDUSTRIAL FACILITIES AS TARGETS

When fire departments prepare for terrorist events, the focus is usually on public buildings or other areas where large numbers of people congregate (fig. 10–1). Prior to September 11, 2001, threats from a terrorist attack on a chemical installation were considered so unlikely that they were not generally included in security plans or in safety and security analyses, except in special circumstances.[2] In a bill introduced to the U.S. House of Representatives on May 10, 2005, the Congress found that because the chemical industry "supplies resources essential to the functioning of other critical infrastructures, the possibility of terrorist and criminal acts on chemical sources (such as industrial facilities) possess a serious threat to public health, safety and welfare, critical infrastructure, national security and the environment."[3] Industrial facilities may be targeted for a variety of reasons. They may be targeted to injure or create fear and panic among a community, or to use the physical and chemical properties of their chemicals to cause an environmental impact. A large incident at an industrial facility may also disrupt supplies that could affect the economy, either locally or nationally. For example, an emergency incident at a gasoline refinery can affect the price of gasoline at the pump. Industrial facilities may be attacked because in many instances they are the critical link between the raw materials and usable products.

Fig. 10–1. **Public buildings may be targets for terrorist events, but fire departments should not direct their planning only to these buildings.** *(Courtesy of M. Barrett)*

In addition to the physical and psychological damage inflicted by a terrorist attack on an industrial facility, the chemicals used in the process will become a weapon. Remember that it does not take a truckload of explosives in an industrial facility to create a disaster. A simple, strategically placed pipe bomb could cause as much impact as a truck bomb, by creating multiple leaks and vessel instabilities that would then instigate a secondary event such as a vapor cloud explosion (VCE). A ready-made chemical weapon is just waiting for the trigger mechanism. Attacking these facilities with explosives can have an impact on a community similar to a weapon of mass destruction (WMD) incident. If we look at the chemicals used and stored at industrial facilities, we realize that many of them have been transported to and from the facility by either road, rail, pipeline, or water; all of these possible additional targets for terrorists. Many port cities and surrounding communities are concerned about ships, especially liquefied natural gas (LNG) and liquefied petroleum gas (LPG) carriers, being attacked in their harbors or rivers, which could cause mass destruction within a large area (fig. 10–2). There has been much debate on this subject, with many supporting and opposing viewpoints. Individual communities and governing bodies must weigh the pros and cons and make informed decisions based on facts, not fear. There must be no knee-jerk reactions to preparedness and response.

Fig. 10–2. **Port cities and surrounding towns are concerned about LPG and LNG carriers traversing their waterways.** *(Courtesy of R. Callis)*

The results of a terrorist attack in the marine environment may include vessels adrift, creating channel or harbor blockage; on-board fires; fuel spills with water-borne fires; chemical releases; and the sinking of ships with the resultant environmental damage, channel or harbor blockage. Fire departments that must respond to marine ports and terminals must assess their capabilities to respond to a marine incident whether it occurs naturally, accidentally, or by a terrorist event.

When responders think about industrial facilities, they usually will only picture those facilities that they normally see in their communities—warehouses, chemical plants, refineries, and the like—but there may be many supporting facilities and infrastructures that are vulnerable, such as pipelines and pump stations in the oil and gas industry (fig. 10–3). The oil and gas industry has onshore oil and gas wells, which may be considered extremely vulnerable areas of the industry to terrorist events. These wells are usually located in remote areas, and providing security to each one may be a financial and physical impossibility. These supporting facilities might be targets of opportunity because of the lack of security and remote locations. It is possible that an explosion at one of these remote locations could be a practice run for a much larger explosion and attack at a more populated area or facility. It should be noted that offshore drilling and production facilities would be a more spectacular target but more difficult to access because of their remote locations.

***Fig. 10–3.* Pipelines may be targets for terrorist activities.** *(Courtesy of Greg Noll)*

To hold explosives in place at a wellhead or pipeline, heavy objects such as sandbags might be used. Fire departments responding to fires at these locations should be aware of this fact and be alert for their presence, indicating the possibility of explosives; either unexploded initial or secondary devices.

PREPARING FOR RESPONSE

As fire departments prepare for terrorist incident response, they must not overlook the "bread-and-butter," or routine, responses. It is the authors' philosophy that how well we prepare for and respond to these types of operations indicates how well we will respond to the major responses, including terrorist events. Fire departments must respond in a disciplined and structured way, adhere to *standard operating procedures* (SOPs), use clear and concise communications, use preincident planning, and have mutual or automatic aid plans in place (fig. 10–4). The effectiveness of our response will depend on how well we have planned and trained for the incident, and how well we have responded to the routine incidents.

Fig. 10–4. Clear, concise communications are a requirement for any response. *(Courtesy of Bloomington, IN F.D. Jeff Barlow, Fire Chief)*

The first step in preparing for a terrorist event at an industrial facility is to identify potential risk sites. How familiar are you with the industrial sites and their process within your response district? A good source of intelligence in this regard is the local emergency planning commission (LEPC). Among the primary responsibilities of the LEPC are to conduct a hazards analysis of hazardous materials (HazMat) facilities and transportation corridors within the community and to receive and manage HazMat facilities' reporting information. Another way to gather intelligence is to visit the facility and speak with plant management. Here are some of the questions that should be asked:

1. What types of processing take place in the facility?
2. What types of chemicals are used, stored, or manufactured at the facility? Obtain the material safety data sheets (MSDS) for the chemicals involved. These sheets include known health hazards of the chemical; the physical and chemical properties of the material; first aid, firefighting and spill control recommendations; protective clothing and equipment requirements; and emergency telephone contact numbers.
3. What are the pressures and temperatures associated with the chemicals?
4. What would be the environmental impact if a leak, fire, or spill occurred? Remember that release scenarios determined for accidental release may not be as severe as a terrorist-related release scenario.
5. What would the economical impact be if a leak, fire or spill occurred?
6. What would be the maximum on-site toxicity, overpressure, or fire scenario?
7. What would be the maximum off-site toxicity, overpressure, or fire scenario?

After identifying potential risk sites, we should look at the worst-case scenarios. When evaluating worst-case scenarios, look at what collateral damage may occur and the consequences of such collateral damage. Upon completing the development of worst-case scenarios, we would then evaluate our resources including mutual or automatic aid, then develop SOPs and emergency response plans (ERPs). We also need to prioritize the likelihood of a terrorist event at a facility. Prioritization can be categorized into high, medium, and low ratings. A high risk would be a facility that has a large asset value or impact severity. The impact could cause severe harm to a large population center or grave economical consequences to the community or nation. A medium risk is one that has lesser consequences than a high risk, and a low risk is one that has negligible consequences. Consultation with local, state, or federal law enforcement officials will be helpful when identifying potential risk sites as well as identifying roles and responsibilities during the incident.

EXERCISES AND DRILLS

Once the industrial facilities have been evaluated and SOPs and preplans have been developed, these SOPs and preplans need to be exercised and drilled. Many times, SOPs and preincident response plans are not exercised and drilled, so when an incident occurs, mass confusion takes place. There are currently various funding sources available, so that large-scale exercises and drills can take place in your community to test the effectiveness of SOPs and preplans. Initially, tabletop drills can be conducted to test the SOP and preplan documents, but full-scale exercises will need to be conducted to actually test the effectiveness and all the components of the documents and plans. Cooperation between all parties involved is essential.

RESPONSE

When responding to an industrial incident, the fire service must be alert to the possibility that the response may be the result of a terrorist act. Remember also that initially a terrorist incident is a local event and should be thought of as such.

Recognition must start with the dispatcher or emergency call taker. These persons must gather information that may give clues that an incident may be a terrorist act. In addition to the standard information that is usually obtained, the following information can provide clues:

- Is the incident at a *target occupancy or target hazard* event?
- Has there been a recent elevation of a terrorist threat level?
- What is the current security threat level?
- Is there a record of previous threats at this location?
- Are multiple casualties involved? Are the reasons known or unknown?
- Are there multiple incidents occurring at the same time? These other incidents could be a decoy to draw first responders away from the targeted facility.
- Are there reports of any unusual odors? Explosions? Hazardous materials?
- Are initial responders down?
- Have any secondary events occurred?
- Is there a simultaneous significant political event?

There are also other clues that indicate that an incident may be terrorism related:

- Does the facility or incident hold the potential for large-scale disruption to the national economy?
- Does the incident have the potential to create large-scale environmental damage?
- Will the incident cause a large-scale financial loss to the private sector?
- Can the incident cause disruption to the regional or local infrastructure?
- Is the facility a symbolic target?
- Does the incident correspond with the anniversary of a significant terrorist or geopolitical event, religious holiday, or national holiday?
- Will the incident be of great media interest?

With industrial occupancies, it may be hard to recognize that the incident may have been caused by a terrorist act due to the chemicals and processes on-site that could cause an explosion if handled improperly.

It is important that the response be effective and organized. The operations by the first responders will set the tone for the operations that follow. As we know, some of the goals of terrorism are to send a message, demoralize the public, and create havoc and insecurity. This may lead to members of the public questioning the government's ability to protect them. If the fire department response is haphazard, chaotic, or ineffective; or the response gives such an appearance, then the effect on the morale of the public can be devastating. A chaotic response will give the terrorists an added victory; an organized, well-planned response will have the opposite effect. For example, consider the London subway bombings of July 7, 2005. The response of the emergency services operating in London was well organized, which contributed to calming the public and reducing panic. As demonstrated by these emergency services, if the response is massive, rapid, and orderly, resulting in a timely control of the incident, then the terrorists' impact will be somewhat lessened.

When responding even to a benign or routine fire or rescue situation, the fire service must be vigilant (fig. 10–5). Even these incidents have the

***Fig. 10–5.* Even when responding to routine situations, firefighters must remain vigilant.** *(Courtesy of Bloomington, IN F.D. Jeff Barlow, Fire Chief).*

potential for primary or secondary devices, release of hazardous chemicals, or an attack on the first responders. Could the response be a diversion so that terrorists can enter the facility during the confusion? One of the tactics used by terrorists is to create a situation that draws onlookers to the scene, then terrorists detonate an explosive device to inflict maximum casualties.

When arriving at the scene, we should always consider that there is a possibility that the incident was the result of a terrorist act. As with any industrial incident, approach from an upwind and uphill direction. Do not commit all resources into the facility proper. Those who do respond into the facility proper should be positioned towards their escape route, so that a quick and orderly retreat can be accomplished if needed. Use staging areas where possible for responding apparatus and staffing. Even though staging areas may have been identified in preincident response plans, if a terrorist act is suspected or known, it may be wise not to use those preidentified staging areas. If operational security breaches have allowed terrorists to learn the location of such preidentified staging areas, then it would be easy for them to plant a secondary device at this location to inflict casualties on responders. Do not preplace signs identifying staging areas for operational security reasons. During suspected terrorist events, have explosive-sniffing dogs search the staging areas and ensure site security by law enforcement personnel. In addition, if a terrorist event is suspected or known, vary your response protocols, including approach direction, to lessen the chance that the department personnel will be targets of secondary devices or attacks while responding.

The incident command post must be located in a safe area, and the immediate and surrounding area must also be searched for the presence of explosives or other dangers. Monitoring the scene for the presence of HazMat or other agents is essential for the safety of the responders. If HazMat or other agents are present, isolation of contaminated persons and emergency decontamination should be started immediately. Symptoms of a terrorist chemical agent exposure must be recognized. Basic terrorism response training is important to ensure that all responders recognize the signs and symptoms. Fire departments should work closely with industrial fire brigades to ensure that the brigade also receives basic terrorism incident response training. The sharing of training resources should be encouraged.

Standard operating procedures for the industrial response should be followed; that is, use on-site plant personnel for intelligence and technical assistance to assist with the fire department operations. These individuals may be able to help in disproving or eliminating the possibility of a terrorist action as the suspected cause of the incident. Perform a thorough size-up and risk–benefit analysis before committing firefighters into the hot zone. There must be no freelancing. Establish isolation zones in accordance with the United States Department of Transportation *Emergency Response Guidebook*, other applicable HazMat and technical response guides, or isolation zones established in the preincident response plans where applicable. A special operations branch or group may have to be established to handle technical HazMat or rescue functions.

Consideration must be given to the possibility that fire protection systems may be unusable due to the event. Preincident response plans should designate alternate water sources, either outside the plant water mains, water tanker shuttles, or in-plant auxiliary sources such as water tanks, cooling tower basins, or cooling water canals.

OPERATIONAL SECURITY

Operational security (OPSEC) is a strategy that should be used by fire departments to protect information that can be used by terrorists or other criminals against a fire department or the citizens that they serve. Some of this information might be in the form of SOPs, preincident response plans, special event operations planning, training manuals, rules and regulations, department policies, and so forth.

How many departments have Knox-Box® or other similar key systems within their communities (fig. 10–6)? Where are the fire department's keys kept? Are they secured on the apparatus or available to anyone who enters the cab of the apparatus?

Fig. 10–6. **Knox-Box**

Operational security extends beyond the fire apparatus and into the fire station. Where are the important documents kept? Are they available to anyone who enters the station, authorized or unauthorized? Do you secure the fire station when you go out on a response? Do you throw away draft documents without shredding them? What is in your trash? Fire departments must now ensure that their premises are secured at all times and that access is limited to authorized persons or visitors. One final thought: We all are proud to be firefighters and love to share our experiences, adventures, and stories with others. Be careful what information you might inadvertently divulge during an innocent conversation. As a part of OPSEC, departments must consider communications security (COMSEC). Who may be listening to your radio communications? There may be hundreds of unofficial listeners to your communications. Do not transmit sensitive information over radio networks. If secure communication frequencies are available, then use them. However, be aware that the terrorists may have planted a bomb that is sensitive to radio transmissions. When radios transmit at the scene, the radio frequency in use may trigger the explosive device.

SECURITY ISSUES

Recently, an industrial fire department responded to a drill at one area of its large complex. Normally, this department would respond from the upwind side of the incident. For this drill they did just that, but on their arrival at the upwind gate, they found it secured and blocked with concrete barriers (fig. 10–7). Without notification to the fire department, the plant had closed this gate for security reasons. Had this been a true emergency and not just a drill, there could have been severe consequences. On another occasion, a group of firefighters on a tour of a ship during a training exercise found that they were unable to exit the ship because the exit had been locked from the outside with a padlock.

Fig. 10–7. **Barriers may be present where access was previously available.**

These two incidents were the direct results of increased security to address the threat of terrorism. The second incident was a consequence of the International Ship and Port Facility Security Code (ISPS). This code was adopted by the International Maritime Organization (IMO) in December 2002. The ISPS code with its added security requirements applies to the following ships and port facilities:

- Passenger ships, including high-speed passenger craft
- Cargo ships, including high-speed passenger craft of 500 and higher gross tonnage
- Mobile offshore drilling units
- Port facilities serving such ships engaged in international voyages[4]

As part of the responsibilities of both the port and shipping companies, a port facility security plan and a ship security plan are developed after a security assessment is conducted. The ISPS sets three levels of security, one of which is maintained at all times. Although security procedures are not uniform in all applications, the fire service should note that in any event, security will be heightened, limiting means of access and egress to port facilities and most ships (fig. 10–8). For port facilities, at the very least, areas and access points previously accessible may now be locked or closed permanently. Even though not required by law, additional security measures will also be in place at other industrial facilities. Concrete barriers; perimeter protection; and fencing, walls, and landscaping that have been placed for security reasons may affect apparatus placement, response routes, and emergency evacuation (fig. 10–9).

Fig. 10–8. **Ships may be secured using unorthodox methods, creating forcible entry problems for firefighters.**

Fig. 10–9. **Concrete barriers may affect apparatus placement, response routes, and emergency evacuation.**

In the past, fire department access to industrial facilities has been fairly easy during nonemergency visits. With increased security, familiarization or inspection visits may be much more difficult. Advance notice may be required, and additional measures such as background checks or identification cards may be requested. It is important that we maintain a working relationship with industry and respect any requests that they may have relating to security while at the same time ensuring that fire protection forces can still operate effectively.

ADDRESSING ADDITIONAL SECURITY MEASURES

How can fire departments prepare to effectively deal with an industrial or marine incident with heightened security measures in place? The following are some of the areas that should be addressed:

- Meet with maritime, port, and industry representatives, including facility managers, port facility officers, and industrial fire brigade chiefs to discuss additional security measures taken. In many instances, only limited information will be divulged because of operational security issues, but specifically ask about fire department access.
- Maintain a working relationship with industrial and port facility security officials, and include the officials' contact information in the fire department's preincident response plans. Include these individuals as part of your planning resources. Information contained in security plans may be useful in developing strategies and tactics.
- Update preincident response plans with accurate information regarding access to the facility. Remember, hastily placed fences or barriers may affect water supply sources or supply points such as hydrants

or fire department connections. If these changes have been made, discuss them with facility management and ensure compliance with local fire department and building codes. Fire department personnel must be vigilant when inspecting or touring facilities to spot any problems with access or other modifications that may affect fire department operations, response times, or means of access or retreat.

- Additional forces may be required to help with the difficult forcible entry and ventilation problems that may be caused by the additional security measures. Call for their response early in the operations. Their forcible entry operations may be necessary for firefighter escape or rapid intervention team entry in case personnel or units become lost or disoriented.
- Test response plans by holding drills and exercises. Tabletop exercises, as well as limited and full-response exercises, should be conducted periodically. During these drills, apparatus access and placement can be verified. Lessons learned should be used to update the department's preincident response plan and the facility's emergency response plan.

For marine responses, the following additional areas will need to be addressed:

- Meet with the crews of ships that enter your area, and discuss additional security measures and how they may affect firefighting efforts. This information may be limited, but establishing a dialog with the ship's master and security officer will enable you to gather as much information as possible. It is important that we remind the ship's crew that we are there to help.
- During a fire, take additional forcible entry tools on board. They may include cutting torches, hydraulic spreaders, power saws with metal-cutting blades, through-the-lock tools such as the K tool, and standard forcible entry tools. On tankers, some areas may contain flammable vapors, so firefighters should consult with the ship's crew before creating an ignition source by using forcible entry tools.
- Additional rapid intervention teams may be needed. Also, dedicated companies may be needed to perform forcible entry at multiple locations onboard the ship that may have been secured. This forcible entry may be necessary for ventilation, providing egress for firefighters who may have become lost or disoriented or for the rapid intervention team's entry.
- During a fire, include the ship security officer in early size-up discussions. Information about the ship's features and entrances that have been secured (and how they have been secured) is critical for a successful operations.

CONCLUSION

With today's increased security issues due to the terrorism threat, it stands to reason that fire department operations will be affected in some way, both at fixed facilities and in modes of transportation. It is up to the fire service to recognize and understand this and acquire the knowledge necessary to modify planning and operations. We can do this by observing conditions in our response areas; educating ourselves relative to changes in laws and codes, preincident planning, and the inherent updating required; and developing dialog and relationships with industry and security personnel. Neglecting to do this will adversely affect our ability to respond to terrorist incidents. We must be ever vigilant to the life-altering changes to our society and the effect on security and emergency response since September 11, 2001.

NOTES

[1] Adams, N. 2003. *Terrorism and Oil.* Tulsa, OK: PennWell, p. 81.

[2] Center for Chemical Process Safety (CCPS) of the American Institute of Chemical Engineers. 2003. *Guidelines for Analyzing and Managing the Security Vulnerabilities of Fixed Chemical Sites.* New York: CCPS, p. 10.

[3] U.S. Congress, HR 2237, *Chemical Security Act of 2005*, sect. 2. Retrieved June 1, 2006, from http://www.theorator.com/bills109/hr2237.html.

[4] International Maritime Organization. 2003. *International Ship and Port Facility Security Code and SOLAS Amendments 2002*. London: IMO, sect. 3.1.

11

Hazardous Materials

INTRODUCTION

Industrial facilities present a significant risk for firefighters to be exposed to hazardous materials (HazMat) during responses to fires and other emergencies. Hazardous materials may range significantly in toxicity and volume depending on the types of processes and production volume of a specific occupancy. This chapter will discuss some of the more common equipment hazards and hazardous materials that responders may encounter during emergencies at industrial facilities, as well as some general response practices to help mitigate these risks (fig. 11–1). Although it is impossible to discuss all hazardous materials found in industrial installations and processes, this chapter's intent is to describe the main issues that are widespread in industrial occupancies while perhaps not so common during a hazardous materials response to a spill or leak on the highway or other nonindustrial occupancies.

Fig. 11–1. **Industrial HazMat response** *(Courtesy of G. Noll)*

COMMON EQUIPMENT HAZARDS

Process and quality control devices

There are a number of devices commonly used in industrial processes that may pose a safety or health risk to personnel responding to emergencies in the associated buildings and process areas. Many times, these types of devices are not considered to be a typical hazardous materials response, in that the device itself and its operation are potentially not harmful. They can be, however, if they are involved in a fire or explosion where the integrity of the device is compromised.

Nuclear instruments

Many industrial processes use devices with radioactive sources to capture process information such as product flow, temperature, density, level, thickness and other related product- or process-related data.

Nuclear instruments are nonintrusive instruments that operate on a simple principle of energy absorption. Gamma radiation is beamed through the vessel or pipe in question. The source has a known energy activity, and a detector measures the activity present on the opposite side of the vessel or pipe. In level measurement, as fluids rise to the level of the beam, the amount of energy at the detector decreases because the product absorbs some of the energy. When product levels fall below the beam, the energy increases because there are fewer products to absorb the energy. Some vessels contain layers of different products. Nuclear level gauges can differentiate between the various strata by referencing the energy absorption capacity of the various products. Similarly, nuclear devices can accurately measure the density of a product such as plastic sheeting or the flow of fluids in a process line.

During a fire or explosion, damage to the instrument could result in a potentially hazardous release of the radiological source. Government regulations require that facilities licensed to use a nuclear device must maintain records on the device and have on staff at least one individual who is trained in the proper use and safety of the device. In responses to major incidents where process equipment or instrumentation are damaged, the incident commander (IC) should ask the facility operations supervisor or the unit maintenance supervisor if any nuclear gauges are present in the involved area. If any nuclear devices—such as level or flow gauges—are present, care should be taken to avoid contact with them or further damaging them during firefighting and overhaul operations. As soon as possible, a *radiological safety officer* (RSO) should inspect the devices to close the shutters and ensure that the device is not leaking. Where extensive structural, equipment, and instrumentation damage has occurred, the IC and RSO should develop a plan to locate and secure all radiological devices.

Nuclear instruments may also play a significant role in confined space rescue operations. Many process vessels and storage silos have nuclear level, flow, or density gauges. Nuclear devices should be secured with shutters closed prior to vessel entry by maintenance and inspections personnel who might subsequently become confined space casualties; however, this is not always the case. The IC should determine if any nuclear devices are present during rescue operations and take appropriate steps, such as shielding and maintaining minimum distances to protect personnel until the device shutter can be closed.

During preincident response planning, the presence of any radioactive sources should be noted to include their specific location, radioactive source, activity, and the contact numbers of personnel authorized to operate them. Because of the proliferation of devices that use radioactive sources, the IC or safety officer should have access to radiological survey instruments or experts during response to such facilities. It would be prudent for the fire department to maintain a contact list of qualified personnel within its jurisdiction who may be available to assist with incidents involving radiological materials. Also, a list of radiological survey equipment found within the jurisdiction should be maintained in the department's response resource lists.

Lasers

A second potentially hazardous technology found throughout industry is the use of lasers. Lasers may be used in process and quality control to conduct a variety of measurement functions. *Simple Class I lasers* may be used to count individual "widgets" and perform simple level or density measurements. More complex level and density measurements such as those involving processes where mist, aerosol, or dust is present may require the higher energy of a *Class II or Class III laser.*

It is logical to think that during a fire or explosion, lasers could be incapacitated by mechanical damage or loss of electricity. Industrial measurement devices are so critical to the operation that they typically have robust construction and their mountings are well secured. However, there is the potential for externally mounted lasers to become dislodged or misaligned in an explosion or impact to the process equipment. In this case, lasers could pose a hazard to responders working in close proximity to the device.

A greater concern is that of the response to an entanglement in an automated assembly line or confined space emergency where process equipment is fitted with laser measurement devices. Although unit supervision and coworkers are often quick to shut down power to conveyer belts, agitators, and other mechanical devices, they may fail to isolate the laser power supply. Responders to any automated process or confined space incident should inquire about the presence and status of laser instrumentation.

COMMON INDUSTRIAL CHEMICALS

This section will discuss industrial chemicals that are manufactured in large volumes and used in a wide variety of industrial processes. Industrial chemicals are used in the manufacture or treatment of other chemicals or in other industrial processes. In 1995, the US chemical industry produced in excess of 750 billion pounds of chemicals.[1] These chemicals can be ranked in production volume and present a valuable opportunity for responders to focus on the most common (highest production volume) chemicals. The 10 highest ranking chemicals in US production volume account for 59% of all US chemical production. These chemicals are discussed in the following sections.

Sulfuric acid

Sulfuric acid is ranked number one in US chemical production.[2] Sulfuric acid is manufactured in a variety of ways—typically involving the generation of sulfur dioxide and subsequent transformation to sulfur trioxide. Finally, the sulfur trioxide is reacted with residual water in a sulfuric acid stream to form more sulfuric acid. Sulfuric acid is an odorless, liquid that is clear to brownish in color. Sulfuric acid is heaver than water, and its vapors are heavier than air.

Sulfuric acid is used in many industrial processes, such as the manufacture of fertilizers, detergents, explosives, petroleum refining, metals processing, dyes and pigments, electroplating, pharmaceuticals, and processing of other acids. Sulfuric acid also plays an important role in the manufacturing of some plastics and resins, as well as the automotive and lead acid battery industries.

Sulfuric acid is a strong corrosive with the capacity to attack metals and animal tissue. Extreme caution should be taken to avoid contact with the liquid or vapors. Sulfuric acid produces an *exothermic* reaction when combined with water. This reaction can result in violent jets of acid if sufficient heat is generated. Water should never be sprayed directly into concentrated sulfuric acid.

Nitrogen

Nitrogen is the second highest ranking chemical in US production volume.[3] Nitrogen is most commonly produced through liquefaction of air and subsequent separation by distillation. A plant that uses this process is normally referred to as an air separation plant (ASP). Nitrogen may also be manufactured by reduction of ammonia. In industrial applications, nitrogen can be found as a liquid or a gas. Liquid nitrogen is lighter than water, and gaseous nitrogen is lighter than air.

Nitrogen is common in almost all industries, including agriculture, refining, automotive, laboratories, petrochemical, chemical, military, aviation, automotive, food processing, and transportation. Nitrogen can be found in piping, cylinders, process vessels, instrument houses, storage tanks, and almost any process container imaginable. Nitrogen is widely used as an inerting or purge gas to reduce a product's exposure to atmospheric oxygen. Typical applications include blanketing of flammable liquids to reduce the potential for ignition, purging of vessels and piping to prohibit rusting, and blanketing of foods to reduce spoilage.

Liquid nitrogen is a cryogenic and therefore poses a danger of freezing and frostbite to personnel. Also, liquid nitrogen can damage other equipment and process lines either through freezing of the container material, for example, rubber hoses, or freezing of the product inside the container, for example, water in a fire hose.

Gaseous nitrogen is a hazard to personnel because it can displace oxygen, thereby creating an oxygen-deficient atmosphere. This is a critical size-up consideration in any confined space rescue. To complicate the situation, many industrial facilities still use nitrogen as a backup to the plant air system. When the plant's air compressors fail, stored liquid nitrogen is gasified and introduced into the plant air system. Even though a utility connection may be labeled "plant air," it could be delivering nitrogen. Before using plant air in a confined space, always consult with plant operations personnel or conduct a quick "go-no-go" check using an oxygen meter.

Nitrogen's predominance in industry combined with its inerting characteristic can also be used to the advantage of responders in snuffing out fires and other oxygen-dependent reactions at the innermost confines of a process system that could not be reached through ordinary means. The use of nitrogen to snuff out a fire should only be performed after agreement with operations personnel. Nitrogen hazards and possible uses in emergencies should be discussed during preincident response planning.

Oxygen

Oxygen is the third highest ranking chemical in US production volume.[4] Commercial production of oxygen is achieved primarily through liquefaction of air and subsequent cryogenic distillation of air liquids in an air separation plant. Liquefaction requires very low temperatures. Oxygen can also be produced from a normal air stream by selective absorption of nitrogen at ambient temperatures. When nitrogen is absorbed in this process, the remaining components, primarily oxygen and argon, are left behind to be separated in the final stage of the process

Oxygen is used in many industries, including steel production, military, medical, aerospace, petrochemical, inorganic chemical, waste treatment, and agricultural chemicals. Oxygen is an important feedstock in the production of some ammonia, acetylene, methyl alcohol, and coal gasification processes. Oxygen can be found in piping, cylinders, process vessels, transport trucks, pipelines, and cryogenic storage tanks.

Industrial oxygen can be found in both liquid and gaseous forms. Liquid oxygen is heavier than water, and gaseous oxygen is heavier than air. Although oxygen itself does not burn, it is an oxidizing agent, and its presence in the gaseous form will vigorously accelerate combustion in combustible and flammable products. Contact of flammable or combustible products with liquid oxygen may result in explosive reactions. Gaseous oxygen also presents an inhalation hazard to personnel who may be exposed to high concentrations. At high concentrations, oxygen can lead to disruption of hearing and vision, dizziness, convulsions, and unconsciousness. Liquid oxygen poses a danger of freezing and frostbite to personnel.

Caution should be used when placing water streams into service in attempts to control fires where liquid oxygen is present. Since liquid oxygen is stored at very low temperatures (boiling point—297.4°F), any water that is introduced to the liquid oxygen will only serve to generate more gaseous oxygen and subsequently increase in the intensity of combustion.

Ethylene

Ethylene ranks as the fourth highest chemical in US production volume.[5]

Ethylene is most commonly manufactured through thermal cracking of hydrocarbon gases in a tube furnace. Ethylene is an olefin, and in typical refining or petrochemical jargon, a process that produces ethylene is called an "olefins unit."

Ethylene can also be produced through dehydration of ethanol or catalytic reaction of synthesis gas, although these methods are not as common as thermal cracking. Under ambient conditions, ethylene is a sweet-smelling, colorless gas. During production, ethylene exists in both gaseous and liquid form. Ethylene is normally stored and transported in liquid form. Liquid ethylene is lighter than water, and ethylene vapors are about the same density as air.

Ethylene is used in a variety of industries, including petrochemical, agricultural, and medical applications. Ethylene is a feedstock for a myriad of plastics and resins, including polyvinyl chloride (PVC), polyethylene, polyester, and polystyrene. Ethylene is also used in production of other common chemicals, including ethylene glycol, aluminum alkyls, acetaldehyde, ethyl alcohol, and various refrigerants.

Ethylene can be found in process piping, pipelines, cylinders, process vessels, pressurized storage vessels, and underground storage caverns. It poses an extreme fire hazard due to its wide flammable range and ability to rapidly vaporize. Ethylene vapors often develop explosive mixtures in air. In high concentrations, ethylene vapors may cause asphyxiation. In low concentrations, ethylene has a narcotic effect and may result in dizziness, headache, nausea, and loss of coordination.

Do not attempt to extinguish an ethylene fire. Any resulting vapor cloud could result in an unconfined vapor cloud explosion. Cool all involved structures and exposures with special attention, keeping streams out of the ethylene furnace (cracker or reactor). The rule in battling ethylene fires is to cool the surrounding structure and process vessels with water—lots of water.

Lime

Lime is the fifth ranking chemical in US production volume.[6] Lime is a general term used to describe several chemicals from the same family, including both calcium oxide and calcium hydroxide. Lime is typically found as a solid in lump or powder form. Lime powder can easily be fluidized. Lime may also be found in a slurry form (liquid).

Calcium oxide is produced in kilns where limestone, or calcium carbonate, is heated. The calcium oxide may then be reacted with water to form calcium hydroxide and subsequently mechanically separated and dried.

Calcium hydroxide is used in many industries, including construction, petrochemical, food, and agriculture. Applications for lime include animal skin and food processing, soil stabilization, plasters, cement, water softening, and disinfecting. Lime is also an important buffering and neutralization agent.

Calcium oxide can be found in the petrochemical, refining, metals, agricultural chemicals, food, and glass industries. Calcium oxide has applications in insecticides, fungicides, sugar refining, leather processing, metal flux, and as a refractory. Calcium oxide also plays a role as a flux compound in steel manufacturing and in pH control in many industries.

Lime is caustic and presents an irritation hazard for body tissues such as the eyes, lungs, and skin. Lime generates heat when mixed with water. Although responses to incidents involving lime are not extremely hazardous, special caution should be taken to control runoff.

Ammonia

Ammonia is the sixth ranking chemical in US production volume.[7] Ammonia is manufactured in a variety of processes that generally require the reaction of hydrogen and nitrogen in the presence of a catalyst. Ammonia is produced in the anhydrous form as a colorless gas and can be easily liquefied under pressure. Ammonia can easily be identified by its sharp irritating odor. Liquid ammonia is lighter than water, and gaseous ammonia is lighter than air.

Ammonia has many uses in the agricultural, petrochemical, chemical, explosives, steel, and aerospace industries. Applications for ammonia include rocket propulsion, bleaching agent, disinfecting, fertilizers, refrigerant, and dyeing. Uses also include synthesis of acids and organic compounds used in dyes, pharmaceuticals, and plastics. Ammonia can be found in process piping, cylinders, process vessels, tank trucks, railcars, barges, ships, and pressurized storage vessels.

Ammonia vapors are corrosive and pose an extreme inhalation hazard. The vapors can be an irritant and corrosive to body tissues including the eyes, skin, digestive, and respiratory tracts. Ammonia is both reactive and flammable. A large volume of water in a finely delivered spray is the rule for handling ammonia emergencies. Special caution should be taken to control runoff.

Phosphoric acid

Phosphoric acid ranks seventh in US chemical production volume.[8] Phosphoric acid is produced in three ways. Two of the production methods involve the reaction of acids with phosphate rock. The third method involves thermal processing of phosphate rock and subsequent hydrating of the phosphoric oxide generated in the thermal process. Phosphoric acid may be found as transparent crystals (solid) or as a viscous liquid that is both colorless and odorless. Liquid phosphoric acid is heaver than water, and phosphoric acid vapor is heavier than air.

Phosphoric acid is used in production of agricultural chemicals, ceramics, glass, pharmaceuticals, metals, food, petrochemicals, refined products, and inorganic chemicals. Common uses of phosphoric acid include manufacturing of soaps, fertilizers, detergents, inorganic phosphates, and soft drinks. Phosphoric acid also has applications in dentistry, as a local anesthetic.

Phosphoric acid can be found in process piping, drums, carboy, process vessels, tank truck, railcar, barges, and atmospheric storage vessels.

Phosphoric acid is both a corrosive and irritant. It is destructive to the eyes, skin, and respiratory tract. As a corrosive, phosphoric acid may also affect emergency operations by damaging process or response equipment. As with any other concentrated acid, phosphoric acid will react violently to the introduction of water.

Unless phosphoric acid is being vaporized and posing a threat to responders or off-site populations, response activities should focus on avoiding contact with the liquid or increasing the liquid volume by addition of water through firefighting streams.

Sodium hydroxide

Sodium hydroxide is the eighth highest ranking chemical in US production volume.[9] Sodium hydroxide is commonly called caustic. While sodium hydroxide is the largest volume caustic produced in the United States, it is not the only caustic material. Sodium hydroxide is produced along with chlorine by electrolysis of sodium chloride brine (saltwater). The electrolysis process produces sodium hydroxide in solution with water, cell liquor, which is subsequently concentrated through repetitive evaporation stages into various liquid concentrations. Sodium hydroxide can also be further concentrated to a solid white bead, pellet, or powder. Sodium hydroxide liquids are clear and heavier than water. Sodium hydroxide liquids give a slick feeling to body tissues such as the skin.

Sodium hydroxide can be found in the petrochemical, rubber, pulp and paper, textile, refining, food, glass, and metals industries. Sodium hydroxide can be found in process piping, plastic drums, fiber drums, glass carboys, vats, process vessels, atmospheric storage vessels barges, tank trucks, railcars, warehouses, and ships.

As a strong alkali, sodium hydroxide can be found in a multitude of processes where acids require neutralization. In addition to neutralization, sodium hydroxide is used in the manufacture of rayon, cellophane, soaps, detergents, aluminum, fibers, and rubber. Sodium hydroxide also plays an important role in electroplating, food processing, ion exchange resins, and etching processes. Removal of fruit and vegetable skins is accomplished by immersing the vegetables in a corrosive bath of sodium hydroxide.

Sodium hydroxide in all forms is highly corrosive. Sodium hydroxide rapidly attacks the skin and eyes. Ingestion of liquid or solid sodium hydroxide can result in tissue damage, scarring, and death. Inhalation of airborne sodium hydroxide particles may result in severe damage to organs in the upper respiratory tract. Exposure may result in severe *pneumonitis.*

Unless sodium hydroxide is posing a threat to responders or the off-site population, response activities should focus on avoiding contact with the product or increasing the liquid volume by addition of water through firefighting streams and controlling any runoff.

Propylene

Propylene is the ninth ranking chemical in US production volume.[10] Propylene can be produced in the same thermal cracking process as ethylene or in subsequent catalytic cracking processes. Under ambient conditions, propylene is a colorless gas. During production, propylene exists in both gaseous and liquid forms. Propylene is normally stored and transported in liquid form. Liquid propylene is lighter than water, and propylene vapors are heavier than air.

Propylene is found in the petrochemical, plastics and resins, and refining industries. Propylene is feedstock for many downstream chemicals such as propylene oxide, polypropylene, pyrolysis gasoline, glycerol, acrylonitrile, and isopropyl alcohol. Propylene can be found in process piping, cylinders, pipelines, process vessels, railcars, barges, and pressurized storage vessels.

Propylene is a simple asphyxiate. Exposure is due to displacement of oxygen and may result in headache, drowsiness, dizziness, vomiting, unconsciousness, or death. Skin contact with liquid propylene or propylene vapor jets can result in freezing and frostbite. Propylene is extremely flammable, and explosive mixtures with air can develop.

As with ethylene, do not attempt to extinguish a propylene fire. Any resulting vapor cloud may result in an unconfined vapor cloud explosion. Cool all involved structures with special attention to keeping streams out of the furnace. Large volumes of water are used to cool exposed process equipment, structures, and vessels.

Chlorine

Chlorine is the 10th ranking chemical in US production volume.[11] Chlorine is coproduced along with sodium hydroxide and hydrogen in the sodium chloride electrolysis process. Chlorine is produced as a greenish-yellow colored gas with a strong, irritating odor. Chlorine is commonly liquefied for storage and use in downstream processes. Liquid chlorine is heavier than water, and gaseous chlorine is heaver than air.

Chlorine is found extensively in the petrochemical, pharmaceutical, agricultural, food processing, and water treatment industries.

Chlorine is used to manufacture various plastics including PVC and neoprene, as well as pesticides, herbicides, fungicides, artificial sweeteners, and fire retardant materials. Although the industry is moving away from halogenated compounds, chlorine can still be found in refrigerants, solvents, and fire-extinguishing agents. In the food industry, chlorine is used extensively as a disinfectant.

Both liquid and gaseous chlorine can be found in process piping, cylinders, process vessels, pressurized storage vessels, railcars, barges, and ships.

Chlorine was used in World War I as a chemical warfare agent. Chlorine gas blisters the skin, eyes, and lungs. Chlorine is not flammable, but as an oxidizer, it can vigorously accelerate combustion in fires involving combustible and flammable products. Liquid chlorine or dense chlorine gas can result in explosive reactions if it comes into contact with flammable or combustible products. Chlorine also exhibits dangerous reactions when in contact with metal powders, ammonia, ethers, hydrogen, and many hydrocarbons. Water spray used to combat fires involving chlorine may generate hydrochloric acid. Caution should be exercised to protect personnel and equipment from this corrosive product as well as in managing runoff.

Response to chlorine emergencies should focus on delivery of copious volumes of water into the resulting vapor cloud. Caution should be taken to avoid water streams coming into contact with liquid chlorine. Control of runoff is essential to preclude off-site migration of the resulting hypochlorous and hydrochloric acids. Respiratory and skin protection is essential for protection from chlorine's acute toxicity and blistering effect.

FIREFIGHTING TACTICS

As with any response, whether a fire or chemical spill, the first thing that needs to be in place is a proper preincident response plan. These plans are even more essential when dealing with a response to a hazardous materials leak or spill. These preincident response plans (discussed in chapter 5) help identify key personnel and equipment needed during a response.

When arriving on the scene of a hazardous materials incident at an industrial facility, you should first try to identify the material in question and determine whether the leak is still active. Unlike responding to a hazardous materials incident on a highway, where there are many unknowns, response to an industrial facility will be very different.

A HazMat response to a highway accident involves many other issues such as the accident that caused the leak, rescue, unfavorable conditions such as location of the incident, and unknowns such as the material leaking, type of leak, access, and many other issues. Many times, during fire department response to a HazMat incident, resources are limited to the department or mutual aid agreements, though sometimes a county agency is available for assistance. In most cases, the type and quantity of the hazardous materials spill is unknown until the fire department units arrive on scene.

Often this will not be the case when responding to an industrial facility that normally handles and processes these types of chemicals. If you have ever responded to both a HazMat incident on the highway and one involving an industrial facility, the difference is noticeable (fig. 11–2). In general, because of government, state, local, and company regulations, these facilities usually have a multitude of resources, or at a minimum, the ability to procure such resources that are normally out of reach for most fire departments. Many industrial facilities have very detailed emergency response plans, a more controlled environment, in-depth information regarding their chemicals, on-site or off-site experts, and access to financial resources for cleanup and spill control.

Fig. 11–2. Industrial process leak *(Reprinted with permission from the* Fire Protection Handbook, *19th Edition, Copyright © 2003, National Fire Protection Association, Quincy, MA 02169)*

Because of various federal and local code regulations such as the Occupational Safety and Health Administration's (OSHA) Process Safety Management (PSM), the Environmental Protection Agency's (EPA) Risk Management Program (RMP), and the International Code Council's Hazardous Materials Management Plan (HMMP), and Hazardous Materials Inventory System (HMIS), most if not all industrial facilities have a very detailed HazMat plan. These plans include the description of the process(es), chemical details and quantities, location or identification of resources (material safety data sheets [MSDS], Chemical Abstract Service [CAS] numbers, etc.), and other useful information (fig. 11–3). These plans also identify potential leak causes and consequences, as well as the likelihood and severity of the events. With this information available to the local responding fire department, the firefighters' ability to make educated decisions is greatly increased, as compared to some of the limited information available during a response to a highway.

Responding to an industrial facility also allows for additional resources and improved conditions. A HazMat incident at an industrial facility offers the ability to use offices, cafeterias, and even dedicated emergency control rooms as a base of operations for the IC. This permits the fire department to access to phones, computers, written and electronic resources, a proper work space, food

***Fig. 11–3.* Reviewing the plan** *(Reprinted with permission from the* Fire Protection Handbook, *19th Edition, Copyright © 2003, National Fire Protection Association, Quincy, MA 02169)*

***Fig. 11–4.* Control room** *(Courtesy of G. Noll)*

and water, and many other useful tools. It also moves the location of the incident management to a clean and well-lighted facility versus standing in the rain or snow in extreme cold or heat. The industrial facility's process or central control room can also provide assistance (fig. 11–4). Control room operators will be able to tell responders the temperature, pressure, and other physical or chemical properties of the material. If the industrial facility is equipped with a gas detection system, information such as the concentration and location of the leak will be easily determined. (See chapter 9 for details.)

With the location of the HazMat incident at the industrial facility, who knows more about their product than the people who manufacture or use as a raw material in their process? Depending on the type of the facility, many locations have a technical or engineering department, research and development department, or a quality assurance department filled with subject matter experts (SMEs) such as chemists, scientists, and engineers. These SMEs can provide the fire department with valuable information about the process, chemicals, and more important, reactions and consequences of the proposed mitigation methods to ensure that you do not escalate the event.

In some cases, off-site experts can be called to provide specialized, technical advice during

Fig. 11–5. Specialist at work*(Reprinted with permission from the* Fire Protection Handbook, *19th Edition, Copyright © 2003, National Fire Protection Association, Quincy, MA 02169)*

an incident. Many times, this is not available to a local fire department, but it may available to the industrial facility. Many industrial facilities have an on-site fire brigade or HazMat team. These teams comprise experts and other personnel specifically trained to handle incidents at their facility or with a specific group of chemicals (fig. 11–5). Some facilities contract out this type of HazMat response to third-party companies that provide these types of services. These companies also have a wide range of experts and equipment to offer.

This brings to the table the issue of cost or financial responsibility. Private industry has more financial means than a public fire department. Because of this, what may seem like a large expense to a fire department might not be an issue with an industrial facility. If, during an event, the need arises for SMEs, heavy equipment, or additional materials, the facility usually has the means to procure these resources (fig. 11–6). If not, a recommendation should be made to the local facility to establish general contract agreements with various vendors to ensure that these resources are always available.

Fig. 11–6. **Cleaning up a spill** *(Courtesy of G. Noll)*

CONCLUSION

The presence of hazardous materials at industrial facilities is a common and daily occurrence. As long as the chemicals remain in their container, pipe, or vessel, they present little to no harm to the public. It is only when they escape their containment that they may threaten both plant personnel and the community. Although, in theory, responses to a HazMat spill is challenging in any setting, a spill or leak at an industrial facility entails many different risks. The volume and physical or chemical state of a material may not normally be found during transportation but only during production. Proper planning and utilization of on-site information, experts, and engineers are essential for controlling and mitigating large hazardous spills at industrial facilities.

NOTES

[1] Chemistry.org . *Production by the US Chemical Industry.* Retrieved August 2006 from http://pubs.acs.org/hotartcl/cenear/960624/prod.html.

[2] Chemistry.org.

[3] Chemistry.org.

[4] Chemistry.org.

[5] Chemistry.org.

[6] Chemistry.org.

[7] Chemistry.org.

[8] Chemistry.org.

[9] Chemistry.org.

[10] Chemistry.org.

[11] Chemistry.org.

12

Petroleum Refining

INTRODUCTION

Refineries are often viewed by outsiders as enormous facilities with complex interrelated processes. Refineries mainly dot the Gulf Coast, but they are also found on both the West and East Coast as well. Anyone who has driven past a refinery, or perhaps has one in the neighborhood, will always remember the large amount of smoke or steam emanating from the complicated network of pipes and vessels. Refineries, in fact, are typically huge facilities with complex interrelated processes; but this does not mean a nontechnical, nonrefinery firefighter cannot understand their fundamental processes and be familiar with their respective hazards. In this chapter, we will discuss the fundamental principles of refinery processes and some important considerations for response to fires in refineries.

DESCRIPTION OF REFINING PROCESSES

Fractional distillation

Refined *hydrocarbons* such as gasoline, kerosene, butane, and other products are extracted from crude oil in various processes throughout a refinery, beginning in the crude unit. Refineries separate these products using the basic principle of *fractional distillation,* which is a way of separating mixtures into their different components. In refineries, the mixture to be separated is crude oil. The principle of fractional distillation is based on the fact that each component, or fraction, in a mixture boils at a different temperature. Remember the term *boiling point* from your hazardous materials (HazMat) training? If you research some of the components of crude oil such as methane, butane, naphthalene, ethylene, and other simple or complex hydrocarbons that form gasoline, diesel, or kerosene in the distillation process, you will see that they all have different boiling points. As crude oil is slowly heated, the first fractions to boil are the lighter fractions because they have lower boiling points.

To illustrate the principle of fractional distillation, imagine a vertical vessel of crude oil and all the necessary equipment to slowly heat it while measuring its temperature (fig. 12–1). As you heat the crude oil mixture, the temperature of the crude oil will slowly rise until it begins to boil. At this point, the first fraction will begin to escape from the crude oil as either gas or vapor. In this illustration, we are using crude oil, which has some liquefied petroleum gas (LPG) components. You will find that as the temperature of your crude mixture is heated to around 31°F, the butane fraction in the crude oil will begin to boil off as gaseous butane. The temperature will remain steady until all the butane boils off. After all the butane is distilled from the crude oil, the temperature of the remaining crude oil mixture will begin to rise again.

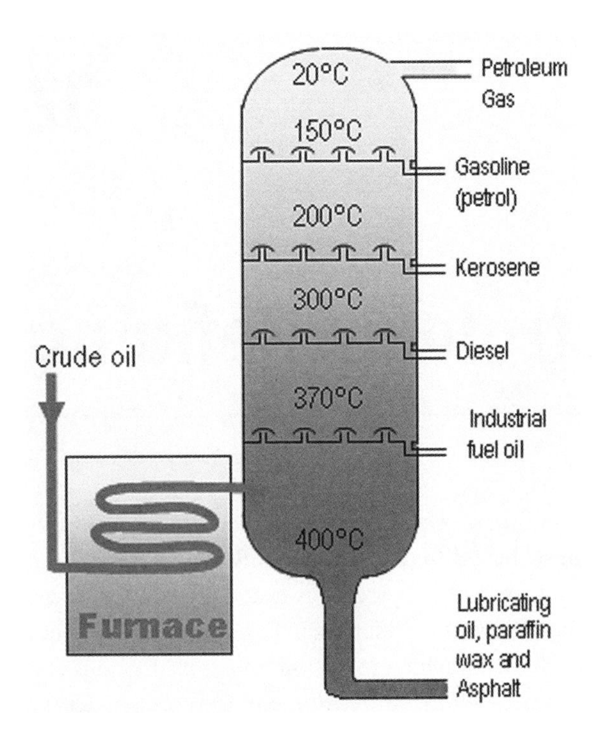

***Fig. 12–1.* Distillation process** *(Courtesy of Theresa Knott, 2004)*

As the temperature rises to approximately 194°F, the crude oil begins to boil again, giving off vapors from a substance known as *straight run gasoline.* In general, about 25% of the crude is straight run gasoline. The straight run gasoline fraction will continue to boil until it has all boiled off, and the temperature in the remaining crude oil mixture will begin to rise again until it reaches the next fraction's boiling point, namely, naphtha. The cycle of boiling off a fraction, temperature rise, boiling, and rising again will continue as the various fractions are distilled from the crude mixture. Finally, the last fraction, residue, remains in the container.

These fractions can be separated by collecting the individual vapors as they boil and cooling the vapors to a point below their boiling point. When a vapor is cooled to a temperature equal to or below its boiling point, it condenses back into liquid.

Generally there are seven primary "cuts" in large-scale fractional distillation operations. Each cut contains raw products within a range of boiling points. For example, the components of straight run gasoline can boil anywhere between 71° and 428°F. The primary cuts are butane and lighter gases, straight run gasoline, naphtha, kerosene, light gas oil, heavy gas oil, and residue. Fractional distillation in a refinery is performed in a tall vessel referred to as a *distillation column.* A distillation column can be separated into two sections, the rectifying section and the stripping section. The rectifying section includes stages where the more volatile cuts are concentrated. The stripping section includes the trays where less volatile cuts are concentrated. There are various types of distillation columns in a refinery, but they all operate on basically the same principle.

How does a refinery collect those vapors and separate them into usable products? First, in a refinery, crude oil is not heated up slowly, which would allow the individual fractions to boil off into individual cuts. Instead, crude oil is pumped into a heater where the crude is heated to approximately 700°F. At this temperature, most of the fractions flash to vapor. The crude stream containing gases, vapors, and some liquids are fed into a distillation column. The column is a tall vertical vessel with horizontal trays strategically distributed throughout it. These trays are filled with small openings. Each opening is fitted with a short vertical pipe, which is in turn fitted with a *bubble cap.*

Remember, there are gases, vapors, and liquids in the distillation column. The gases rise in the column and are drawn off through a large pipe at the top of the column—these are the *overhead gases.* The heavy liquids fall to the bottom of the column, where they are removed through another large pipe. These heavy liquids are the distillation "bottoms." The remaining vapors are forced upward through a series of trays. In each tray, the vapors are bubbled through hydrocarbon liquids, which collect on the top of the trays. As the vapors bubble through the trays, they are slowly cooled and begin to condense back into liquids. Since the trays in the lower section of the column are closer to the heater, vapors passing through them are hotter. Only the heavier fractions condense in the lower trays. The lighter fractions continue rising to the higher trays, where they are condensed by the cooler liquids. As the condensing liquids accumulate on each tray, some of the liquid overflows to lower trays through an opening called a *downcomer.* In the lower trays, the downcomer liquids are reheated by the hotter liquids, allowing any lighter fractions to flash off and rise again to the trays above. Eventually, fractions become trapped in a series of trays where they can no longer be vaporized.

At various points in the distillation column, the individual fractions are drawn off through pipes. Each cut point where individual fractions are drawn off represents products from several trays within the fractionation column. These cuts are further processed and separated into continuously purer products in refining process downstream of the crude unit.

Cracking

As you may have noticed in the discussion of fractional distillation, there were several fractions that are probably not familiar. Light gas oil, heavy gas oil, and residue all have their own uses, but they are not as valuable in the market as other, lighter products such as gasoline and *olefins,* which feed the petrochemical industry. *Cracking* is a process refineries use to increase the volume of more useful products like gasoline from less desirable products like residue.

Cracking is possible because of one fundamental characteristic of hydrocarbons is that they are all made of hydrogen and carbon atoms. In cracking, the larger hydrocarbon molecules are broken apart and rearranged into several smaller hydrocarbon molecules. Cracking can be accomplished in two ways, thermal cracking or catalytic cracking.

In thermal cracking, a product such as residual or “bottoms” is heated to temperatures above 900°F. At this high temperature, the long chains in the hydrocarbon molecule are broken, allowing smaller hydrocarbons to form in a special cooling process. Thermal cracking can be enhanced through various methods such as the addition of hydrogen to the cracked gases or performing the cracking process under a vacuum.

Catalytic cracking takes place in the presence of a catalyst. *Catalysts* are chemicals that promote chemical reactions in other materials without being affected themselves. There are many common materials that are used as industrial catalysts, such as platinum and silver. In the case of the catalytic cracking of hydrocarbons, the catalyst is normally a porous *zeolite* material. The cracking process causes deposits to form on the surface of the catalyst, which must be removed in a separate process known as *regeneration.* Zeolite is similar to the material found in a household water softener, which also must be regenerated to remove deposits.

Some of the hydrocarbons produced in the cracking process are not useful in fuel blending. These materials need to be processed further to be useful as motor fuels.

Alkylation

In the *alkylation* process, smaller unused hydrocarbons are joined to form larger hydrocarbons under relatively low temperatures, approximately 40°F. In fact, you can think of the alkylation process as the opposite of the cracking process. The products generated in the alkylation process are used primarily in fuel blending. Feed stocks for the alkylation process are by-products of the cracking process, primarily *propylene* and *butylene.* These products are chemically forced to combine with *isobutane* to form a new product called alkylate. *Alkylate* is actually a mixture of several chemicals, all of which have a high octane rating, burn relatively cleanly, and have good antiknock properties.

In the alkylation reactor, butylene and propylene “react” in the presence of an acid catalyst with isobutane. The catalyst can be either sulfuric acid or hydrofluoric acid. In addition to alkylate, the reaction produces a number of other chemicals. After reaction, these products are sent to a *settler,* where the dense liquids (mostly acid) are separated from the lighter liquids (mostly hydrocarbons). The hydrocarbon stream exits the settler from the top, and the acid exits from the bottom. The hydrocarbon stream is then moved to another vessel, where it is washed with *caustic* to neutralize any remaining acid. Finally, after neutralization the hydrocarbons are fed into a series of fractionators to separate these new hydrocarbons. The settler bottoms (acid) are recycled to the reactor, where they are reused in the alkylation process.

Alkylation units usually produce alkylate, propane, isobutane and normal butane. The isobutane is recycled to the reactor for further alkylation.

From the chemicals present, it is clear that alkylation units present a multitude of hazards including fire or explosion and materials that are both toxic and corrosive. Flammable hydrocarbon and strong acids are present in both liquid and gaseous forms. Caustic, another corrosive, is also present.

REFINERY TYPES

Processes found in modern refineries designed to separate petroleum (crude oil) into its various fractions are all based on these three simple principles: fractional distillation, cracking, and alkylation (fig. 12–2). A number of processes found in refineries use variations of these principles, but the processes are fundamentally the same as already described. This is not to say that there aren't other important processes that are integral to the basic refinery processes. In fact, there are many processes that support the primary goal of separating crude into various fractions. There are processes that remove contaminants to allow safer, more effective, and economical operation of the cracking and distillations processes. There are processes that clean and purify the fractions after they are separated into usable products or for feedstock into downstream processes. There are also many processes that support the refinery's smooth and continuous operation, such as regeneration of catalyst, production of steam, provision of cooling water, and neutralization of hazardous wastes.

***Fig. 12–2.* View of process units** *(Image by Leonard G. Barton, 2005)*

Refineries can be broken into three basic categories that relate to the complexity and number of processes within the refining facility (fig. 12–3). These definitions were developed for economic analysis in the highly competitive refining industry. For our purposes, they can be useful in establishing a mental picture of the processes found in a particular refinery, the products in these units, and the types of products stored in their tank farms.

Simple refineries

Simple refineries typically have a crude distillation unit, a *hydrotreater*, and a catalytic reformer.

Complex refineries

Complex refineries have all the components of a simple refinery plus an alkylation unit, a catalytic cracker, and a gas plant. The gas plant may include desulfurization and liquidation of natural gas for sale and distribution.

Very complex refineries

Very complex refineries have all the components of a complex refinery plus an olefins unit and a *coker* or other residual reduction unit.

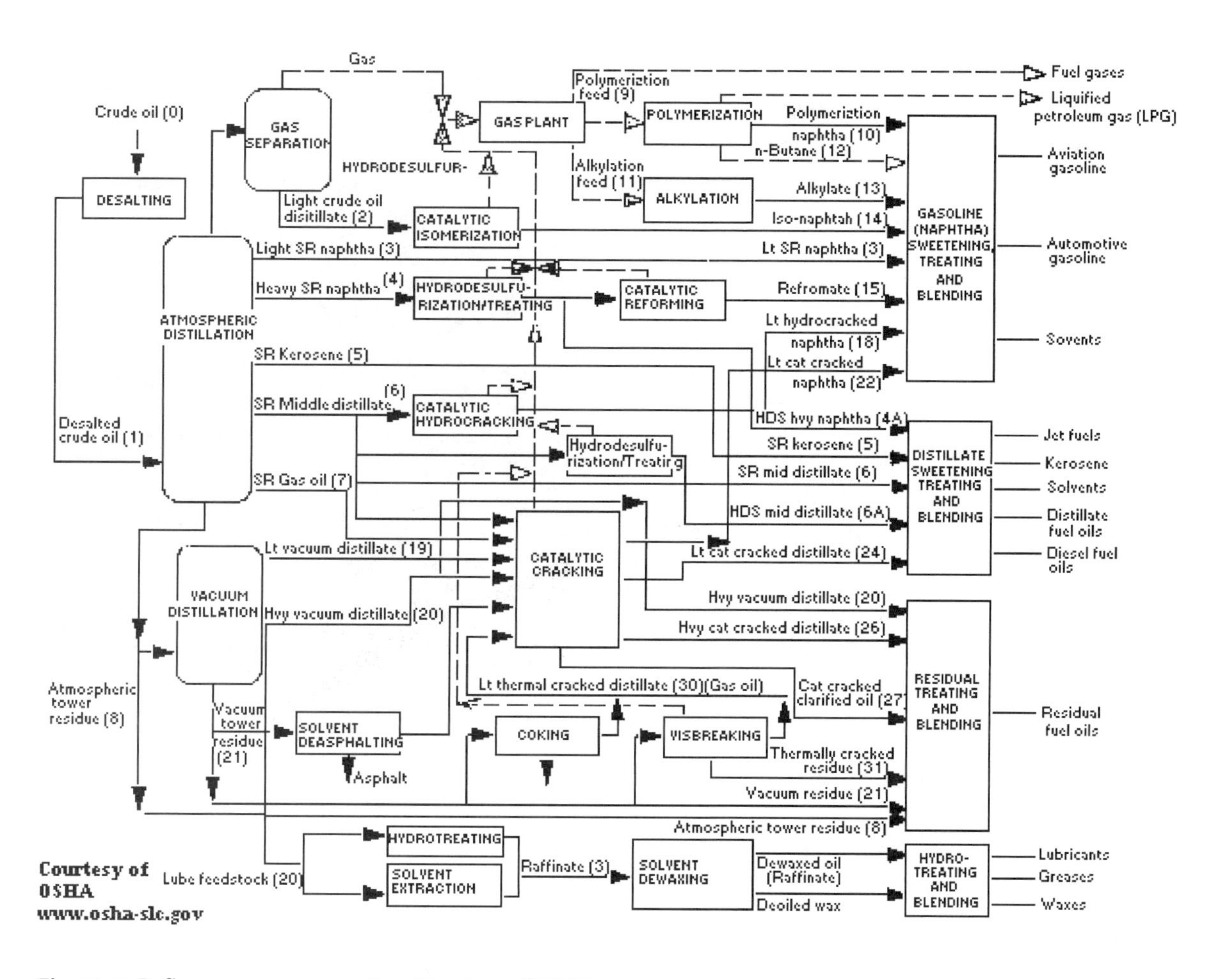

***Fig. 12–3.* Refinery process overview** *(Courtesy of OSHA)*

COMMON HAZARDS

Large complex facilities such as refineries contain a multitude of hazards. The most common hazards consist of toxic gases, vapor cloud explosion (VCE), boiling liquid expanding vapor explosion (BLEVE), jet fires, and pool fires resulting in an unplanned release of flammable or combustible liquids or gases (fig. 12–4). These releases may be slow or large with respect to the amount of mass or volume released, and they may be instantaneous or delayed over time. They may involve small leaks (less than 1 inch in diameter) or large, catastrophic releases such as vessel failures. These hazards exist throughout the plant and are common to all refineries.

***Fig. 12–4.* Process fire (Courtesy of Dave Cochran)**

Toxic gases

Refineries typically contain huge volumes of toxic gases that may be released during a process emergency. Hydrogen sulfide and sulfur dioxide are some of the most common toxic gases found in refineries. Toxic levels of hydrogen sulfide are present in many grades of crude oil and are commonly found in various plant areas such as the distillation unit and sulfur recovery unit. When dealing with hydrogen sulfide, the material safety data sheets (MSDS) and other chemical references should be used to aid in the development of both plans and actual responses. Additionally, several refining processes use toxic and corrosive materials such as sulfuric acid and hydrofluoric acid in their processes. Chemical references should also be used to determine appropriate action during a release.

Vapor cloud explosion (VCE)

A VCE is one of the most dangerous hazards associated with these types of facilities (fig. 12–5). Normally, a response to this type of incident is after the fact and very rarely during or before. VCE is an explosion occurring outdoors that produces damaging overpressures. It is initiated by the unplanned release of a large quantity of flammable vaporizing liquid or high-pressure gas from a storage tank, system or process vessel. These overpressures can damage and destroy pipes and pipe racks, process vessels, buildings, and tanks. (See chapter 13 for more details pertaining to VCEs).

***Fig. 12–5.* Explosion and fire** *(Courtesy of Dave Cochran)*

Boiling liquid expanding vapor explosion (BLEVE)

A BLEVE is the sudden release of a large mass of superheated liquid into the atmosphere. The main cause is usually related to external heating via flame impingement on a shell of a vessel above its liquid level, thus weakening the container and leading to a sudden shell rupture. Although a pressure-relieving valve may be installed, it does not protect against this mode of vessel failure since the shell failure is likely to occur at a pressure below the set pressure of the relief valve. Also, the rate of internal pressure increase and the rate of the relief valve discharge might not provide the necessary relief of the internal pressure. Note that a BLEVE can occur due to any mechanism that results in a sudden failure of containment, including impact by an object, corrosion, or internal heating. The sudden containment failure allows the superheated liquid to flash and expand its volume.

Jet fires

When flammable gases or liquids leaking from a pressurized container such as process equipment or piping are ignited, a flame is formed that takes the shape and characteristics of a cutting torch. This condition is referred to as a *jet fire,* an intense fire that often impinges on adjacent equipment and structures. A jet flame, as it relates to a jet fire, is a function of the diameter of the hole and the velocity or speed of the exiting fuel. (See chapter 13 for more details about jet fires.)

Pool fires

Pool fires are simply spill fires that have accumulated on the ground or solid surfaces to form pools of burning liquids. Pool fires tend to be localized and are mainly a concern for establishing the potential for spreading the fires to other noninvolved areas. The primary effects of pool fires are thermal radiation. For example, a small spill with ignition under a pressure vessel, while initially minor, can lead to a BLEVE if action is not appropriate or is delayed. In this case, fireproofing or drainage is an important consideration in the prevention of escalation of pool fires. (See chapter 13 for more details about pool fires.)

FIREFIGHTING TACTICS

Firefighting in refineries is very different from house or business-related fires. Refinery fires are more complex and require a large amount of preplanning with both your department and the plant's fire brigade. During preincident response planning, the municipal fire officer should determine the organizational structure of any existing fire brigade or emergency response team and develop a working relationship with the team leaders. (See chapter 3 for more details.)

One of the more difficult concepts for a municipal fire department incident commander (IC) to appreciate in responding to refinery emergencies is that he or she is not in full control of the response personnel and strategic decisions. Furthermore, initial responses may only involve exposure cooling versus actual suppression and extinguishment activities, much as it is done during a municipal response. This, however, is the case in safe and successful responses to emergencies in process units such as those found in a refinery. In process emergencies, there are a number of critical considerations and actions for which the fire department IC is not properly trained, staffed, or prepared to undertake. Upon arrival, actions that seem obvious to veteran fire department personnel may not be the appropriate action to take. Large fires in a particular section of the unit would seem to be the logical place to start firefighting operations; however, smaller fires impacting critical process equipment may in fact be the priority. Before undertaking any operations, the fire department IC should consult with the unit process supervisor or plant IC.

Because of their intimate understanding of process operations within a specific unit, the unit process supervisors will normally assume command of emergency operations within the affected process unit. Similarly, the plant superintendent or plant manager typically assumes the role of overall IC due to his or her detailed understanding of the refinery's integrated processes and support systems. This leaves the fire department IC to do what he or she does best—command fire and rescue activities in support of the response strategy. Sometimes municipal fire departments will respond to industrial facilities where they are the only response organizations, but typically in a refinery this is not the case. (See chapter 4 for more details.)

In refinery emergencies, initial exclusion zones are typically easier to establish, delineate, and maintain than in a municipal setting. Evacuation zones are commonly established in a refinery to coincide with unit battery limits or boundaries. Each process area can be delineated according to the physical location of equipment required to operate the unit and the property on which that equipment sits. The outer limits of these units are referred to as the unit *battery limits* and normally are identified by the valves required to isolate the unit from the rest of the refinery. Anything inside the battery limits is *inside battery limits (ISBL)*, and anything outside of the battery limits is *outside battery limits (OSBL)*. Normally during unit emergencies, the entire unit is evacuated of all nonessential personnel. If emergency conditions such as a flammable or toxic vapor cloud extend into adjacent units, then all nonessential personnel in those units are also evacuated. This makes a simple task of defining initial exclusion zones. Initial exclusion zones for responders should simply mirror any existing evacuation zones until the fire department IC and unit process supervisor or refinery IC jointly establish a more refined exclusion zone for responders.

A refined exclusion zone can be established to minimize the distances traveled by personnel operating in the contamination reduction and support zones after consulting with the unit process supervisor or refinery IC on issues such as:

- Condition and stability of the process
- Existing process hazards
- Air-monitoring results
- PPE requirements
- Required response actions

Nonessential refining and response personnel should remain outside the affected units even if the exclusion zone is minimized for response personnel operating at the scene of the emergency.

Often it is safer to continue process operations during an emergency than to stop the process. Safely shutting down a unit may take hours or even days. The personnel and time required to shut down the entire unit can be more effectively utilized in controlling the process and focusing on stabilizing, isolating, and removing inventory from the affected sections of the process unit. Liquid flow in piping, process equipment, and process vessels may be serving to cool the equipment and bring the process under control. The fire department IC typically will not have sufficient understanding of all necessary considerations to make these decisions. Further, the fire department IC typically would not have a full appreciation for how the process unit integrates with other units in the plant and how it is affecting the safe operations of the entire refinery.

If it is safe to continue running the unit, then this option is preferred until the process is sufficiently stabilized to bring the unit down in a safe and controlled manner. If the process operations supervisor or refinery IC determines that the process must be shut down immediately, emergency shut down (ESD) procedures may be implemented. (See chapter 9 for ESD details.) During ESD operations, many elements of the process may not be operated under less than optimum conditions. Flammable and toxic products may be vented to atmosphere or be dumped to flares. Pressures and temperatures in process equipment may rise or fall, thereby affecting any existing fire or leak conditions. Inventories in pipelines, process vessels, and process equipment may stop, increase, or reverse flow, changing the existing fire or leak conditions. Relief valves may open, creating more points of product discharge and expanding emergency conditions beyond the existing exclusion zone.

Fire suppression and detection systems

The fire department IC should determine the presence of any existing fixed fire protection systems, such as water spray and foam systems (chapter 8) or gas detection or ESD systems (chapter 9). These systems can be useful in containing or extinguishing refinery fires. If these systems are present but not activated, they may be activated manually to augment or replace cooling streams. Containment of runoff and effect on total water flow should be considered when activating these systems. Conversely, accumulation of water in the process area, total water flow, and the systems present effectiveness of should be reviewed periodically to determine if the system(s) should be turned off.

During preincident response planning, any fixed systems should be investigated to identify what areas of the unit are covered by specific systems, and what type of support is required from the fire department to ensure that these systems are functioning properly during a fire. The fire department may require special connections or adapters to connect to these systems. Also, preincident response plans should identify the types of extinguishing agents for which these systems were designed and the minimum flow rates required to accomplish extinguishment according to design. Fixed fire and gas detection systems are also important (chapter 9). These systems can tell responders what is going on in and around the incident area with respect to fires and gas leaks. Any specialized training required to support these systems should be identified.

Process area containment

During process emergencies, flammable and toxic liquids may accumulate in the process area and adjacent ditches. Extreme caution should always be taken to avoid overflowing a process containment area. There are a number of process area containment features that should be considered during responses to process fires and emergencies.

Many refinery units have curbed areas around the process equipment to contain any spilled materials. These curbed areas comprise large concrete pads surrounded by concrete curbs. The concrete pads are sloped to allow spilled product and washdown water to collect in process trenches. During a major fire or spill, the entire curbed area may fill with product and firefighting water. This condition is referred to as *pooling*. During pooling, flammable liquids that float on the water surface are spread as the curbed area is filled with firefighting or storm water. When selecting positions for portable monitors and hose teams, the IC should recognize that the area involved in the fire may grow as the result of pooling within the curbed area(s).

Process trenches collect liquids from the curbed area and route them to process sumps where solids and liquids are separated. Often a process sump will also separate liquids into water and oil phases. From the process sump, liquids are pumped off for treatment either in other sections of the unit or at the wastewater treatment plant. Process sumps may become involved in fire. In a major fire this is not a huge concern, because these sumps do not include equipment that will cause a catastrophic event. However, if fire has not extended to the sump upon arrival, caution should be exercised because a covered sump may contain flammable vapors in the explosive range.

Process sewers are similar to process trenches in that they collect accumulated liquids and route them to process sumps or other treatment or collection areas. The difference is that process sewers are underground pipes rather than open trenches.

One effective tactic to minimize the potential for overflowing process curbs, trenches, and sumps is to foam down areas where burning liquids have accumulated, allowing excess foam to flow into process trenches and sumps. Where the potential exists for fires to extend to covered sumps or process sewers, preemptive foam application should be considered to minimize accumulation of explosive atmospheres.

Process trenches and sewers can become an avenue for extension of fire to other sections of the process unit. Process trenches and sewers should be monitored periodically to ensure that fire is not migrating through them and flammable liquids are properly blanketed with foam.

During prefire inspections, process engineers should be able to provide information on the drainage capacity of process areas and any data on pooling in the process areas. Process engineers in the wastewater treatment plant should be questioned regarding the effect of foam solutions on the wastewater treatment processes. Some treatment processes cannot tolerate foaming agents, and certain foam solutions may be toxic to the microorganisms in biotreatment systems.

Vessels found in the process area will also typically have secondary containment systems to collect spilling product in the event of a vessel failure. These systems are simply concrete or soil dikes surrounding the vessels. During a process unit fire involving these vessels, caution should be taken to prevent overflow of burning or toxic materials from these containments. When applying cooling streams to these areas ICs should establish a system to monitor and control dike levels.

Process equipment such as fractionators, columns, and reactors create a fire hazard because of the large volume of hydrocarbons present in this equipment. A leak with an ensuing fire can be devastating. Even more disastrous would be a leak without immediate ignition leading to a VCE. Process equipment can range in sizes up to 300 feet in height and 30 feet in diameter, or for horizontal vessels, more than 100 feet in length. Process equipment and vessels can be under vacuum, at normal atmospheric pressure, or pressurized. These vessels are constructed of a variety of components ranging from various types of steel, valves, gauges, measuring equipment, insulation, and pipes. Vessels are designed, constructed, and tested according to various American Society of Mechanical Engineers (ASME) standards. Typical process vessels may appear to extend fully to the ground; however, upon closer examination, these vessels are actually sitting on a steel "skirt" that is vulnerable to flame impingement. Passive fire protection such as fireproofing is normally used on critical process vessels.

The use of a ladder truck may be beneficial in these types of incidents, to reach the fire at higher elevations (fig. 12–6). Cooling can be from the ground as well, or from truck-mounted "deck guns" or monitors. Note that many of these vessels are insulated; therefore, if a fire occurs internally, cooling from the outside may have little or no effect at all. It may become necessary to get water inside via a manhole or other access point. Gaining this access may require climbing the man-ladder on the side of the vessel. Although a known tactic in industrial firefighting, this needs to be evaluated on a case-by-case basis. Vessels under pressure require rapid attention due to the BLEVE potential versus vessels that are at or near atmospheric pressure. Many of these pieces of equipment are known to possess pyrophoric materials as well. Caution should be exercised when operating in the presence of pyrophoric materials.

***Fig. 12–6.* Elevated streams** *(Courtesy of Dave Cochran)*

One guaranteed condition in responding to a refinery unit fire is that there will be numerous pipes running in every direction. Pipes of various sizes are used to transport product, catalysts, process wastes, process-cooling water, fire water, and a host of other fluids throughout the unit. These pipes will be found in pipe racks on the ground, elevated above the road and process equipment, running in the horizontal position, in the vertical position, and making loops. All these pipes serve a specific purpose and may be of significant concern to responders when fire impinges on them. Always notify the process unit supervisor when fire is located in or near piping. Hazards in this area can include VCEs and pool fires. (A more detailed discussion on pipe- and pipeline-related response is in chapter 33.)

Valves are located everywhere in continuous process plants such as refineries. There are a multitude of valve types, components, materials, and construction. Valves are used to control process flows; isolate pumps, vessels, and other process equipment; and for other common uses. Valves can range from 1 to greater than 48 inches in diameter and can be operated manually or motorized in their operation, either locally or remotely. Valves create a weak point in a pipe or vessel. Because of this weak point, combined with the large number of valves in refineries, these tend to be the locations of the majority of leaks, spills, and fires. Valves can be a source of both fires and/or explosions.

Caution should always be exercised in closing valves. The isolation of valves may cause a rise in the pressure of interrelated equipment or an increase in the volume of fire in other sections of the unit. Always consult with operations personnel prior to opening or closing any valve. When actually closing the valve, pay close attention to any change in sounds within the piping or movement of the piping or equipment. Spray and pool fires and toxic releases are common around valves because valves are the weak points. During these types of incidents, industrial tactics such as multiperson hoselines, foam lines, and cooling monitors may be needed. If your department has not been training on fighting valve-related fires, it may be best to allow the plant's fire brigade to handle this task.

The most common fires in refineries are pump fires. Like valves, pumps are very common in process industries. Pumps range in types, capacities, construction, and materials. They range in capacity from 100 to more than 7,000 gpm and can be used to pump liquids such as cooling, fire, and other utility water; crude oil; refined products such as gasoline and diesel; and other production liquids. The main hazards associated with pumps are spray fires. These can cause impingement on surrounding equipment and rapidly spread fires to other pumps or equipment in the area. Toxic releases, spray, and pool fires are common around pumps. Pump and pump seal fires are commonly referred to as three-dimensional fires. Although there are several methods you can use to combat pump-related fires, the most common involve using dry-chemical extinguishers or Hydro-Chem® technology. The Hydro-Chem nozzle combines the effectiveness of an automatic water/foam nozzle, while incorporating the delivery of a dry-chemical in the same device. This gives you the advantage of dry-chemical, but at further distances not normally found with using on a dry-chemical extinguisher. Another method for pump-related fires involves similar tactics as those used in valve fires, that is, isolating the pump and allowing the fuel to be consumed while keeping the area cool.

Compressors are similar to pumps, in that they move the gaseous form of product and can often be isolated remotely by motor control centers and motor-operated valves. Unlike pumps, compressors found in refineries are typically huge pieces of equipment that move enormous volumes of product (gas). The difference in pumps and compressors is that compressors compress gas to increase pressure by reducing their volume. The compression process also may increase its temperature. Although related, compressors and pumps increase pressure on a fluid (liquid or gas) and can both transport the fluid through a pipe. Since gases are compressible (liquids are not considered compressible), the main job of a compressor is to reduce the volume, while a pump's main job is to raise the liquid pressure to overcome head losses in order to pump the liquid to another location. The compressor's suction and discharge valves are typically larger than can be quickly closed by hand. Compressors may be driven by steam turbines or large electric motors. Special attention should be focused on keeping compressor structures from failing. Because of the large capacity of these machines, failure of their structures may significantly increase the volume of fire. Fortunately, compressor structures are usually well designed and constructed of heavy concrete castings. Gas compressors are prone to vane failures on the impellers and can cause large catastrophic loses. Caution should be used during cooling operations so as to not cool the compressor in such a way that it may cause the compressor to fail.

A heat exchanger is a device built for efficient heat transfer from one fluid to another, whether the fluids are separated by a solid wall so that they never mix, or the fluids are directly contacted. They are widely used in refrigeration, air conditioning, space heating, power production, and chemical processing. Heat exchangers may be classified according to their flow arrangement. The most commonly used in a refinery is a shell and tube heat exchanger. A shell and tube heat exchanger consists of a series of finned tubes, through which one of the fluids runs. The second fluid runs over the finned tubes to be heated or cooled. Heat exchangers contain heated gases or liquids, and when leaking, pressurized fires may occur. Because of their process, heat exchangers can contain large volumes of flammable liquids. When processes are isolated, heat exchanges can become closed containers subject to pressure failure or BLEVE under fire conditions, as well as spray fires, and they can be the sources of large pool fires. Even when a particular piece of process equipment is not directly involved in the fire, due to process-piping connections, it may represent the weakest link in an interrelated system that is impacted by the fire and fail without warning. Heat exchangers, because of the amount of liquid holdup, can be prone to VCE, BLEVE, and pool fires.

Knockout drums are vessels that collect or knock out condensed liquids in a system that is designed to transport gases. Condensed liquids are collected in the bottom of the knockout drum and are pumped off to other processes. Gases continue through the drum to their intended destination. During a process upset such as a fire, the knockout drum may fill with liquids due to excessive flow or carryover from upstream process equipment. This is a significant process problem because liquid may then carry over into the process intended for gases. This is especially critical if the process is a furnace, incinerator, or flare. When the liquids reach the furnace, incinerator, or flare, they may flash to gas and ignite, potentially causing an explosion. Liquid carryover in flares can result in burning liquids being spewed from the flare tip and extending fire to more sections or the refinery.

Caution should be exercised when applying cooling streams near furnaces or incinerators even after they are shut down. Any water introduced to the combustion chamber of a furnace or incinerator can result in a steam explosion.

Many process units have storage tanks within the unit to store enough feed stock or product for daily operations. These tanks are typically referred to as *day tanks.* Additional tankage is located in the tank farm, port, marina, or other large storage depot. During preincident response planning, the presence of any *day tanks* should be noted. (Storage tanks are discussed in greater detail in chapter 31.)

All responders should presume respiratory hazards are present and use self-contained breathing apparatus (SCBA) until instructed otherwise by the refinery operations personnel; after removing SCBA, they should continuously monitor the atmosphere. Skin contact with products and runoff water should be avoided in all cases, and decontamination procedures for personnel and equipment should be specified by the plant IC or safety officer.

Generally, the preferred method of controlling process equipment fires is to isolate the equipment remotely and remove inventory of any remaining product to *flares,* incinerators, storage tanks, or uninvolved process equipment. Another less desirable option is to allow the fire to burn out under controlled conditions or bleed the product to the process sewer. Remote isolation can be accomplished by deenergizing pumps at the motor control center or using motor-operated valves. Deenergizing pumps and compressors may not fully stop the flow of product, but this procedure should significantly reduce the flow of any escaping product.

Refinery fires require large flows for long periods to cool exposures and attain control. The typical refinery fire is not extinguished; fuel is isolated, and the fire is allowed to burn out while huge quantities of water are applied to cool exposed process equipment and structures. It is not uncommon to find industrial fire apparatus capable of flowing in excess of 4,000 gpm (fig. 12–7). Many refineries have fire-water systems that exceed 20,000 gpm flow capacity (fig. 12–8). Before the emergency response team arrives, the unit operations personnel may have established more than 8,000 gpm in cooling water from fixed monitors and deluge systems.

Fig. 12–7. **Large-capacity industrial pumper**

Fig. 12–8. **Large flow** *(Courtesy of Dave Cochran)*

During preincident response planning, consideration should be given to water flow requirements and sources of fire water. Large-diameter hose (LDH) supply lines may be laid from adjacent process units and remote locations within the refinery (fig. 12–9). Supply hose routes should be carefully selected to avoid trapping apparatus or creating road blocks. Hose should be laid to the side of the roads to avoid snaking. Consideration should be given to using culverts and areas under pipe racks to avoid obstructing roadways. When it is necessary to lay hose across roads, keep in mind that in addition to other fire apparatus, specialized equipment such as vacuum trucks and cranes may be required to access the process unit. The typical LDH hose bridge may not support this heavier equipment.

Fig. 12–9. **Large-diameter hoses** *(Courtesy of Dave Cochran)*

One of the first activities undertaken by responders at a refinery fires should be the placement of effective cooling streams onto involved equipment and structural exposures (fig. 12–10). This defensive measure is critical in minimizing structural collapse and containing the spread of fire. Pooling liquid fires, liquid jet fires, and gas fires that are not affecting the structural integrity of vessels, structural supports, or critical valves are secondary considerations. The rule of thumb is "wet steel is cool steel." If cooling water does not vaporize on the surface of process equipment or steel structural members, then the equipment is cooled below the boiling point of water and is cool enough to maintain its structural integrity. Here are some good, general rules of thumb for dealing with refinery fires:

1. Maintain communications with a process operations representative.
2. When possible, extinguish pooling fires and let jet fires burn.
3. Use fixed systems, fixed monitors, and portable monitors as the primary means of cooling.
4. For flame impingement, apply cooling streams to vessels, the base of columns, vessel supports, pipe rack supports, piping, and any involved process equipment supports.

5. In process areas, be careful of your footing, especially in water that has pooled. Often fires and explosions have damaged or displaced trench covers, platform grating, and manhole covers.
6. Handlines are for blocking small valves and overhaul.
7. Avoid overflowing containment systems such as curbs, dikes, and walls.
8. Isolation valves should never be operated without concurrence of the process operations supervisor or his or her representative.
9. Any increase in sound volume or pitch, flame height, or vibration should be carefully evaluated. Immediate evacuation of the area may be necessary.
10. Constantly monitor process stability upstream and downstream of the immediate emergency area.

***Fig. 12–10.* Multiple streams** *(Courtesy of Dave Cochran)*

CONCLUSION

Refineries are typically huge facilities with complex interrelated processes containing a multitude of hazards. This is why a detailed incident plan and close cooperation with local plant personnel is essential. Training and site familiarization are extremely important in these types of facilities. Upon arrival, establishing cooling streams and isolation of the process unit, pipe, or pump should be a primary concern. Establishing a close working relationship with the local plant fire brigade will also provide a better working environment, increasing the likelihood of successful control and extinguishment.

13

Chemical and Petrochemical Facilities

INTRODUCTION

Chemicals and petrochemicals are used in the production of agricultural, plastics, rubber, paint and coatings, automotive equipment, pharmaceutical, electronics, foams, and many other products. The facilities that manufacture chemicals and petrochemicals (fig. 13–1) share many common esthetic features with each other, as well as with the refining facilities (fig. 13–2). Therefore there is a tendency for those who are not familiar with the industry to refer to any facility with tall silver vessels and smoke stacks as either a chemical plant or a refinery. Knowing the basic differences in these industries is helpful to the responder in understanding the general hazards associated with emergencies in these facilities.

Fig. 13–1. **Purified terephthalic acid plant**

Fig. 13–2. **Oil refinery** *(Courtesy of G. Noll)*

The term *chemicals* is very broad, and to describe chemical production and the processes used to manufacture them is even more challenging. The water you swim is a chemical solution, but we would not refer to a swimming pool as a chemical facility. A chemical can be defined as a substance that is produced by various processes and reactions in an attempt to manufacture other substances or products. In this context, *chemical* is typically used in reference to the production of an *inorganic or organic* substance. An example of an inorganic substance is phosphorus or chlorine, while an example of an organic chemical is a hydrocarbon such as crude oil; gasoline; or polymers such as polyethylene, polyvinyl chloride (PVC), or nylon.

Petrochemicals include a wide spectrum of petroleum-based chemicals that are used in the processing or manufacture of numerous *downstream* products. Defining the term *petrochemical* can become problematic due to its varied application in our society and perceptions of those not familiar with the interrelation of the refining, chemical, and petrochemical industries. *Petrochemical* has been used differently by various industries and organizations to define broad groups of chemicals that contain hydrocarbons (fig. 13–3). These chemical processes produce chemicals that are the direct feedstock for manufacturing petrochemicals in adjacent processes. For example, chlorine does not contain hydrocarbons, but chlor-alkali plants are commonly found in petrochemical plants where chlorine is used in the manufacture of vinyl chloride, which is a petrochemical. Even in the industry there are certain petrochemicals like ethylene and propylene that can be manufactured in a petrochemical facility or refinery. Generally, it is accepted that refined products used as fuels such as gasoline and kerosene are not considered to be petrochemicals. Likewise, petroleum-based lubricants such as oils and greases are not considered to be petrochemicals. Fuels may, however, contain petrochemical additives to enhance combustion, increase engine performance, and reduce emissions.

***Fig. 13–3.* Chemical Plant** *(Courtesy of G. Noll)*

In general, the biggest difference between hazards at a chemical plant versus a petrochemical plant is toxicity and flammability. Responses to chemical plant emergencies tend to involve very toxic and less flammable materials, while petrochemical emergencies include both toxic and flammable hazards. This is not always the case, and both chemical and petrochemical plants can include toxic, flammable, and explosive materials. In developing chapters for this book, inorganic chemicals were not allocated a specific chapter of their own; therefore, for the purpose of this chapter, we will discuss the primary petrochemicals as well as interrelated chemical processes.

DESCRIPTION OF PROCESSES AND THEIR HAZARDS

There are a number of chemicals that are manufactured in huge volumes by chemical and petrochemical plants that the consumer will never see in their original form. These products serve as the building blocks for more familiar products such as plastics, textiles, detergents, and countless other consumer products. This group of chemicals is known as *base* or *commodity* chemicals. For the purpose of this chapter, we will restrict our discussion to specific base chemicals.

The process unit that produces *ethylene, propylene,* and associated petrochemicals is known as an *olefins* unit, because the primary products are olefins. Olefins are unsaturated open-chain hydrocarbons.

Olefin units are typically divided into two sections—the hot section and the cold or recovery section. The hot section includes the furnace and quenching process. The recovery or cold section includes acid removal, chilling, compression, *demethanizer, deethanizer, depropanizer, debutanizer,* acetylene recovery, acetylene *hydrogeneration,* ethylene fractionation, and propylene fractionation. The cold section is referred to as such because product recovery is conducted under *cryogenic* conditions.

In the furnace, olefins are *cracked* from naphtha, gas condensate, ethane–propane mix, or other light liquid hydrocarbons usually in the presence of steam. These liquids are heated in tubes within the furnace to temperatures between 1,380 and 1,740°F, where the high temperature cracks the hydrocarbons into various products including ethylene, propylene, pyrolysis gasoline (pygas) and other hydrocarbons. The cracked products are very reactive and are quickly quenched with quench oil and water to stop their further cracking and to prevent reaction between the hot olefins in the stream. Olefins furnaces may also be referred to as crackers or reactors. An unwanted material forms on the inner walls of olefins furnace tubes during production of olefins. This material is called *coke,* and if left unchecked, it would build up, causing uneven heating and restriction of flow in the tubes. The tubes in olefins furnaces require decoking on a regular schedule—typically every 7 to 10 days. To maintain a constant flow of olefins to the cold section during the decoking process, most olefins units have multiple furnaces constructed adjacent to each other. Typically, there will be five to eight furnaces, with a different furnace going through decoking every day. During a furnace emergency, the operations department will be trying to deinventory the furnace to the flare. Normally the *flare*(s) cannot deinventory all affected equipment simultaneously, and adjacent furnaces will remain online until operations can stabilize or deinventory the most critical equipment.

In the cold section, the product stream coming from the hot section is cleaned and separated into individual petrochemicals. The feed from the quench area has some acids in it, which are removed, and the stream is then compressed and chilled to form a cryogenic product. This cryogenic stream is then sent through a series of towers to separate methane, ethane, propane, and a mixture of olefins. This is an efficient process because the individual components are separated through a fractionation process, where they are simply allowed to flash back to gas in cascading order based on their boiling points.

If you can remember the difference between the hot section and the cold section, you've got a good leap on tactical considerations in the olefins unit. In the hot section, there will normally be gaseous hydrocarbons and an ignition source. A positive factor regarding olefin furnace emergencies is that they have almost always peaked in intensity by the time you arrive. Do not let this detail lull you into a false sense of security. If a boiler tube fails, you will have a fire burning inside of a furnace, and you can let it burn. If the tube failure blows the wall off one of the furnaces, however, you will have a fire impinging on other furnaces. Avoid putting water in the furnace(s). It will be hot for days, and cooling does not provide any advantages; at worst, you will create a steam explosion by putting water in the furnace. Any escaping product in or downstream of the furnaces will likely be gaseous flammable vapors. Due to the presence of operating furnaces, these vapors will most likely have ignited prior to arrival. Focus should be on cooling structural components and critical valves. In the event that ignition has not occurred, evacuation of nonessential personnel and vapor cloud management should be a primary concern.

Product releases upstream of the furnaces will be liquids with very low flash points. Pooling liquids should be blanketed with foam whether they are burning or not. Process area trenches and process sewer systems should be blanketed with foam to minimize generation of flammable vapors.

In the cold section, almost everything is cryogenic. If you have a pooling liquid and you put water on it, you will increase the rate of vaporization. Use foam blankets (low or high expansion) to reduce the rate of vaporization and water fog to dilute and disperse vapor clouds. To prevent fugitive vapors from reaching ignition sources, portable or fixed monitors performing vapor disbursement techniques should be used.

Ethylene could be considered the king of base chemicals because it is the primary feedstock for more products known to consumers than any other petrochemical. Ethylene is a feedstock for common products such as PVC, polyethylene plastic, and antifreeze. Ethylene occurs naturally in the tissue of certain fruits, flowers, and vegetables. Ethylene for commercial purposes is produced in an olefins unit.

Ethylene is normally a gas, but it is found as a liquid in the cold section of the olefins unit and associated processes. Liquid ethylene is a cryogenic that flashes to vapor rapidly but can pool if the leak rate is sufficient. Ethylene can be found in manufacturing processes for the following substances:

- Ethylene oxide and glycol (ethylene glycol)
- Ethylene dichloride and vinyl chloride monomer
- Polyethylene
- Ethylbenzene and polystyrene
- Polyvinyl acetate
- Ethanol (industrial)

Ethylene is extremely volatile. This volatility, coupled with its high flammability, creates a very intense fire when ignited. If ethylene in the atmosphere has not ignited, there is a potential for an unconfined vapor cloud explosion (VCE). Copious volumes of water should be deployed to cool exposed vessels, piping, and process structures.

Propylene ranks in the top 10 chemicals manufactured worldwide.[1] The primary use of propylene is in the production of polypropylene plastic. Polypropylene has many applications and can be found in consumer products including food packaging, textiles, laboratory equipment, rope, and automotive parts.

Propylene is a volatile gas, but it is normally stored and transported in process piping as a liquid. Liquid propylene is a cryogenic and easily pools while vaporizing when leaks occur. Propylene has a pungent odor that has been described as that of acetylene or garlic.

Fires involving propylene require large amounts of water due to the high heat output of propylene. In process and storage fires, cooling streams should be placed into service where flames impinge on vessels, process equipment piping, and supporting structures. Where propylene gas has not ignited, water spray should be used to disperse the vapor cloud. Caution should be taken to avoid casualties associated with VCEs. Master streams should be set up and allowed to operate unmanned where possible.

Ethylene oxide ranks 26th in worldwide chemical production.[2] As an industrial chemical ethylene oxide is the precursor to ethylene glycol (antifreeze). Ethylene oxide is used in the manufacture of *surfactants,* detergents, adhesives, and pharmaceuticals. Ethylene oxide is also a sterilizing agent and a defoliant.

Ethylene oxide is derived from the reaction of ethylene and oxygen in the presence of a silver catalyst in the tubes of a fixed bed catalytic reactor. The reaction is *exothermic* and occurs at temperature ranges of 400 to 550°F under pressures of 150 to 400 psi. Ethylene oxide is carcinogenic and toxic, and it exhibits an extremely wide flammable range (3%–100%) and is known to undergo violent (explosive) *polymerization.* Due to the potential for violent polymerization, ethylene oxide processes and storage facilities usually include an emergency venting system where ethylene oxide is either vented to atmosphere or vented to a water *sparge* system.

Atmospheric venting events are extremely hazardous due to the toxic and flammable characteristics of ethylene oxide. Emergency ethylene oxide venting is normally released to the atmosphere from a tall vent stack in the process structure away from normally populated areas. Still, ethylene oxide gas is considerably heavier than air, and the vapors fall to ground level as they disperse in the atmosphere. An atmospheric vent system should have a preventing alarm system to alert personnel to the impending release of ethylene oxide.

Water sparging involves piping the ethylene oxide into a water reservoir. Ethylene oxide is fully miscible with water. In the sparging process, ethylene oxide is mixed with water thereby reducing the flash point of the resulting solution. When ethylene oxide is diluted in water, it is reduced to a 2% concentration, the product has a flash point of 37°F; at 1% dilution in water, the flash point rises to 88°F. In most cases, emergency sparging processes are designed for durability and not precision. As a result, the dilution typically is not fully effective, and ethylene oxide vapors may be present in the water reservoir area.

Responses to an operating ethylene oxide unit should always consider the possibility of an emergency dump of ethylene oxide into the immediate area. As noted previously, the feedstocks for ethylene oxide include ethylene and oxygen. Any response to an ethylene oxide process unit could involve these feedstocks as individual hazards. Ethylene is highly flammable and can generate explosive vapor clouds. Oxygen is a cryogenic and supports combustion.

Fires in ethylene oxide are intense and require huge volumes of water. A positive factor regarding fires in ethylene oxide process units is that you don't need to be as cautious concerning cooling streams. With the exception avoiding pooling ethylene and oxygen liquids, there is very little damage resulting from the stray water stream.

Also, there will normally be no concern for overflowing dikes and containment areas since the primary chemicals are gases. Caution should be taken when dealing with liquids in the glycol reaction area, because mild acids may be used as a catalyst.

Methanol is the 19th highest volume chemical manufactured worldwide.[3] Methanol is an important base chemical used in the productions of acetic acid, formaldehyde, dimethyl ether (aerosol propellant), and a host of downstream products including plastics, plywood additives, paints, explosives, and textiles. Methanol can also be used as a motor fuel, fuel additive, solvent, and antifreeze. Methanol is transported by pipeline, ship, barge, railcar, and tank truck.

Methanol is produced in a series of steps starting with steam-methane reforming (SMR). In the SMR process, methane (natural gas) and steam are reacted in the presence of a nickel catalyst to form synthesis gas (syngas). Syngas is a mixture of carbon monoxide, carbon dioxide, and hydrogen. Syngas is subsequently synthesized into raw methanol in the methanol reformer. The raw methanol is then refined by distillation.

Methanol has a flash point of 52°F and a boiling point of 418°F; therefore, in storage at room temperature, it is normally found as a liquid giving off flammable vapors. However, in the processes area, it will most likely be found as vapor or a liquid that is heated above its boiling point. Under these conditions, if it has not already vaporized, it will quickly flash to vapor—presenting the possibility of a vapor cloud explosion.

Methanol flames are almost undetectable, and caution should be taken when approaching any methanol spill to avoid the possibility of unseen flames. The first indicator of a methanol pool fire will be the smoke and flame produced by Class A materials within the pool area, such as insulation, grass, wood products, and plastic materials. Fire may also be detected by discolorations of painted surfaces.

Methanol is a polar solvent, and it is miscible in water. This creates a challenge when trying to maintain a foam blanket but also presents an opportunity for dilution of small to moderate spills with copious amounts of water. As previously mentioned, the methanol process requires generation of carbon monoxide and hydrogen in the SMR process. Any gas leaks in the process area should therefore be considered both toxic and flammable.

Chlor-alkali facilities produce chlorine and sodium hydroxide (lye or caustic) through an electrolytic process. In addition to these chemicals, the chlor-alkali process produces hydrogen as a by-product. In the chlorine cell, an electric current passes across electrodes (an anode and a cathode). Concentrated brine (saltwater) in the cell is separated by the electrical current into its individual components: chlorine, sodium hydroxide, and hydrogen. The chlorine and sodium hydroxide are separated within the cell by a membrane with chlorine gas, hydrogen gas, and cell liquor (liquid sodium hydroxide) coming out of different nozzles (openings) on the cell. Many chlorine cells are connected together to form a large circuit with electricity flowing through the individual cells. The circuit may be referred to as a train or bridge and can reach several hundred feet in length. When individual cells are removed, the electricity required to power a chlorine unit is equal to that consumed by a small town. Most often the energized electrical conductors connecting the cells are large copper plates (busbars), which are not insulated and are open to the atmosphere. These busbars may be located in trenches below the cells or running overhead. Electrical currents in the cell buildings are so large that they generate their own magnetic fields. The magnetic fields can be on metallic objects such as belt buckles, steel tools, and steel shell fire extinguishers. When responding to emergencies in a cell building, special precautions should be taken to avoid contact with energized process equipment and electrical busses. Overloads originating in chlor-alkali plants have been known to trip the electric grids in adjacent plants and even entire regions. Responses to chlor-alkali process units must always include strategies and work techniques to avoid electrocution. Personnel working in chlor-alkali units normally wear special *dielectric* footwear for additional protection against electrocution.

There are three types of chlorine cells: diaphragm, mercury, and membrane. The primary differences between these cells play a role in the health and safety aspects of emergency response.

The membrane of a diaphragm cell is made of asbestos. The cathodes are stainless steel, and the anodes are graphite blocks. During anode replacement, the graphite blocks are secured into place within the cell chambers by molten lead poured around the base of the blocks. The lead cools to provide both a structural mounting and a conductor of electricity. In addition to other precautions mentioned in this chapter, special precautions to avoid exposure to asbestos fibers and lead vapors must be taken in both the production and cell maintenance areas of diaphragm cell units. Diaphragm cell use is declining worldwide due to the health and environmental issues associated with asbestos and lead.

The mercury cell produces a high-concentration caustic because it uses mercury as the cathode. In this process, mercury, a heavy liquid, pools on the floor of a large cell. The graphite anode is mounted above the mercury pool. This process uses large quantities of mercury. Mercury is toxic, and exposure should be avoided when responding to a mercury cell unit. Mercury cell use is declining due to the heath and environmental issues associated with mercury.

Most chlor-alkali facilities being constructed today use ion-exchange membrane technology because of the savings in capital construction and operating costs. Ion-exchange membranes are constructed of synthetic materials that do not contain asbestos. These cells provide more efficient disassociation of chlorine and sodium hydroxide from brine than do diaphragm or mercury cells. Hazards associated with membrane cells are chlorine, hydrogen, sodium hydroxide, and electricity.

Cell liquor is collected and sent to the caustic unit, where it is concentrated for captive use or for sale as various concentrations of sodium hydroxide including 30%, 50%, and dry powder or pellets. Sodium hydroxide has many industrial uses, including feedstock in glass manufacturing, paper production, detergents and soaps, chemical peeling of vegetables and fruits, and neutralization of acids and textiles. Sodium hydroxide is not flammable, but it attacks human flesh, and concentrated sodium hydroxide can quickly cause chemical burns. Sodium hydroxide in very small quantities can damage the eyes and cause blindness. Any response to a chlor-alkali or caustic unit should include precautions for skin and eye contact. Equipment and protective clothing must be carefully decontaminated and examined for chemical attack. Sodium hydroxide is extremely aggressive in attacking leather materials. Leather and similar animal products cannot be decontaminated.

Chlorine gas is toxic and presents a significant inhalation hazard. Chlorine gas was used in World War I as a poisonous gas. Any response to a chlor-alkali unit or process involving chlorine must be conducted using self-contained breathing apparatus (SCBA). Even if chlorine is not present initially, there is always the possibility of a chlorine release. Chlorine gas (cell gas) produced in the cell is collected and purified. In most cases, chlorine gas is compressed to liquid (liquefied) and stored for sale or use in downstream processes. Liquid chlorine can be found in storage tanks, pipes, cylinders, railcars, ships, and barges. Anytime chlorine liquid escapes, it will rapidly begin to vaporize. Under ambient temperatures and pressures, liquid chlorine expands by a ratio of 450 to 1 when vaporizing to form chlorine gas. Chlorine vapors are denser than air and will collect in low areas such as trenches and pits.

Cell gas is not liquefied in some modern petrochemical plants. In these processes, the cell gas is purified and immediately consumed in the gaseous form by other processes within the manufacturing complex. An example is the modern ethylene dichloride (EDC) process, in which ethylene dichloride is produced directly from cell gas. Since the chlorine is not liquefied, the potential for a catastrophic release from the chlor-alkali plant is significantly reduced. Careful survey of processes downstream of the chlor-alkali unit, including rail and barge terminals, should be conducted during preemergency response planning to determine if alternate means of importing chlorine are available. Chlorine-dependent processes may have alternate means of receiving chlorine when the chlor-alkali unit is shut down for major maintenance. In this scenario, railcars or barges of liquid chlorine may be brought in as a source (fig. 13–4). The process would include a liquid chlorine gasifier.

Fig. 13–4. **Rail loading** *(Courtesy of G. Noll)*

Responses to chlor-alkali units also present an opportunity to encounter hydrogen fires. Hydrogen is easily ignited; in fact, it can ignite from the heat generated when it is released to atmosphere from a leaking vessel or pipe. Hydrogen is present in the cell building and associated processing areas. Hydrogen burns with a colorless flame and can be difficult to detect. The first indication of a hydrogen fire may be discoloration of nearby painted surfaces or other materials that have ignited from the hydrogen fire. Caution should be exercised to avoid walking into a hydrogen fire. Walking with a broom extended in front or throwing sand will help to identify the outer edge of a hydrogen fire.

Precautions against chemical burns should be taken when responding to chlor-alkali units. Also, consider the probability of highly corrosive runoff from water streams, and plan for runoff containment in any preincident emergency planning.

OTHER COMMON HAZARDS

Chemical and petrochemical facilities are very similar to refineries and contain many of the same multitude of hazards. The most common of these hazards, such as toxic gases, VCE, and BLEVE, begin in much of the same way, via unplanned releases. Of these types of hazards, the VCE is the most destructive of the group and fortunately the least frequent.

A VCE is one of the most dangerous hazards associated with these types of facilities. Normally, a response to this type of incident is after the fact and very rarely during or before. VCE is defined by FM Global as "an explosion occurring outdoors which produces damaging overpressures. It is initiated by the unplanned release of a large quantity of flammable vaporizing liquid or high pressure gas from a storage tank or system, process vessel, pipeline, or transport vessel."[4]

Generally speaking, for a VCE with damaging overpressure development to occur, several factors must be present. First, the material released must be flammable and processed or held under suitable conditions of pressure or temperature. Examples of such materials are liquefied gases under pressure (e.g., propane, butane), ordinary flammable liquids at high temperatures and/or pressures (e.g., cyclohexane, naphtha), and nonliquefied reactive flammable gases (e.g., ethylene, acetylene). A cloud of substantial size also must form prior to ignition.

With most common flammable materials, should ignition occur instantly with a release of the material, a large vapor cloud fire (fireball or flash fire) may occur, causing extensive localized heat radiation damage; however, significant blast pressures causing widespread damage will likely not occur.

Should the cloud be allowed to form over a period of time within a process area and subsequently ignite, blast overpressures away from the cloud center can create *deflagration* pressures or exceed those pressures developed from detonation of high explosives and result in extensive damage over a wide area. Ignition delays of 1 to 5 minutes are considered most probable, although major incidents with ignition delays as low as a few seconds and higher than 60 minutes have occurred.

A sufficient amount of the cloud must be within the flammable range of the material to cause extensive overpressure. The percent of the vapor cloud in each region varies, depending on many factors including type and amount of the material released, pressure at release, size of release opening, direction of release, degree of outdoor confinement of the cloud, wind speed, atmospheric stability, and other environmental effects. The cloud will move over time, changing the flammable regions. For example, a continuous release over a long period of time will generally have a rich region near the source, a lean region at the cloud leading edge, and a flammable region in between. A puff release (essentially instantaneous release) will usually have a rich region at the leading edge with flammable regions following. Important factors that must be present for an ignited vapor cloud to produce overpressure are outdoor confinement and turbulence generation. Research testing, incident investigation, and computer modeling have demonstrated that the greater the horizontal and vertical confinement and the more turbulence in the gas cloud, the greater the potential for overpressure development.

Turbulence can be caused by two primary mechanisms. First, repeated obstacles in the center of a cloud can accelerate gas mixing, which in turn can increase flame speeds within the cloud. This will strongly influence pressure development due to flame instabilities. Second, turbulence can be directly initiated from a high-pressure release.

Wide-open spaces do not easily promote VCE events unless the area presents unusual conditions of confinement (such as long, narrow ravines) or repeated obstacles (such as dense forests, large railroad staging yards, or inside close process buildings). However, a cloud released in an open area may be of sufficient size, and winds may be of suitable velocity and direction to disperse the cloud into a congested process area at great distances from the actual release. While the ignition of the cloud could occur anywhere in the cloud (even in the open space, e.g., by a vehicle), the apparent explosion epicenter will be the area where the cloud is confined and where obstacles exist that can cause transition from a cloud fire to a cloud explosion. Remaining portions of the cloud outside the congested area will not contribute to blast effects, although radiant heat effects will occur.

FIREFIGHTING TACTICS

As with all responses, petrochemical response strategies must consider the availability of resources and the ability of responders to deploy these resources. As will be mentioned several times in subsequent chapters, incident preplanning and identification of credible scenarios is critical in these facilities. Just as in preincident response planning for refinery fires, it is critical in petrochemical preplanning efforts to identify personnel and resources. One of the most common and useful resources in industrial responses is water streams.

Water streams can be used alone or in combination with other agents to control or extinguish fires; disperse, redirect, and dilute vapor clouds; and dilute or decontaminate many chemicals. When initially placing streams into service or when there are insufficient water resources to apply water to all exposures, the priorities in the following paragraphs should be considered.

As discussed in chapter 12, the general rules of thumb still apply in chemical and petrochemical plants. If cooling water does not vaporize on the surface of process equipment or steel structural members, then the equipment is cooled below the boiling point of water and is sufficiently cool to maintain its structural integrity. As mentioned previously, when applying cooling streams during a fire situation, priority must be given to exposed vessels, supporting structures, piping, and valves, especially where there is a radiant heat impact. If there is not enough water to cool everything you need to cool, then prioritize in this order:

1. Columns and vessels
2. Structures supporting large vessels
3. Piping
4. Critical valves

In the event of direct flame impingement, such as jet fires or heavy radiant heating of process equipment and structures, water streams should be prioritized as follows:

1. Rescue
2. Cooling for pressure vessels and the base of columns
3. Cooling for atmospheric storage vessels
4. Cooling for the structural supports of major process equipment
5. Cooling the structural members of pipe racks
6. Cooling of pipes
7. Cooling of critical valves

When vapor clouds are present, special precautions must be taken to protect personnel exposed to a potential explosion. Strategic decisions regarding vapor clouds are complicated by the fact that they may be toxic as well as flammable. Consider the possibility of a toxic and flammable vapor cloud that poses an overpressure threat to a building where plant employees have sheltered in place from the toxic effects of the vapor. On the one hand, if employees are allowed to remain in the building, they are threatened by the possible collapse of the building if the vapor cloud explodes. On the other hand, if they are evacuated from the building, there is the possibility of overexposure to the toxic effect of the vapor. Either way the situation is problematic.

In the event of vapor clouds, water resources should be used to minimize the potential for ignition and mitigate the effect of a vapor cloud explosion. When attempting to disperse a vapor cloud using water streams, the streams should be deployed as close as possible to the source of the vapor.

When the source of the vapor is pooling liquids, water streams should not be allowed to agitate the pool. Any stream impacting the pool will result in greater evolution of vapors. When pooling liquids are cryogenic, any water applications will heat the liquid and only serve to increase the vaporization of liquids. Aspirated foam should be used to blanket pooling liquids.

The use of water streams to knock down or disperse vapor clouds will generate additional water runoff, which must be properly managed. Proper management of vapor clouds is particularly critical when the cloud threatens occupied building, process control buildings, or process areas; or when flammable vapor clouds are encroaching on ignition sources. Vapor clouds that migrate off-site into populated areas, rail corridors, highways, and places of potential human occupancy are of the highest priority.

Historically, *three-dimensional fires* have been described as fires where burning liquids cascade down through a process structure, showering burning liquids onto platforms and equipment below the source of the fuel. Three-dimensional fires spread out horizontally from their source as burning liquids splash off vessels, equipment, and structural components located below the fuel source. Control of three-dimensional fires is difficult at best. First, if you attempt to extinguish from the bottom up, burning liquids will continue to fall onto the recently extinguished liquids, thereby reigniting them. Second, if you attempt to extinguish from the top down, the flames from lower-level fires will continue to rise through the structure, reigniting fuels previously extinguished in the upper sections of the process structure. The objective is to achieve control one level at a time and put controlling streams in vertical positions between the burning levels and any extinguished sections of the process.

Foam may be effective on the ground level where falling liquids begin to pool, but it has limited effect above ground level where fuel is airborne, falling through grading, and splashing off of process equipment. Since falling liquids exhibit little or no outward pressure, pressurized water streams can easily move them in lateral (sideways) directions. Therefore, water streams from portable monitors and handlines can redirect falling fuels in three-dimensional fires away from process equipment, piping, and structural supports. Depending on the leak location, in elevated process structures, fuel and flames may be projected completely outside of a process structure. Caution should be exercised not to impact other personnel and lines when redirecting fuel and flames.

There is an argument for starting at the bottom of a process fire and working your way up when controlling leaks. The exact sequence will be dictated by the fire conditions present. When flammable liquids are present, burning or not, a foam blanket should always be applied to the ground level. This will suppress any existing fire and help to control any fires that might develop while teams are operating in the process structure. Ground-level pool fires should be controlled, and a foam team should remain on the ground level to maintain the foam blanket.

As teams reach fires in upper levels of the process structure, they can approach block valves by redirecting flames and fuel from jet fires associated with the valve, as well as any three-dimensional fires falling from levels above. This may take several hose teams. Another common obstacle to approaching equipment in elevated sections of the process area is the presence of flames coming from fires on lower levels. If fires in lower levels cannot be controlled, their fuel and flames can be redirected with portable monitors to allow teams working above to penetrate further into the process area and possibly access control valves or set up portable monitors.

One piece of equipment that may prove to be beneficial is called a Hydro-Chem nozzle developed by Williams Fire and Hazard Control. The Hydro-Chem nozzle combines the effectiveness of an automatic water nozzle and the delivery of a dry-chemical in the same device. This unique nozzle, which can be mounted on a handline or on a monitor nozzle, projects dry-chemical agent at roughly 20 pounds per second through the center of the water (or foam/water) stream resulting in a highly effective range and reach, normally not possible with standard dry-chemical hand or wheeled extinguishers. Using this nozzle, when used in conjunction with additional foam lines for ground fires, enables firefighters to attack and extinguish three-dimensional fires by maintaining a safe distance, without the need to get too close to the origin of the fire.

Jet fires generally must be redirected from the level of the leak, because grating interferes with the water stream dynamics required to offset pressures associated with jet fires. Grating may be helpful to teams operating below, where the grating acts as a liquid deflector, thereby converting liquid jet fires into less threatening three-dimensional fires. In this situation, caution should be taken to avoid the effect of intense heat from the jet fire on the grating. The grating may fall onto personnel during the fire or fail to support equipment and personnel during later operations.

Pool fires are simply spill fires that have accumulated on the ground or solid surfaces to form pools of burning liquids. Small pool fires may be extinguished with dry-chemical, aqueous film-forming foam (AFFF), or CO_2 fire extinguishers. Larger pool fires may be extinguished with foam.

Some products can be easily extinguished with unaspirated foam solution.

It should be noted that a significant number of petrochemicals will resist all attempts to extinguish them with foam. Many petrochemicals aggressively attack the components of foam, or generate vapors so rapidly and with such pressure that foam cannot suppress the vapors quickly enough to extinguish the fire. In all cases, it is necessary to blanket the pooling liquids with a compatible aspirated foam or similar agent after suppression to minimize the generation of vapors.

When flammable gases or liquids leaking from a pressurized container such as process equipment or piping are ignited, the flames form a jetting pattern. This condition is referred to as a jet fire. Jet fires are intense and often impinge on adjacent equipment and structures. The source of jet fire fuels should be isolated, and the jet fire itself should be allowed to burn out whenever possible. If a jet fire involving a gas is extinguished, the continuing gas leak will create a potentially explosive atmosphere. Likewise, a jet fire with a liquid source will continue to create vapors, especially where the leak is creating an aerosol spray. The toxicity of the combustion products should also be considered when determining if extinguishment is appropriate. Several techniques can be used to allow a jet fire to burn while process operators isolate and deinventory the source.

Cooling water can be applied to all areas of flame impingement and areas affected by heat. In this case, be careful to consider flame impact in upper levels of the process structure and the effect of burning liquids as the cascade through the process structure below the fire level. When flammable liquids are involved, you must consider places where flammable liquids may accumulate on lower levels and on the process area floor. Careful examination of process trenches and sewers must be made to preclude extension of fires into other sections of the process area.

Where jet fires impinge on adjacent process equipment or piping, the flames and jetting fuel can be redirected by close proximity application of tight water cone patterns from portable monitors and handlines. This technique is very effective, but caution must be used to ensure that the redirected flames and fuels do not adversely affect other equipment or personnel working on other levels of the process structure.

Chemical facilities typically contain huge volumes of toxic gases that may be released during a process emergency. Toxic releases can involve a multitude of chemicals, typically in large quantities. When dealing with toxic releases, the material safety data sheets (MSDS) and other chemical references should be used to aid in the development of both plans and actual responses. (See chapter 11 for more details.) Chemical references should also be used to determine appropriate action during a release.

Many petrochemical processes require special chemicals to initiate or control the process chemistry. Polyolefins, such as polypropylene and polyethylene, require organic peroxides for their polymerization processes. Organic peroxides used in these processes are liquids that react explosively with water. These organic peroxide spills are controlled by application of light inert solids to blanket them and reduce the ability of moisture in the air to react with them. Control products include *vermiculite,* which is often found in large bags near organic peroxide storage tanks. Preincident response plans should identify any chemicals that require special control measures, as well as the location of special extinguishing agents.

No amount of postresponse critique or incident investigation can replace the potential effect of a thorough preincident response planning. Never miss an opportunity to visit plants in your jurisdiction. If an invitation to visit these facilities is not presented, then initiate one yourself. Become familiar with their processes, the chemicals they use, the systems available in their facilities, and especially the names and faces of the personnel whom you will work with when responding to emergencies in their facilities.

CONCLUSION

Similar to refineries, petrochemical plants are large, complex facilities that manufacture a wide range of chemicals and petrochemicals and share many common esthetic features with refining facilities. A thorough plan, identifying credible scenarios and hazard identification, will help to ensure a proper response to chemical facilities. Close cooperation and joint training exercises will help close the gap between your local fire department and the industrial fire brigade. It is also important to remember that many of these fires might not involve rapid intervention and suppression but rather other measures such as cooling exposures and isolating the involved equipment. For a quick reference, use the general rules of thumb from chapter 12 and prioritize your cooling efforts

NOTES

[1] Chemistry.org. *Production by the US Chemical Industry.* Retrieved August 2006 from http://pubs.acs.org/hotartcl/cenear/960624/prod.html.

[2] Chemistry.org.

[3] Chemistry.org.

[4] FM Global. 2001. *Guidelines for Evaluating the Effects of Vapor Cloud Explosions Using a TNT Equivalency Method.* Johnston, RI: Factory Mutual Insurance Company, p. 4.

14

Nuclear Facilities

INTRODUCTION

The nuclear power industry is highly regulated by the United States (US) in all aspects of the production, storage, and waste of nuclear energy. In 1954, Obninsk, in the Union of Soviet Socialist Republics (USSR), opened the first nuclear power plant in the world to generate electrical power. The Shippingport Reactor in Pennsylvania was the first nuclear generator in the United States to produce electrical power, built in 1957. Other forms of nuclear research and technology had taken place since 1938, with the first nuclear fission experiments being conducted in Berlin by Otto Hahn, Lise Meitner, and Fritz Strassman. Since 1957, the nuclear industry has grown in capacity through the 1980s.[1]

Nuclear power plants are highly complex and strictly secure facilities and require special consideration by the fire service. The facilities often occupy large areas of land and are located near bodies of waters such as rivers and lakes, because of the large demand of water for the production of steam and cooling water. Therefore, this chapter will mainly focus on regulated reactors and discuss the basics of power and research or test reactors with respect to basic process description, response planning, and tactics. These reactors are regulated by the US Nuclear Regulatory Commission (NRC). This chapter will not discuss storage or waste sites, as these are nonprocess-related areas.[2]

DESCRIPTION OF PROCESS

Nuclear energy is produced by atomic fission. In physics, *fission* is a nuclear process, meaning, basically, that it occurs when the nucleus of an atom splits into two or more smaller nuclei (fig. 14–1). This process releases additional by-products such as free neutrons and photons, which produces large amounts of energy. The heavier the nucleus, the more energy that is released. In the case of uranium, the splitting into fragments may release more than 100 million volts of energy. The free neutrons may trigger additional breakups in the form of a chain reaction. If this is the case, then the chain reaction may be controlled, releasing energy that can be used to boil water, which in turn produces steam to drive a turbine. If this chain reaction cannot be controlled or begins to run away, the end result is infamously known as a *meltdown*.[3]

***Fig. 14–1.* Nuclear fission** *(Courtesy of Alex Wellerstein, 2005)*

The first step in producing electrical energy from a nuclear reactor is to find a suitable fuel source. Uranium is currently the fuel of choice for all nuclear power plants. Uranium enrichment is the first critical step in transforming natural uranium into a nuclear fuel source to produce electricity. Currently, the only uranium enrichment facility in the United States, located in Paducah, Kentucky, is operated by the United States Enrichment Corporation (USEC), locally known as the Paducah Gaseous Diffusion Plant (PGDP). At the PGDP facility, natural uranium is enriched from a raw product of less than 1% fissionable uranium to almost 5% fissionable uranium, commonly referred to as U-235. This is the minimum required level needed by nuclear power plants to produce viable electricity. USEC uses a process called gaseous diffusion technology to produce the 5% U-235. In simple terms, uranium hexafluoride (UF6), a solid at room temperature, is heated to more than 135°F, at which point it becomes a gas. USEC uses this gas to begin the gaseous diffusion process. Basically, the process separates the lighter U-235 isotopes from the heavier U-238 isotopes. The gas is forced through a series of porous membranes with very small openings, visible only by a microscope. As the gas flows through the diffusion process, the U-235, because it is lighter, is able to pass through the microscopic barrier more easily than the U-238. Similar to a reduction or continuous distillation process, the two different isotopes are separated, and the level of U-235 increases, becoming a usable fuel source.[4] After uranium is processed and transformed into a usable fuel source, it is transported to one of two regulated types of reactors: research and test reactors or power reactors.

Research and test reactors

Research and test reactors, often called *nonpower* reactors, consist mainly of nuclear reactors used for research, development, and training purposes. They do not produce power for sale or consumption but produce it primarily for scientific purposes. This level of research activity permits the contribution of nuclear science in everyday life. Advances in all forms of science, including chemistry, physics, medicine, biology, and archeology can be traced back to nuclear research. These facilities are regulated by the NRC for licensing, inspection, security, and safety. These types of reactors are commonly found at leading technical universities such as Ohio State University, Worcester Polytechnic Institute, and Massachusetts Institute of Technology. Some of these research reactors may be found at other sites, also, such as the US Geological Survey in Denver, Colorado, the National Institute of Standards and Technology in Gaithersburg, Maryland, and the Armed Forces Radiobiological Research Institute, Bethesda, Maryland. Most research and test reactors operate at reduced power and are currently licensed to operate only at these levels. For example, the University of Missouri–Rolla (UMR) has had a research reactor in operation since 1961. Its current license permits operating at a level of only 200 kilowatts of thermal power or about the same heat release output of a burning small recliner chair. The NRC currently regulates about 50 of these research and test reactors. Of that, only about 30 are actually operating. The remaining reactors are in some stage of decommissioning or awaiting removal of radioactive material.[5]

Power reactors

As mentioned earlier, the NRC also regulates commercial nuclear power plants that generate electricity for sale. Although there are several types of power reactors, only pressurized water reactors (PWRs) or boiling water reactors (BWRs) are in operation in the United States. According to the NRC, there are currently 104 reactors licensed to operate nuclear power plants in the United States. The breakdown includes 69 PWRs and 35 BWRs, which produce almost 20% of the energy in the United States.[6] (Additional information on power generation is also in chapter 30.)

Pressurized water reactors (PWRs)

Pressurized water reactors, which account for more than 65% of the operating units, involve using the principle of generating steam to rotate an electric generator. The PWR uses ordinary water to accomplish three tasks: primary coolant, secondary coolant, and neutron moderator (fig. 14–2). A *neutron moderator* is a way to reduce the velocity of fast neutrons to stabilize and sustain the chain reaction mentioned earlier. PWRs belong to a family called *light water reactors*. The basic operating principle of a PWR is based on the nuclear fuel in the reactor vessel undergoing a chain reaction. The reactor vessel and the steam generator are located in the familiar concrete pillar or dome-type structure. Oftentimes, the large hourglass-shaped concrete structure is thought to be the nuclear reactor, but in fact these are very large cooling towers. The main objective is to produce heat in the reactor building for the production of steam. In the nuclear reactor vessel, control rods are raised and lowered in a very controlled environment. This process heats the water in the primary coolant loop, which is a separate loop of pipe and is always fed into the reactor vessel. The hot water that is produced by this reaction is pumped to a steam generator, which permits the primary coolant to heat up the secondary coolant. This is done in a large vessel-type structure where the primary coolant, in a separate loop, can come in contact with the secondary coolant, in another separate loop, via the steam generator, which is acting as a heat exchanger. This is very similar to the radiator of a car or truck. It is important to note that the primary and secondary loops do not mix or come in contact with one another due to the potential of the primary coolant to become radioactive. Once the secondary coolant is hot enough to turn into steam, it can be used for generating electricity via a *turbine*. The turbine is connected to an electric generator by a shaft. The spinning turbine shaft inside the generator creates an electrical current. The secondary coolant then is cooled down in a *condenser* before being fed into the steam generator again.[7]

Boiling water reactors (BWRs)

Boiling water reactors account for the other 35% of power reactors in the United States and involve using the same principle of generating

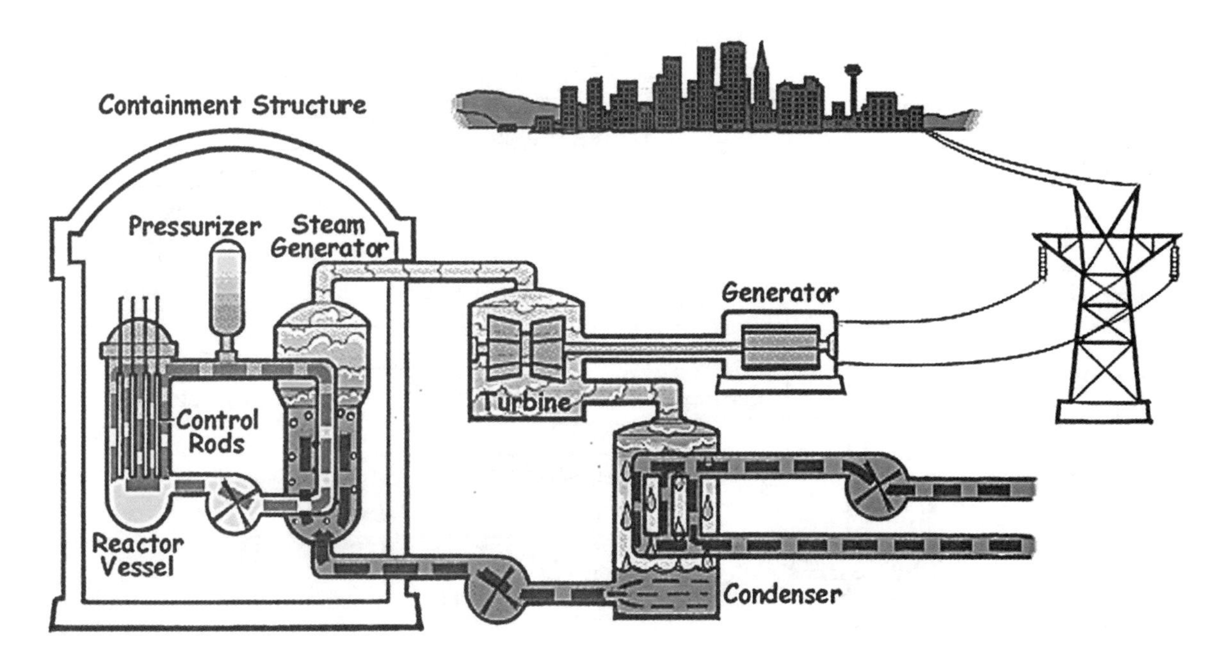

Fig. 14–2. **Pressurized water reactor** *(Courtesy of NRC)*

steam to rotate an electric generator. Like the PWR, the BWR is a light water reactor, and it has many other similarities to the PWR, except that in a BWR, the steam going to the turbine is produced in the reactor core rather than in a steam generator or heat exchanger (fig. 14–3). That is, there are no primary or secondary coolant lines, but rather a single discharge line on top of the reactor vessel. In a typical BWR, the reactor core creates heat, and a single loop both delivers steam to the turbine and returns water to the reactor core to cool it. The cooling water is force-circulated by electrically powered pumps. Emergency cooling water is supplied by other pumps, which can be powered by on-site diesel generators. In the BWR, the single circuit in which the water flows is at a lower pressure than in a PWR, so it boils in the core at about 545°F (285°C). The reactor is designed to operate with about 15% of the water in the top part of the core as steam, resulting in less moderation (slowing down of fast-moving neutrons), lower neutron efficiency, and lower power density than in the bottom part of the core. The BWR design is very safe because it is essentially a PWR that has been designed to operate permanently under fault conditions. Boiling water in the primary loop of a PWR would be considered a very dangerous coolant leak and would result in immediate shutdown.[8] The BWR does not use a steam generator or secondary coolant line, and therefore the design is less complex.

COMMON HAZARDS

Although the obvious hazard associated with these facilities is a problem in the reactor building, other hazards exist as well. These hazards include fires and incidents relating to power generators (chapter 30), warehouses (chapter 17), laboratory or research and development (chapter 23), offices, and other common occupancies. These incidents are minor when compared to the serious hazard at nuclear facilities. Because of the nature, complexity, and severity of an incident involving the reactor building or associated component or process, this type of hazard per se will not be covered in this chapter, but rather how to prepare, plan, respond, and interact with other agencies during these types of incidents.

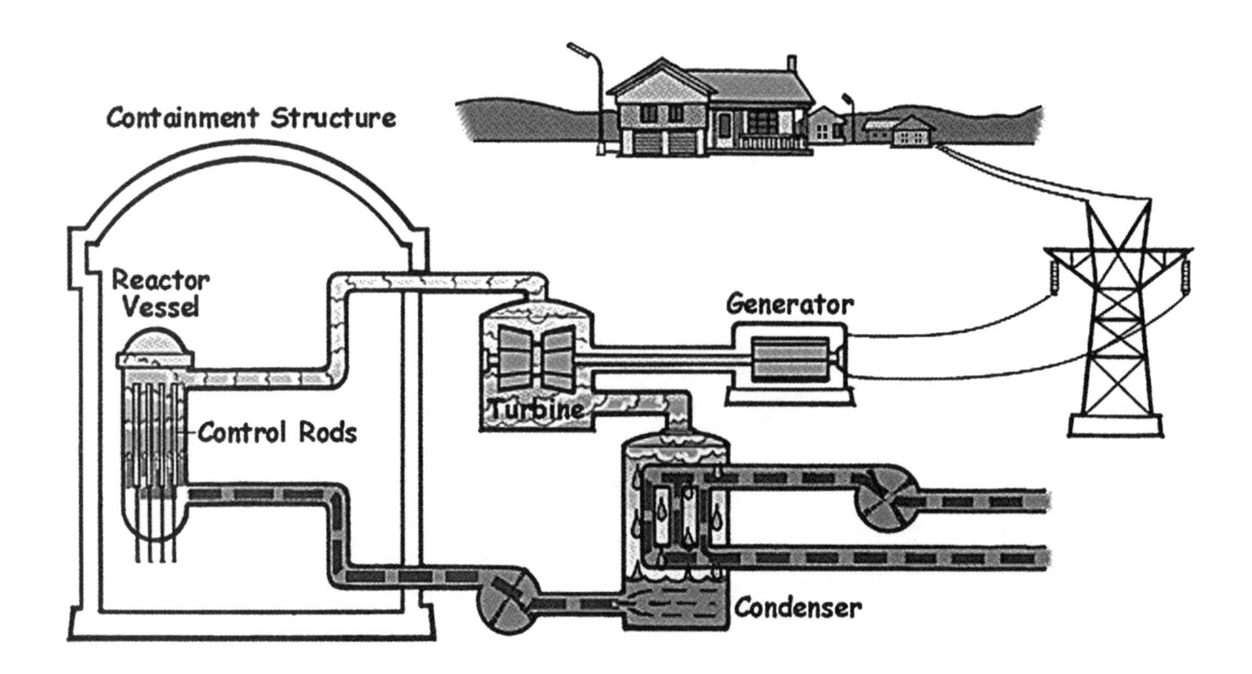

Fig. 14–3. **Boiling water reactor** *(Courtesy of NRC)*

COMMUNITY PLANNING

As mentioned and described throughout this book, the need for proper planning and identification of hazards is essential for these major types of industrial occupancies. Chapter 5 addresses the common methods for preplanning an industrial facility. Although community planning is mentioned briefly in chapter 5, this type of planning is very important because it involves both state and federal agencies, as well as the local entities. Also, these facilities have been previously identified as a potential target for terrorism (chapter 10). Most residents are also aware of planning efforts. Residents within a radius of approximately 10 miles from a nuclear power plant receive emergency information materials annually. This information is commonly distributed via brochures, phone books, calendars, and utility bills. Some cities might have a community awareness or preparedness offices. These offices offer a great source of information and contacts for state and other governmental resources. These materials contain educational information on radiation, instructions for evacuation and sheltering, special arrangements for the handicapped, and contacts for additional information. Included in the information packet will be details pertaining to potassium iodide (KI). In January 2001, a rule change to the NRC emergency planning regulations included the consideration of the use of potassium iodide. If taken properly, KI will help reduce the dose of radiation to the thyroid gland from exposure to radioactive iodine and reduce the risk of thyroid cancer. The Food and Drug Administration (FDA) has issued guidance on the dosage and effectiveness of potassium iodide. The NRC has supplied KI tablets to states and local governments requesting it for the population located within a 10-mile zone. If necessary, KI can be to be used to supplement evacuation or sheltering, not to take the place of these actions. If radioactive iodine is taken into the body after consumption of KI, it will be rapidly excreted from the body.

Part of these local preparedness plans include an alerting system, similar to the common tornado early warning system is many communities. The Alert and Notification System (ANS) is in place to notify the public within a 10-mile radius of a nuclear power plant in the event of a declared emergency. This system will be activated within approximately 15 minutes of a decision by government agencies of the need to take protective actions. This system typically uses sirens, tone-alert radios, route alerting, or a combination of these methods. If you receive an alert, tune your radio or television to an Emergency Alert System (EAS) station located in your emergency information pack for details.[9]

The NRC and the Federal Emergency Management Agency (FEMA) have responsibilities in dealing with these types of events. These organizations review the state and local plans, resources, and other regulatory requirements. State and local government officials have the overall responsibility for deciding and implementing the appropriate preparedness planning and protective actions for the public during a nuclear power plant radiological emergency. They are responsible for notifying the public to take protective actions, such as evacuation, sheltering, or taking potassium iodide pills. State and local officials base their decisions on the protective action recommendations by the nuclear power plant operator and their own radiological or health organization.

The NRC provides oversight and guidance of the protective action decision by the state and local government officials. Neither the nuclear power plant operator nor the NRC can order the public to take protective actions. If your department is located within these designated 10-mile zones, or if the potential exists for response to these areas because of local or state mutual agreements, close cooperation is required.[10] Because of potential terrorist threats to these facilities, it may be necessary to arrange for your plant visit weeks if not months in advance.

Depending on the circumstances in the event of a nuclear power plant radiological emergency, the public would be evacuated or sheltered. Under most conditions, evacuation is preferred. However, under some conditions, sheltering provides protection that is equal to or greater than evacuation, considering weather, competing events, fast-breaking or short-term release, and traffic conditions. Depending on the type of building and construction used, sheltering can result in a reduction of radiation dose of up to 80% compared to a dose received outdoors and unsheltered. It is possible that your department could be involved in these evacuation or sheltering operations. Your station could be a designated evacuation shelter. As part of your planning phase, food and water rations need to include not only those who would be potentially sheltered but also responders.[11]

There are different types of planned emergencies at nuclear facilities. To help in classifying these emergencies, a classification system has been established to aid in determining the level of response needed and the associated risk to life. Emergency Classification (EC) is a set of plant conditions that indicate a level of risk to the public. Both nuclear power plants as well as research and test reactors use the following four emergency classifications in order of increasing severity. The vast majority of events reported to the NRC are routine in nature and do not require activation of the incident response program. The classification comprises these components:[12]

- **Notification of unusual event.** Under this category, events are in process or have occurred which indicate potential degradation in the level of safety of the plant. No release of radioactive material requiring off-site response or monitoring is expected unless further degradation occurs.
- **Alert.** If an alert is declared, events are in process or have occurred that involve an actual or potential substantial degradation in the level of safety of the plant.
- **Site area emergency.** A site area emergency involves events in process or which have occurred that will result in actual or likely major failures of plant functions needed for protection of the public.
- **General emergency.** A general emergency involves actual or imminent substantial core damage or melting of reactor fuel with the potential for loss of containment integrity.

Immediately upon becoming aware that an incident has occurred that may result in a radiation dose that exceeds federal government protective action guides, the responsible nuclear power plant personnel should evaluate plant conditions, then recommend actions to the state and local governmental agencies for the protection of the population from exposure to radiation. Nuclear power plant representatives are required to give the recommendations to the state or local government within 15 minutes and to the NRC Operations Center as soon as possible, within 1 hour of determining that an accidental radiation release could affect public health and safety. The emergency plans, part of the National Response Plan (NRP) (fig. 14–4), The Nuclear/Radiological Incident Annex for all NRC-licensed facilities is required to provide reasonable assurance that adequate measures can and will be taken in the event of an emergency. For nuclear power plants, both on-site and off-site emergency response plans are required. This is because a severe accident at a nuclear power plant could reasonably be expected to affect people located some distance away from the power plant.[13]

Fig. 14–4. **National Response Plan** *(Courtesy of NRC)*

In response to an event at an NRC-licensed facility that could threaten public health and safety or the environment, NRC activates its incident response program at its Headquarters Operations Center and one of its four Regional Incident Response Centers (Region I in King of Prussia, PA; Region II in Atlanta, GA; Region III in Lisle, IL; and Region IV in Arlington, TX). NRC's highest priority is to provide expert consultation, support, and assistance to state and local public safety officials responding to the event. Once the NRC incident response program is activated, teams of specialists are assembled at the Headquarters Operations Center and Regional Incident Response Center to obtain and evaluate event information and to assess the potential impact of the event on public health and safety and the environment. Other involved agencies include Department of Defense (DoD), Department of Energy (DOE), Department of Homeland Security (DHS), Environmental Protection Agency (EPA), National Aeronautical and Space Administration (NASA), as well as the NRC. Other governmental agencies that provide assistance include the Department of Agriculture, Department of Commerce, and the Department of Transportation. At the Operations Center, scientists and engineers analyze the event and evaluate possible recovery strategies.

Meanwhile, experts from other governmental agencies evaluate the effectiveness of protective actions that have been recommended by the local nuclear operator or licensee and implemented by state and local officials to minimize the impact on public health and safety and the environment. Communications with the news media, state, other federal agencies, the Congress, and the White House are coordinated through the Headquarters Operations Center. As described in the Nuclear/Radiological Incident Annex to the NRP is the coordinating agency for events occurring at NRC-licensed facilities and for radioactive materials either licensed by NRC or under NRC's Agreement States Program. As coordinating agency, NRC has technical leadership for the federal government's response to the event. If the severity of an event rises to the level of general emergency, or is terrorist-related, DHS will normally declare an Incident of National Significance under the NRP. In this case, DHS would take on the role of coordinating the overall federal response to the event. If event conditions warrant, NRC will immediately dispatch a site team, consisting of technical experts and a site team director, from the regional office to the site. An executive team will be assembled in the Headquarters Operations Center to lead the response. The executive team is typically headed by the chair of the NRC or a commissioner acting as chair. Once the site team is in place, authority to manage event-related activities is turned over to that team. The site team serves as the NRC's eyes and ears on-site, allowing a firsthand assessment of the situation and face-to-face communications with all participants. The Headquarters Operations Center provides round-the-clock logistical and technical support throughout the response.[14]

Neither the nuclear power plant operator nor the NRC can order the public to take protective actions. The plant operator recommends to the appropriate state and local governments what protective actions to take. Once the local emergency response organization, whether a city or county fire department, has been activated, it will establish a local emergency operations center to coordinate decisions and implementation of protective actions with other government organizations. The main function of the fire department, along with following response plans (chapter 5) and the incident command system (ICS; see chapter 4), is to fill the role of support. Response to these incidents is not a rapid intervention mode of response but more of a long-term support role.

FIREFIGHTING TACTICS

As mentioned earlier, response to a nuclear power facility for a nuclear-related event might never occur. The frequencies are very low for these types of events. In the event that one does occur, your role would not be a rapid type of response but more likely a supportive role with either community evacuation or shelter assistance. But there are other, more conventional types of emergencies that can occur at nuclear facilities. These emergencies can include office, cafeteria, warehouse, or other ordinary combustible occupancy fires, cable tray or tunnel fires, or events dealing with spills of hazardous materials or flammable/combustible liquids.

Fires involving offices, cafeterias, warehouses, or similar business-type occupancies, while not normally challenging, could be at nuclear facilities. Access due to security concerns, sensitivity of information, and other proprietary concerns could hamper firefighting efforts. The proper procedure, protocol, preplanning, and facility communication can help lessen the chances of problems.

Cable fires, whether overhead in bundles on trays or in tunnels, can spread to other areas of the facility. If a fire occurs in the cable trays, fires can spread to the control room, causing the loss of plant operations and control. If this happens, it would create a very dangerous situation. Therefore, every effort should be made to protect, isolate, and contain cable tray fires to prevent control rooms' motor control centers from becoming affected by the fire. Further, cable tunnel fires are extremely dangerous. These involve entry into areas that have limited access, may interfere with radio communication, and will be much like a basement fire—hot. (See chapter 19 for more on cable tray fires.)

Incidents involving hazardous materials or flammable or combustible liquids are also challenging. More than likely, nuisance and minor spills will be handled by on-site teams. If you are called in to support or provide equipment and/or personnel, remember to follow your preincident response plans. (See chapters 11, 12, 13, and 30 for more information about these types of responses.)

CONCLUSION

Nuclear power plants are highly complex and strictly secure facilities that require special consideration. These reactors are regulated by the US Nuclear Regulatory Commission (NRC). The NRC and the Federal Emergency Management Agency (FEMA) have the responsibility to deal with the types of events related to regulation by the state and local government. The state and local agencies have the overall responsibility of deciding and implementing the appropriate protective actions for the public during a nuclear power plant radiological emergency and for activating the National Response Plan (NRP). These types of incidents require a great deal of detailed planning and coordination with federal, state, and local agencies. This type of planning will also involve large sections of the community. It is important to remember that the function of the local fire department is to have a long-term supportive role rather than a first responder or rapid response agency. Responses to conventional type emergencies can even be in limited capacities. It is best to keep clear and open communication lines with local plant management.

NOTES

[1] Free Software Foundation. *Nuclear Power.* Retrieved July 2, 2006, from http://en.wikipedia.org/wiki/nuclear_power.

[2] United States Nuclear Regulatory Commission (NRC). Retrieved June 4, 2006, from http://www.nrc.gov/reactors.html.

[3] Free Software Foundation. *Nuclear Power.*

[4] United States Enrichment Corporation (USEC). Retrieved July 1, 2006, from http://www.usec.com/v2001_02/html/aboutusec_quickfacts.asp.

[5] NRC.

[6] NRC.

[7] NRC.

[8] NRC.

[9] United States Department of Homeland Security (DHS). December 2004. National Response Plan. Washington, DC: DHS.

[10] DHS, *National Response Plan.*

[11] NRC.

[12] Department of Homeland Security: "National Response Plan." December 2004

[13] NRC.

[14] NRC.

15

Bulk Grain Storage and Processing Occupancies

INTRODUCTION

Bulk grain occupancies range in size from small local facilities typically found in rural areas and farms to large industrial complexes (fig. 15–1). Bulk grain facilities can be separated into two broad classifications, storage and processing. Storage consists of grain elevators intended to store and ship grains and seeds. Processing involves industrial grinding (sometimes called pulverizing or milling) processes by which combustible and some noncombustible materials are reduced in size via various processes to a very small particle; this reduction in particle size from the grinding or milling process causes combustible and some normally noncombustible materials to become a serious fire or explosion hazard. The explosion hazard arises when the finely reduced particles are suspended in air, where the dust concentrations can reach the *lower explosive limit* (LEL).[1]

Dust hazards can also be present in nonmilling or grinding operations such as bulk grain storage as typically found in grain elevators (fig. 15–2). There are other dust-generating occupancies, but for the purpose of this chapter, we will discuss dust hazards associated with bulk grain processing and storage. Other occupancies that are dust generating, such as paper manufacturing, woodworking facilities, as well as plastic manufacturing and molding are discussed in other chapters of this book.

Fig. 15–1. **Commercial grain elevator** *(Courtesy of Randall Merriott, 2004)*

Fig. 15–2. **Aerial view of grain elevator** *(Reprinted with permission from the* **Fire Protection Handbook**, *19th Edition, Copyright © 2003, National Fire Protection Association, Quincy, MA 02169)*

DESCRIPTION OF PROCESS

Storage

The raw materials associated with grain operations that are capable of burning or creating explosive atmospheres are those that consist mainly of starches or carbohydrates, protein, and fiber. While in their whole-kernel state, these grains are relatively stable and are not readily combustible, but outside contamination from moisture, insects, or fungi can cause spontaneous combustion, a product of microbiological spoilage.

The raw materials are typically delivered via truck, railcar, or barge (fig. 15–3). The locations for these types of occupancies are usually in rural areas, although many large cities have grain elevators located near rivers or large rail-switching yards. The size of the facility will also dictate the frequency and quantity of deliveries. The more deliveries, the more likely dust can accumulate, thus creating the hazards associated with dust explosions.

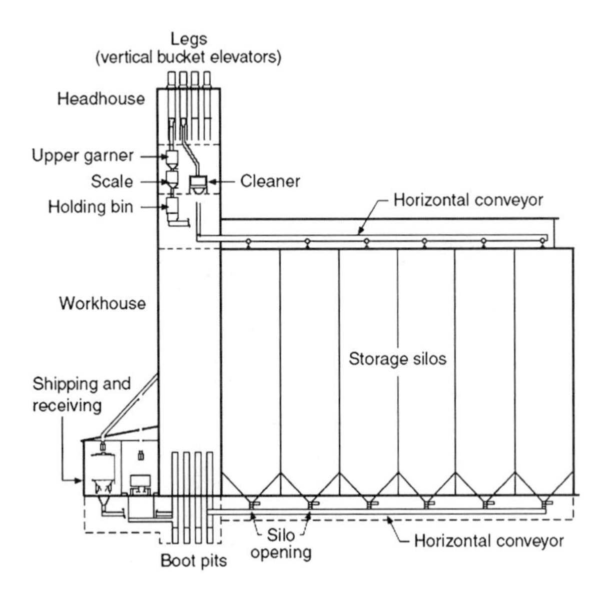

Fig. 15–3. **Grain elevator overview** *(Courtesy of NFPA)*

The raw materials or grain that is delivered via trucks is offloaded by one of two methods, either bottom unloading or rear dumping. *Bottom unloading* of grain consists of a truck with a load of grain driving over a special bin below grade level at the grain storage facility. The receiving bin or receiving pit design is similar to the way "quick oil change" shops are laid out, in which you drive you car over the opening in the floor, and below is the mechanic who accesses this area via steps to be able to gain access to the underside of your car. In a similar fashion, the truck drives over a pit in the receiving area and opens the bottom of the grain hauler. A series of screens are used to filter out large pieces of non-grain-related materials such as stones, husks, and other items that could cause sparks or lead to spontaneous heating and combustion. From the receiving pits, the materials are transferred via a conveyor belt to the *boot pit* and elevator bucket or *legs*.

There are several methods to transfer grain, including belt conveyors, chain conveyors, screw conveyors, and pneumatics (to convey finely ground grains). The most common conveyor found in grain processing is the belt conveyor. The belt conveyor is commonly used to transport grain horizontally from point to point. The belt conveyor also presents a fire hazard, in addition to those created by the dust. The hazard mainly comes from the rubber material used to manufacture the belt, although flame-resistant belts are available. In addition, magnetic separators are installed over the belt conveyor to ensure that the metal does not enter the storage process. Small metal fragments from farm equipment or other outside containments could lead to spark generation from friction, thus introducing an ignition source for a fire or explosion. These magnetic separators are typically installed at the beginning of the conveyor, near the receiving bin or pit and at the elevator legs, prior to being loading into the bucket.

The belt conveyor dumps the grain into the elevator boot, where the movement from horizontal to vertical takes place. From the elevator boot, the grain is transferred to the bucket elevator, which is sometimes called the *elevator leg* or simply *leg* (fig. 15–4). The basic principle of gravity is used in this application to transfer the grain to the *headhouse*. The grain elevator is designed to transport the grain to the highest elevation point; then gravity will force down the grain through various garners, weighing scales, and cleaners. The bucket elevator is one of the most hazardous pieces of equipment from an explosions hazard perspective. The dust concentration levels will most likely always be above the LEL during all times of operations.

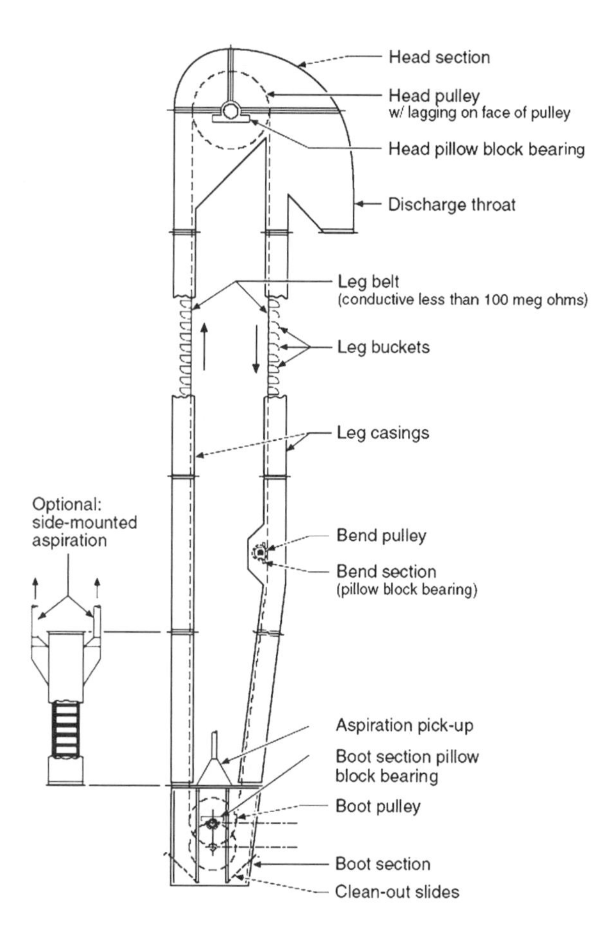

Fig. 15–4. **Overview of bucket elevator**
(Courtesy of NFPA)

From the bucket elevator, the grain is transferred to the headhouse, which is the tallest part of the entire facility. Inside the headhouse is the elevator head, where the grain is transferred from mechanical transportation to forced gravity. From the headhouse, the grain is forced via gravity through screens or *garners,* weighing hoppers or scales, and cleaners; then to spouts; and finally dumped onto a horizontal conveyor belt. Further, automatic grain samplers are used to monitor the grain quality, moisture content, and other quality assurance items. From the horizontal conveyor belt, a movable hopper is used to slide back and forth along the top of the grain storage bins or silos and pours the grain into individual silos.

For grain retrieval and loading onto railcars or into trucks, the process is basically the same. From the storage silos, the grain is released, again using gravity, from the bottom via a silo opening or loading spouts onto the same horizontal belt conveyor used to load the bucket elevator. Once on the belt conveyor, it travels again to the elevator boot and into the bucket elevator and back to the headhouse. In the headhouse, the grain is weighed and transferred by gravity through spouts to shipping bins, usually located in the *workhouse.* The workhouse is located underneath the headhouse and above the elevator boot. The workhouse contains storage bins used to hold in process grain for loading and shipping, as well as additional cleaning and bagging for local packaging. From the shipping bins, grain is then loaded via gravity to the loading spouts for truck, rail, or onto a belt conveyor that leads to a marine terminal for distribution onto barges or other marine vessels.

All along this process of storing the grain is the dust control process. At each point where grain is transferred and handled, dust is generated and suspended in air. A properly designed dust collection system used in conjunction with good preventive maintenance and periodic housekeeping will help reduce fugitive emissions. The dust collectors are located in the baghouse, which can be on the roof of the silos or at ground level. The primary purpose of the dust collection system is to remove the dust being generated by the storage process of grain. The dust collection system comprises a series of ducts, spouts, and collection points that work like a big vacuum cleaner. Negative pressure is generated by the remotely located blower, and dust is forced into the collection points by the lower air pressure. Dust collection systems are also fire and explosion hazards and should be evaluated to ensure proper operation.

Processing

Now that we have described the process relating to storage and retrieval of grains, we will focus on the grinding and milling operations that could take place at the same site or in a different location. Although products such as wheat, corn, sugar, rice, and flour present an explosion hazard during storage and storage handling, perhaps it should be noted that the grinding and milling operations present a more hazardous situation due to the generation of smaller and more refined dust particulates. In processing grains, an explosion or fire may originate in the process equipment (equipment explosion hazard) or in the ambient environment in the room containing the process equipment (room explosion hazard). It is important to differentiate the processes used to reduce the size of materials via grinding or milling from other operations or processes that use abrasive disks and wheels to shape or smooth wood or metals. Although the processes used to mold, smooth, or shape wood and metal generate dust and could produce levels that might cause an explosion, these processes are not likely to sustain dust concentrations at the *minimum explosion concentration* (MEC). Additionally, these processes could be either wet or dry. Therefore, in this section we will only consider processes that generate large quantities of dust above the LEL for dry processes (because those that involve only occasional dust production or involve wet processing are a lesser risk or hazard).

Most grinding or milling operations are done by machines called mills. There are six basic types of mills:[2]

- *Tumbling mills* consist of a cylindrical or conical shell with steel or porcelain balls inside. As the shell revolves on its horizontal axis, the balls are tossed around, grinding the material against the vessel shell, balls, or themselves. The size of the balls will dictate the size of the particle (fig. 15–5).
- *Ring-roller mills* consist of a grinding ring or plate sliding between a series of rollers. The orientation of the ring or plate could be either horizontal or vertical, and the differences in the ring or plate and the roller grind the grain (fig. 15–6).
- *Roller mills* are similar to ring-roller mills, except that the grain is ground between several rollers revolving in the opposite direction and at different speeds. A blade at the discharge end is used to remove the fine particles. These rollers are jagged, and the degree of jaggedness determines the final particle size (fig. 15–7).
- *Hammer mills* are more popular types of mills. They use hammers that are attached via a hinge or fixed onto a shaft that rotates. As the shaft rotates, the grain is hammered against the vessel shell wall. The final particle size is determined by the rotating speed of the shaft, the spacing between the hammers, the feed rate, and the size of the discharge opening (fig. 15–8).
- *Attrition mills* use grinding plates (steel or stone) that rotate at very fast speeds. One disk may be stationary, or both may rotate in opposite directions in either the horizontal or vertical position. The grain is introduced at the axis or center and is discharged at the ends of the rotating grinding plates.
- *Jet mills* use a process that does not involve grinding against a hard surface, but instead a gas to convey the grain at high velocity in opposite directions of the gas stream. The high turbulence forces the grain particles to collide and grind against themselves.

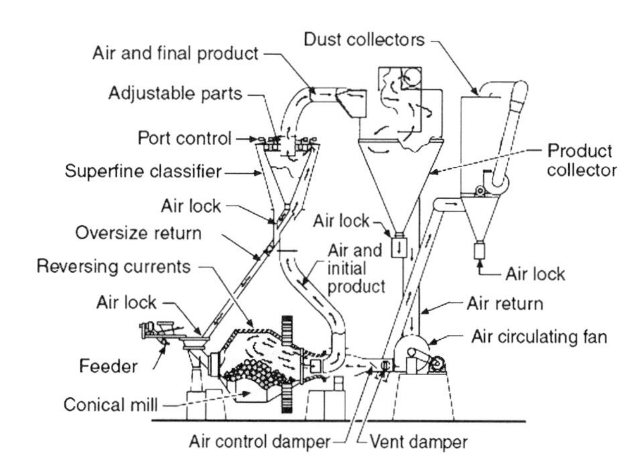

***Fig. 15–5.* Tumbling mill** *(Reprinted with permission from the* **Fire Protection Handbook**, *19th Edition, Copyright © 2003, National Fire Protection Association, Quincy, MA 02169)*

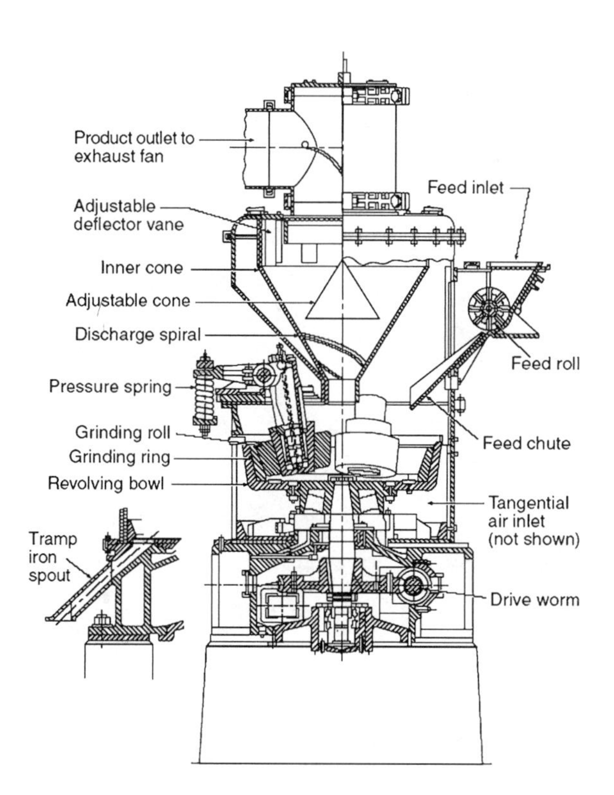

***Fig. 15–6.* Ring-roller mill** *(Reprinted with permission from the* **Fire Protection Handbook**, *19th Edition, Copyright © 2003, National Fire Protection Association, Quincy, MA 02169)*

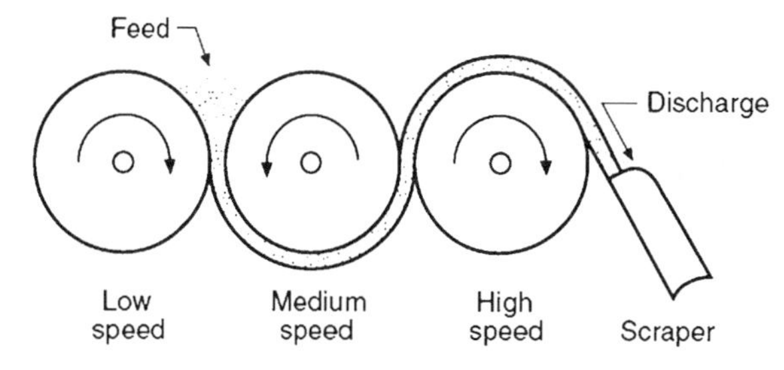

***Fig. 15–7.* Roller mill (Reprinted with permission from the *Fire Protection Handbook*, 19th Edition, Copyright © 2003, National Fire Protection Association, Quincy, MA 02169)**

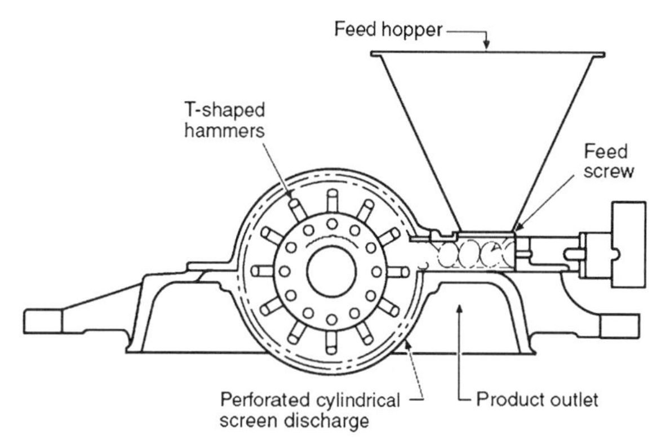

***Fig. 15–8.* Hammer mill** *(Reprinted with permission from the* **Fire Protection Handbook**, *19th Edition, Copyright © 2003, National Fire Protection Association, Quincy, MA 02169)*

The grinding process ranges from a simple batch mill, in which an operator grinds 55-pound bags of grain for a small grocery store bakery or local restaurant, up to large-scale grinding operations using large silos or railcars to feed the mills, similar to those you find at large bakeries, food-manufacturing facilities, or other food-based occupancies.

Although many different types of milling or grinding equipment exist, the hazard is the same, independent of the type or equipment used. The hazard of grinding grain comes from the fact that the process produces fine particles of materials mixed with air from the process or the environment—a mixture that could be flammable or explosive.

COMMON HAZARDS

Fire Hazard

Although these occupancies primarily have an explosion hazard and therefore tend to attract the most attention, fires can occur as well. Fires are likely to occur in all areas of the grain facility, unlike dust explosions, which tend to be limited to areas containing dust. Fires often precede a dust explosion. In general, grain facilities are of noncombustible construction, usually concrete, but older facilities might contain some wood construction. Typically, due to the age of the majority of these facilities, sprinkler protection and fire detection are not on the premises.

Fires can occur in equipment such as motors, drives, bearings, and electrical equipment, as well as from poor housekeeping and smoking. Areas that contain these pieces of equipment can be found in dryers, motor control centers or rooms, elevators, storage bins, milling areas, and offices. Fires can also occur in maintenance-related areas or storage rooms.

To help control these sources of ignition, the electrical wiring and equipment should be suitable for this environment. All electrical wiring and equipment should comply with the National Fire Protection Code (NFPA 70), National Electric Code (NEC) articles 500, 502, and 504. These installed components should be rated for use in Class II locations and the appropriate group to ensure safe operation and shutdown of faulty equipment. This means that firefighters should only use intrinsically safe equipment in these facilities.

As previously mentioned, the conveyor belt is another fire hazard. Conveyor belts can range in length up to 100 feet or more and operate in a series of several sections. In the event of a fire, the belt can break, and the delay in shutting down the conveyor motors could force the belt to form a pile in the conveyor tunnel. If ignition occurs from overheating or friction from the motors and rollers, a large fire that is fed from a large pile of burning rubber conveyor belt in a tunnel could ensue, which would prove to be very challenging from a firefighter tactical viewpoint.

The grain dryer also exhibits a potential and severe fire hazard (fig. 15–9). If ignition of the grain or grain dust and other accumulated materials in the area occurs, this could again prove to be a very challenging fire. Especially when you consider the large size of these dryers (up to 80 feet) and the gas-fired units providing more than 20,000,000 Btu per hour, these type of files require careful strategically planning and tactical operations.[3]

Fig. 15–9.* Grain dryer** *(Reprinted with permission from the* **Fire Protection Handbook, 19th Edition, Copyright © 2003, National Fire Protection Association, Quincy, MA 02169)*

The grain itself is a fire hazard. The hazards associated with igniting grain must be prevented, whether loose or spilled grain or the grain in the silos. Preventive measures should be in place to monitor open-flame operations (cutting, welding, brazing) via a hot work program and strict no-smoking policy inside the facility. Additional preventive measures should include adequate housekeeping, weekly and monthly testing of the fire protection and detection system, and maintenance of the dust collection system and all safety interlocks on the conveyor belts and bucket elevators.

Fires could also occur in the truck and rail loading and unloading areas. These fires can be as a result of overheated brakes, engine fires, electrical shorts, and the like. Many times, these areas are used for excess storage and tend not to be included in the housekeeping programs because the area is considered to be outside and therefore not a problem. Consideration should be given to ensure proper housekeeping, removal of oil spills, and other safety precautions.

Explosion Hazard

Dust is certainly one of the major hazards in the grain industry and certainly gets the most attention. As previous mentioned, in processing or storage of grains, an explosion (or fire) may originate in the process equipment (equipment explosion hazard) or in the ambient environment in the room containing the process equipment (room explosion hazard).

This is a good time to explain the anatomy of a dust explosion (fig. 15–10). Similar to the fire triangle (heat, fuel, and oxygen), for an explosion to occur, there must be present fuel (grain dust), oxygen (air), ignition sources (open flame, sparks, or hot surfaces), as well as a fourth element or component, confinement. Confinement could be intrinsic to the equipment, such as the inside of a vessel or bin, or confinement caused by the room containing the process equipment.

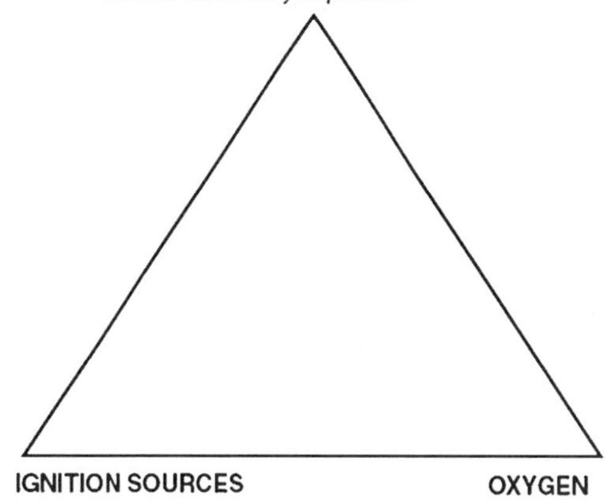

Fig. 15–10.* Explosion triangle** *(Reprinted with permission from the* **Fire Protection Handbook, 19th Edition, Copyright © 2003, National Fire Protection Association, Quincy, MA 02169)*

For an explosion to occur, this fourth element is required to confine the rapidly expanding heated gases of combustion within the enclosure until the pressures generated by the expanding gas exceed the strength of the enclosure. Similar to fire and the fire triangle, an explosion does not always occur when all four components are present. Instead, luckily, only under ideal situations will a dust explosion occur. This situation is a multifaceted combination of dust particle size, dust-to-air concentration, and the amount of energy from the ignition source.[4] Additional factors to be considered are the moisture content of the dust and the chemical composition of the dust.

Process equipment that either dry, weigh (bin scales), or grind grain can present an explosion hazard. As the grain is being handled and processed, dust is generated, and if not properly removed, can accumulate and provide the necessary fuel for an explosion. The hazard is the same for rooms containing grain processing-equipment. If the equipment and the dust collection systems are not functioning properly, fugitive dust emissions may accumulate in the room. It is common to find dust accumulations on window sills, structural building members, the equipment itself, and light fixtures. In areas such as the conveyor corridor, where the conveyor is open to the air, the dust collection system is essential to maintain a relatively dust-free environment.

In addition to explosion hazards, firefighters may be enveloped by the grain particles themselves. Many a firefighter or rescuer has sunk into the grain while trying to assist a victim. Finally, many hazards are presented by the equipment itself. As discussed in other chapters, the mechanisms of process equipment present an extreme danger. Electrical power supplies, equipment fuel, and gases also present potential hazards.

FIREFIGHTING TACTICS

As with all firefighting tactics, a well-thought-out and tested emergency response plan is critical. Preincident response planning must be carried out in cases where a building is identified as having the potential to develop into a major incident. In the event of a fire at such a building, it would be advantageous to have information readily available that relates to the structure, fire safety provisions, location of water sources, presence of hazardous materials, and any other information that may be relevant.

When responding to fires at a grain-storage or grain-processing facility, caution should be used when approaching the area. It should first be determined what is burning and where. As a precaution, all operations should cease, and the dust collection system should be permitted to operate. This will aid in the removal of dust and, potentially, smoke and heat. The bucket elevator and conveyors should shut down, as well as the grain dryers, scales, and other grain-processing equipment.

A fire in the conveyor tunnel should be approached similarly to a basement fire. These tunnels are below finish grade, extend over a long area, and usually have limited access. The area inside the tunnel is very limited, given that the majority of the space is occupied by the conveyor system, dust collection points, and other miscellaneous equipment. Proper entry procedures should be used when entering this area. Due to the construction and location, it is likely that most radios will not work in the tunnel. A second form of communication using runners or rope will have to be utilized. Positive-pressure ventilation should be set up as soon as possible to aid in the removal of heat and smoke. As with a basement fire, this fire can and will be hot. A large pile of conveyor belts burning has the same characteristics as a pile of burning tires; so large amounts of water would be needed, as well as a good Class B foam to aid in extinguishment. During operations, the level and accumulation of dust should be noted. Agitation of the dust into the air could form an explosive atmosphere and further complicate the situation. Wetting down the area is the best precautionary method to prevent the dust from being suspended in the air.

If responding to a report or confirmed explosion at these types of facilities, approach the area with caution; secondary explosions can occur, and falling debris presents another hazard. Staging should be out of the danger zone and far enough away such that secondary explosions would not affect response operations. Many times, the explosion(s) have consumed the majority of the fuel, and thus postexplosion fires are not a problem. There are occasions, though, when either the fire is the cause of the explosion, or the fuel load remaining is substantial enough to cause secondary fires. These fires should be treated as previously mentioned with the noted exception that the facility and structure might not be safe for entry.

CONCLUSION

Responding to bulk grain and storage occupancies where both fire and explosion hazards exist can be very dangerous. It is important to follow your department's response plans and guidance from the local plant personnel. Secure a good water supply, check for secondary explosion hazards, and use caution when entering these types of facilities. Remember that when entering the tunnels, radio communications may become inoperable; therefore, a second form of communicating would need to be established.

NOTES

[1] National Fire Protection Association (NFPA). 2003. *Fire Protection Handbook,* 19th ed. Quincy, MA: NFPA, p.6–372.

[2] NFPA. 1990. *Industrial Fire Hazards Handbook,* 3rd ed. Quincy, MA: NFPA, p. 837.

[3] NFPA, *Fire Protection Handbook, p. 6–368.*

[4] Zalosh, R. G. 2003. *Industrial Fire Protection Engineering.* West Sussex, UK: Wiley, p. 196.

16

Textiles

INTRODUCTION

Every day, we come into contact with textiles. Textiles are in most every aspect of our lives; our clothing, furniture, carpets, curtains and automotive upholstery are examples of textile materials, that is, materials produced with fibers. Almost all fibers are combustible. This combustibility creates problems in all areas of fiber manufacture, use, and storage. This chapter will examine the manufacture of textiles and the fire hazards associated with this manufacture. The basic fire hazards associated with textile manufacture are combustibility of the raw materials, ignition of raw materials, fire spreading because of the open flow of material between individual processes, and the congestion of the machinery and stock.[1]

Textiles are made from a basic component called *fibers*.[2] These fibers can either be natural or man-made. Cotton is the most common fiber used in textile manufacturing. Cotton arrives at the manufacturing plant in large bales. Plants that produce a blend of natural and synthetic fibers must also rely on additional raw materials, depending on which final products are produced, for example, polyester blends. The production processes vary based on the nature of the raw materials used. The combustibility of the raw materials also vary. For example, wool is more difficult to ignite, burns slower, and is easier to extinguish than cotton, all else being equal. Fully synthetic fibers are acrylics or thermoplastic, such as nylon or polyester. The acrylic fibers will char, *intumesce,* and burn when exposed to heat, whereas thermoplastics will not char but will shrink away from heat sources and melt, forming holes near the heat source and/or *ablate.*

Bales of cotton can either be the domestic, compressed, or imported types. The domestic type is less densely packed and so will be more affected by fire or water.

Synthetic fibers are also packed in bales, in the form of cut filaments. These bales are not as easy to ignite as cotton fibers, but will burn readily once ignited by the heat from combustible wrappings or cartons.

DESCRIPTION OF PROCESS

The manufacture of textiles consists of the opening of raw materials, picking, carding, combing, drawing, roving, spinning, weaving, and finishing.[3]

Opening

Bales of raw materials are brought into opening rooms in the first step of the manufacturing process. Foreign materials such as dirt, leaf matter, stones, and bailing wire may be present within the bales, especially with the cotton bales. The cotton bales are opened and allowed to sit to adjust to the humidity of the area. Various machines are used in the opening process, all with the same goal of opening and cleaning the cotton stock. This is done by beating, tearing, and shaking the fibers.

The cotton is then blended using various grades of fibers from different bales to ensure uniformity in the final product. The cleaning operations are performed mechanically and pneumatically, while the fibers are separated using spiked rollers. Light impurities are sucked away, and heavy waste is collected in traps. Initially, the cotton proceeds through blending feeders, which are slow-moving machines. From the blending machines, the cotton proceeds through vertical openers where the cotton is pneumatically driven upward through the machine, while arms on a vertically rotating steel beater throw the cotton outward against grid bars. The foreign matter is passed through grids, where it is then removed.

Other machines used to clean the cotton are horizontal cleaners, centrif-air machines, gyrators, and magnetic separators. Magnetic separators remove ferrous metal from the raw material. This metal could cause friction sparks as well as physically damage the machines.

Picker room

After leaving the opening room, the cotton is usually conveyed pneumatically to the picker room. Here it is further cleaned and aligned. This is done by using beaters, screens, and rollers. Upon completion of this process, the cotton is a flat sheet of stock.

Carding

The cotton is transferred from the picking room to the card room for further cleaning and straightening of the fibers by hand, wheeled rack, conveyor, or ductwork. Carding machines separate and lay the cotton fibers parallel to each other. The final product leaving the card room will be a soft, fleecy, ropelike strand of cotton known as a *sliver.*

Combing

The combing process is optional, depending on the end-product requirements; combing produces a fine, high-quality product. The combing process is similar to carding, but is more exact and removes all the short fibers under a preset length.

Drawing and roving

Drawing further straightens the cotton fibers, reduces the size of the sliver, accomplishes a degree of blending, and parallels the fibers further. Roving further reduces the size of the sliver, puts in enough twist to give the yarn the required strength, and winds the product on bobbins. The product is called *roving* at this stage.

Spinning

Spinning reduces the roving to the final weight required in the finished yarn. Ring and open-end spinning frames are the two main types of spinning frames.

Warping

In preparing the thread for the weaving process, the end threads are tied into continuous lengths that are threaded through guides and wound in parallel on the section beam. These beams, known as *warper beams,* may be stored in a high-piled rack configuration consisting of cantilever beams.

Slashing

In the slashing process, the warp is run through a sizing solution of starch and other additives to treat and coat the yarns to prevent them from breaking due to tension forces or rubbing and chafing during the weaving process.

Weaving

The culmination of the preceding processes is weaving. It is at this stage that the finished product is woven into the cloth as we know it by using a *loom*. The yarn is converted to cloth on the loom by two interlacing sets of yarn, one longitudinal and one running the width of the loom. In some instances, the finished cloth may be taken up on a lower floor, rather than at the loom. This may create unprotected floor openings.

Cloth inspection and cleaning

The inspection and cleaning process follows weaving. At this stage the cloth is inspected for deficiencies, and the fabrics are made ready for shipment to a finishing plant. Loose ends are either cut or singed off.

Finishing

At the finishing area in the mill or at a separate plant, the cloth may be bleached, dyed, printed, have backings or coatings applied, or receive other processes to achieve the desired effect for the finished cloth. The finishing work usually involves dip tanks, spreaders, rollers, and various types of heating and drying processes, each with their inherent hazards.

Knitting and carpeting

Knitting and carpeting are other processes involved in the textile industry. Both of these processes involve many of the preceding processes, but in the case of knit products, the final fabric is made on a knitting machine instead of a loom.

Carpet forming uses a tufting machine in which the basic yarns are interlaced. The finishing process requires that the carpet material is backed with jute or similar material and a form of latex. This is followed by a heat process that cures the latex. Direct heat or steam drums are used in the heating process.

COMMON HAZARDS

The basic fire hazards in textile manufacturing are the congestion of machinery and stock, the combustibility of the stock with its ease of ignition, and the rapid fire spread that is associated with the flow of material between processes. Textile mills typically create dust and lint-laden atmospheres, and therefore housekeeping must be maintained to minimize the dust and lint buildup. Machinery must be kept in top working order to eliminate errant sources of electricity that may create sparks. Areas that create dust conditions must be carefully monitored and cleaned to prevent dust accumulations that create fire hazards.

In the opening rooms, the storage of the bales of raw materials, which are stacked to heights up to 20 feet, can create storage and fire hazards. If a bale is brought in that already contains a smoldering fire, this hidden fire can flash over the surface area of the loosely packed bale. It can rapidly spread to other bales. If the fire proportions are such that the ties holding the bales together are affected, then the bale will break open, causing the fire to spread, and the stacks of bales to fall. Hidden fire that does not flash may be hard to detect, and it may create fire hazards later on in the process by allowing fire to take hold in the dust or lint buildup with a process machine.

Fork lifts or other lift trucks are used to move the bales of raw materials. These vehicles in themselves create fire hazards unless they are properly maintained. Fuel used in these vehicles must be stored in appropriate storage areas that are well ventilated and away from the material storage or process areas. In the case of a fire in an unoccupied mill building, these types of vehicles may be stored on the premises. In this case, the fuel supplies on the vehicles must be factored in during our initial size-up. Liquefied petroleum gas (LPG) used as a fuel can create serious hazards to firefighters if the tanks are exposed to the heat of the fire or direct flames.

As stated previously, the bales of raw materials may contain foreign objects. If these objects are not adequately removed during the manufacturing process, there is a risk of sparks in the machinery, causing potential sources of fire. This, combined with the dust and lint buildup, creates a fire hazard. If raw material wraps around moving parts of the process machinery, then friction can develop, also creating a source of ignition. Jammed stock within a machine can cause friction and heat buildup, creating a source of ignition and fuel supply.

Good housekeeping practices and keeping machinery in top operating condition will reduce fire hazards in mill buildings.[4] Constant awareness and vigilance is necessary to keep loose cotton from around and under materials. Machinery should be on a schedule of regular preventive maintenance. Electrical arcs from wiring that has been improperly maintained or from faults in electrically driven motors and traveling hoists can be a further source of fires. The area above suspended ceilings is another void where dust and lint can also accumulate.

In the finishing areas, dust explosions can occur from the starch used in the sizing solution if it is improperly handled in its dry state. (In recent years, polyvinyl alcohol is replacing starch.) In the inspection and cleaning areas, fire hazards are created when using torches used for the removal of loose ends on the cloth rolls. Air-conditioning systems throughout the mill must also be maintained free of dust accumulations. Filters must be kept clean and replaced according to manufacturers' recommendations. Air-conditioning ducts should be well supported and dust-tight, so that if a fire does occur in a duct, it will not spread to other combustibles outside of the duct itself. Under-the-floor air return systems can accumulate lint and dust and can spread fires in the event an ignition source is present.

Hazardous materials

Most of the hazardous materials encountered are the result of using solvents in such areas such as:

- Dry cleaning of synthetic fiber knit fabrics and woven and wool fabrics
- Dyeing operations
- Some finishing operations for impregnating or coating textile fibers

Solvents are also used to clean machinery such as rollers and spinning machines. Fire departments should have copies of the material safety data sheets (MSDS) for any hazardous materials present in textile occupancies within their response areas. These MSDS should be attached to the preincident response plan for the specific occupancy.

FIREFIGHTING TACTICS

Textile mill fires are known to be some of the most spectacular fires that firefighters will encounter. The large areas required for the production and the combustibility of the materials will cause rapid fire spread. In the New England area, textile manufacturing was the primary industry in many of the old cities. The buildings that housed this industry were large, heavy timber construction, also known as mill construction (fig. 16–1). Over the years, the solvents, oils, and greases penetrated the floors and added to the fire hazard. While it is true that this type of construction was designed to limit fire spread, once it is ignited, it is difficult to extinguish.

Fig. 16–1. **Large, heavy timber construction, also known as mill construction typically housed textile mills.**

Many mills will have fire detection and suppression systems in place that will aid the fire department by early detection, notification, and limiting the fire spread. However, these systems must be properly maintained to be effective. Manufacturing, finishing, and storage areas should have automatic sprinkler protection installed.

Warehouses will have sprinkler systems as well as smoke- and heat-venting systems. When arriving at fires where a sprinkler system has activated, consideration must be given to the additional weight added to the loads on the structure because of the cotton bales absorbing the sprinkler water. Cotton bales may also swell with water and exert pressures on exterior walls, causing partial collapse if adequate clearance was not provided between the bales and the outer walls. Ensure that for sprinkler-equipped buildings the fire department connection is supplied by the first-in engine company. Scuppers and drains should be used to assist with water runoff, but they may be blocked or clogged, allowing the water to accumulate on the floors and be absorbed by the stock present. In fires involving the opening room, it may be necessary to use the equipment available, for example, forklifts to bring the smoldering bales out of the facility, so that they can be opened further and the contents spread out for final extinguishment and overhaul. An area in the parking lot should be designated for this purpose. Be aware of the wind conditions present to ensure that flying embers do not start fires elsewhere. Before and while opening the bales outside, a hose stream should be applied to reduce the possibility of flying embers. Self Contained Breathing Apparatus (SCBA) should be used by firefighters. It would be helpful to designate in your preincident response plans the areas that can be used for this purpose or for apparatus placement. For small fires in the cotton bales, water extinguishers may be helpful in the extinguishment of small fires. Remember, a thorough overhaul and breaking down of the bale to ensure complete extinguishment is required.

For buildings equipped with heating, ventilation, and air-conditioning (HVAC) systems, it may be possible for the fire department to use the system to remove or channel smoke from the fire area. Initially, though, the system should be shut down until it is assured that it will not spread the fire.

Some of the individual machines may have additional automatic detection and extinguishing systems such as dry-chemical or carbon dioxide (CO_2). These systems should be indicated on preincident response plans if present. If one of these systems activates because of fire, then a thorough inspection of the fire area and surroundings must be made to ensure that the fire has been completely extinguished and no fire extension has taken place. Fires in the machines powered by electricity should be treated the same as any live electrical fire. The main power must be shut down, and lockout/tagout should be performed. If lockout/tagout is not possible, a firefighter with a radio must be stationed at the power switch to ensure that accidental activation of the power source does not take place. Dry-chemical or CO_2 are the preferred extinguishing mediums for electrical component fires. When the fire has extended to materials or lint, water may be used after the power has been disconnected.

Class A foams are very useful when fighting fires in textile-manufacturing or storage areas. Class A foams are ideal for penetrating bales of raw materials or piles of textile scraps. These types of foams have the advantage of a reduced knockdown time as well as a reduced overhaul time. With its reduced knockdown capability, Class A foams significantly reduce the volume of water used, in turn reducing the amount of water absorbed by the materials involved in fire.

When the fire is large enough to warrant an exterior attack, place all apparatus and personnel outside the collapse zone. Large fires in mill buildings have been known to have partial or full collapse early in the fire. In many instances, older mill building complexes have the buildings interconnected on many floors (fig. 16–2). This was intended to make it easier to move stock throughout the complex during the production process. These openings, if not properly protected, will spread the fire rapidly throughout the complex. Older mill buildings have been known to have inoperable fire doors, which contributed to rapid fire spread at some of the spectacular fires witnessed in the past.

Fig. 16–2. Older mill buildings may have interconnected buildings or sections. Note the different construction of the building section.

CONCLUSION

The textile industry, just as many other industries mentioned in this book, is not immune from the potential for fires. The dust and lint, the machinery used, and the combustibility of the materials set the stage for fires both large and small. Fire departments should be aware of the process used and the inherent fire hazards they possess so they can develop an effective fire prevention program and effectively preplan a response. Remember that we will need to call for additional resources early and ensure that we have adequate water supplies available to fight the fire. Keep in mind that some of the most spectacular fires witnessed in recent years involved textile mill buildings.

CASE STUDY

In July 2004 fire crews from a large municipal fire department responded to a fire in a multibuilding warehouse complex housing a mattress-manufacturing company. Lightning was the cause of the fire when it struck a rooftop air-conditioning unit at the rear of the complex. The initial responders began an interior attack by stretching a hoseline through the front entrance and continuing through to the rear in an attempt to hold the fire in the rear of the fire building. Large stocks of mattresses and boxsprings lined the walls of their entry route, and rapidly deteriorating conditions caused the interior firefighting forces to be withdrawn. An exterior attack was instituted. Six elevated master streams were used in conjunction with exterior hoselines and ground monitors to bring the fire under control in approximately 3½ hours.[5]

Random Thoughts

- Early recognition of deteriorating conditions and hazards such as falling raw or finished stock is paramount to the safety of operating personnel. A dynamic risk assessment must be conducted by the incident commander, sector officers, and company officers immediately upon arrival and should be continued throughout the operations. Interior fire crews must be withdrawn when conditions are such that the safety of firefighting crews is doubtful. All firefighters on the scene should maintain situational awareness and report any perceived or actual unsafe conditions.

- The weight of the raw and finished stock when it has absorbed water must be considered. Many of these structures are old, and floor connections to the walls may be compromised by rot and connection deterioration. When textiles absorb water, their weight increases considerably, and bales of textile materials can bulge, pushing out exterior walls. Water runoff should be monitored to determine a rough estimate of the water accumulating within a structure.
- Collapse zones must be established and maintained during exterior fire attack (defensive) operations. Rope off collapse zones so that firefighters will not unintentionally enter the zone. A collapse zone should be at least the full height of the affected wall; some have argued that because debris can bounce when it hits the ground, a collapse zone of one and a half times the height of the wall is safer.[6] Elevated master stream devices, including the aerial tip, basket, boom, and the truck itself, should be outside the collapse zone. This includes the arc of a wall falling at a 90° collapse.[7] If it is absolutely necessary to operate a master stream inside a collapse zone, the portable deluge nozzle or aerial stream should be secured to direct the stream effectively and safely, then left unattended.[8]

NOTES

[1] National Fire Protection Association (NFPA). 1990. *Industrial Fire Hazards Handbook, 3rd ed.* Quincy, MA: NFPA, p. 397.

[2] NFPA. 2003. *Fire Protection Handbook,* 19th ed. vol. II. Quincy, MA: NFPA, pp. 8–61.

[3] NFPA, *Industrial Fire Hazards Handbook*, p. 400.

[4] NFPA, *Industrial Fire Hazards Handbook,* p. 406.

[5] This case study is based on information contained in Peter Matthews, Burning Beds, *Fire Rescue,* November 2004, pp. 90–94.

[6] Avillo, A. L. 2002. *Fireground Strategies.* Tulsa, OK: PennWell, p. 423.

[7] Dunn, V. *Safety and Survival on the Fireground.* Saddle Brook, NJ: Fire Engineering, 1992, p. 225

[8] Dunn, V. *Collapse of Burning Buildings, A Guide to Fireground Safety.* New York: Fire Engineering, 1988, p. 66

17

Storage and Warehouses

INTRODUCTION

Storage and warehouses have been in existence for centuries and have been the center of some of the largest fires in recorded history. This is mainly attributed to the simple fact that the very essence or purpose of a warehouse is to store large quantities of about anything in a single, easy-to-access location. From a fire department or loss-control perspective, this accumulation of *commodities* creates a large fire load, which if ignited, can be very challenging to control, let alone try to extinguish. Warehouses come in a variety of construction characteristics and can range from old heavy timber construction with no or minimal fire protection features used in the late 1890s and early 1900s to new, state-of-the-art precast, tilt-up concrete construction using some of the current advances in fire protection technology (fig. 17–1). Warehouses can even be found in underground limestone mines, old buildings that have been converted to warehouses, and portable buildings and containers.

This chapter will discuss design characteristics and hazards associated with various storage occupancies, building construction, storage arrangements, and fire protection features. It will also discuss some tips on firefighting tactics for general warehouse occupancies. Details on fire protection and detection systems, water supply, and other specific hazards and details are discussed in other chapters of this book.

Fig. 17–1. **Warehouse**

DESCRIPTION OF OCCUPANCY

Warehouses are buildings that are used to store all types of goods or merchandise (*commodities*). Warehouses can be separate and purpose-built facilities, or part of a multiple-use occupancy (fig. 17–2). Warehouses are located in all types of areas from urban, suburban, and rural to wharfs, industrial areas, and airports. They range in size from just 50,000 ft^2 up to more than 1,000,000 ft^2 and are constructed in a multitude of ways, ranging from heavy timber to precast tilt-up concrete (fig. 17–3). These facilities also include warehouses that are used to store fresh foods at low temperatures, in the range of 40°F, to other types of warehouses that store frozen foods at or below –18°F. Many warehouses also contain flammable and combustible liquid storage. These flammable and combustible liquids typically contain such goods as motor oil in cases, cooking oils in boxes, paints and finishes in cans, or various chemicals used in production facilities on- or off-site (fig. 17–4).

Fig. 17–2. **Warehouse**

Fig. 17–3. **Warehouse**

Fig. 17–4. **Dry goods storage** *(Courtesy of Safeway Inc.)*

Other types of warehouses exist in unusual locations. Many old limestone or granite quarries have been converted into storage for all types of commodities, including having been converted into a cooler or freezer type storage. Because of the varying size, construction, location, and commodities stored in warehouses, they present a large and unusual fire hazard.

COMMODITY OVERVIEW

First, the chapter will discuss the types of goods or merchandised that are typically stored in a warehouse. To determine fire protection features, the type of goods and their storage arrangement must first be determined. In the remaining part of this chapter, we will refer to storage of goods or merchandise as *commodities,* and we will also discuss commodity classifications. A commodity is defined as the basic product itself (e.g., can of soda), the packaging (e.g., 12 soda cans in a cardboard box or case), and its container (e.g., 50 cases of soda, wrapped in plastic on a wood or plastic pallet). When considering the classification of a commodity and their relative fire hazard, you must take into account more than just the product itself. You must also consider the package material that it is stored in (on an individual level) and how the entire pallet or bin is loaded and packaged. In determining fire hazards and hazard classification within a storage occupancies, you actually establish its commodity classification. Evaluating the classification of a commodity can be accomplished either by generic material classification, in which you simply match the commodity to previous testing records, or via large-scale testing using a certified and approved testing laboratory.

Generic commodity classification information can be found in the National Fire Protection Association (NFPA) Standard, NFPA 230, *Standard for the Fire Protection of Storage* (2003) or FM Global Property Loss Prevention Data Sheet *8–1, Commodity Classification.* These standards use a classification system that contains seven categories of commodities. They are Class I, II, III, IV, and

Group A, B, C Plastics. Class I commodities are the least flammable, and Group A plastics are considered to be the most flammable. Many special hazards, such as flammable and combustible liquids and aerosols present many challenges. Flammable and combustible liquid commodity classification use the NFPA's standard schedule of Class I A, B, C, Class II, and Class III A, B. Aerosols typically follow the Department of Transportation (DOT) guidelines for classification.

Many commodities are wrapped in plastic for weatherproofing and transportation. This form of packaging is called *encapsulation.* Encapsulation is a packaging method consisting of plastic sheeting completely wrapping and enclosing a commodity, usually on wood pallets.

A Class I commodity is a noncombustible product or good stored in an ordinary combustible package such as a cardboard box or by itself with little or no additional packaging. Examples include glass bottles with nonflammable liquids, bags of cement, canned foods, meats, metals (pots, pans, or wire), and dry cell batteries.[1]

Class II commodities consist of Class I commodities stored in wooden crates or boxes, double- or triple-walled corrugated boxes, or equivalent material on wood pallets. Examples include beer or wine (not to exceed 20% alcohol) in wood crates or barrels, large washing machines in triple-walled cardboard boxes, fiberglass insulation, combustible free-flowing powdered products like sugar or flour, fragile glass such as lightbulbs, or dinner/china plates in rigid, triple-walled corrugated cardboard boxes.[2]

Class III commodities are products made from wood, paper, or natural cloth stored on wood pallets. This does include Classes I and II products that contain less than 5% plastic either by weight or volume. For example, a metal bicycle frame with plastic handles, pedals, seats, and tires is considered to be a Class II commodity because the amount of plastic is considered to be negligible, less than 5%. The metal bicycle frame and plastic handle alone would be considered as Class I. Other examples of Class III Commodities include books, magazines, shoes, luggage, furniture, mattresses, tobacco products, and refrigerators (due to the plastic interior).[3]

Class IV commodities are Class I, II, or III products containing themselves or in their packaging no more than 25% by volume or 15% by weight of expanded or unexpanded plastic stored in cardboard boxes. Examples of Class IV commodities include cameras, crayons, plastic beverage bottles, wax-coated paper cups, electrical wire on plastic spools, pharmaceuticals (pills and capsules) in plastic bottles, and synthetic fiber padding.[4]

At times, it is extremely difficult to determine the classification when it comes to classification of plastic commodities. Often, commodity classification will have already been done by a fire protection engineer, insurance underwriter, loss control engineer, or a local research laboratory. Perhaps for an *authority having jurisdiction* (AHJ), the best procedure to follow is to seek the advice of these individuals or agencies. Plastic is difficult because of the large variety of plastic types, the complexity associated with their nomenclature, and the burning characteristics relating to difference additives used to manufacturer plastics. Accurate commodity classification of plastics is crucial because the *heat release rate* of a burning plastic can be three to five times greater than ordinary combustibles. For example, ordinary combustibles such as wood or paper have a *heat of combustion* that have a range in the 6,000 to 8,000 Btu/lb range, while plastics generally have a heat of combustion between 12,000 to 20,000 Btu/lb. Additionally, the burning rate of a commodity is dependent on many characteristics but in general displays a higher maximum burning rate over ordinary combustibles.[5]

Plastics are broken into two large groups: *thermoplastic* and *thermoset.* Thermoplastics become soft when heated and hardened when cooled. Thermosets set into a permanent shape from heat and applied pressure, and reheating will not soften these materials. Plastic materials are manufactured into two basic forms: expanded and unexpanded. Expanded plastics are low density and are commonly referred to as foam, such as foam coffee cups (polystyrene) or foam padding and sheeting (polyethylene), whereas unexpanded plastics are considered to be more solid and of a higher density, such as plastics totes or toys.

Group A plastics have a higher heat of combustion than ordinary combustibles and a much higher burning rate as well. Group A plastics are considered to be the most flammable and include items that contain thermoplastic polystyrene or acrylonitrile-butadiene-styrene (ABS) such as certain toys, souvenir cups, or similar. Paraffin and beeswax also fall into this category.[6]

Group B plastics have heat of combustion rates that are less than Group A plastics but more than ordinary combustibles. These include products made with thermosetting polyesters and thermoplastics such as polyethylene, acrylics, and nylon.[7]

Group C plastics are products that incorporate plastic materials having a heat of combustion and burning rate similar to ordinary combustibles. These products contain thermoplastic fluorocarbons and thermosets such as phenolics and silicones. Group C plastics can be treated and protected as Class III commodities. Caution should be used if these materials are combined with other materials or packaging that would change the burning characteristics of the commodity.[8]

The question might arise of what to do if you have a mixture of commodities in a warehouse. If the commodities cannot be separated by a recognized or approved method as described in the standards, the worse-case or most hazardous commodity applies. This means that if you have a warehouse that is housing a Class I and a Class III or Group A plastic in the same area or rack, fire protection and detection should follow the requirements for the Class III or Group A plastic, which ever commodity is the most hazardous.

STORAGE ARRANGEMENTS

Now that the commodity classification is determined, we will next discuss the storage configurations. To determine fire protection features required after identifying the commodity classification is to establish the storage arrangements. Storage configurations may vary from location to location. In general, storage arrangements can be classified as palletized, solid-pile, shelf or bin-box, and rack storage (single-row, double-row, or multirow racks).

Palletized storage is commodity storage on pallets that are then stacked on top of each other, often up to four pallets high. The stacking height is dependent on the commodity and its ability to be stable under a load. Pallets are constructed of wood, metal, or plastic, and are loaded onto forklifts to move and stack commodities.

Rack storage is storage in warehouse racks that is constructed using metal members in a vertical, horizontal, and diagonal orientation (fig. 17–5). The racks can have open or closed shelves to aid in the storage of commodities. Closed shelves are considered to be more of fire hazard than open-shelve racks. This is attributed to the simple fact that the solid shelves block and shield the burning commodity from overhead sprinkler ceiling protection. Rack storage can be fixed or portable. Loading of the racks varies from lift trucks, stackers, and manual (by hand), or automated robotic storage systems. Within the rack storage arrangement, there are three different types: single-row racks, double-row racks, and multiple-row racks. Racks have flue spaces (fig. 17–6), which are the open spaces between rows of storage. In rack storage, the longitudinal flue spaces are perpendicular to the direction of the loading (space in between racks), and transverse flue spaces are parallel to the direction of loading (between loads or racks side by side). In solid-pile and palletized storage, flue spaces may run in either direction (fig. 17–7).

Single-row racks have no *longitudinal flue spaces* or open space in the back of the rack. These

***Fig. 17–5.* Example of rack storage** *(Reprinted with permission from the* **Fire Protection Handbook**, *19th Edition, Copyright © 2003, National Fire Protection Association, Quincy, MA 02169)*

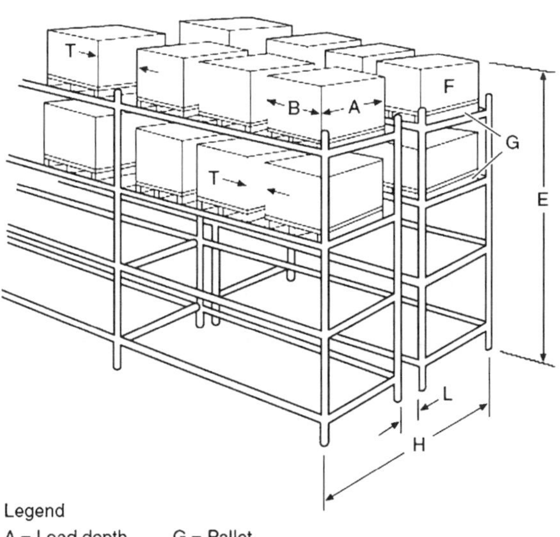

***Fig. 17–6.* Flue spaces** *(Reprinted with permission from the* **Fire Protection Handbook**, *19th Edition, Copyright © 2003, National Fire Protection Association, Quincy, MA 02169)*

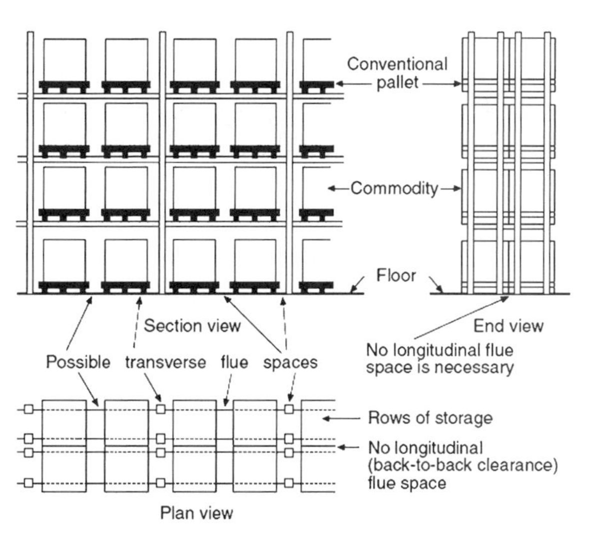

***Fig. 17–7.* Flue spaces and double-row racks** *(Reprinted with permission from the* **Fire Protection Handbook**, *19th Edition, Copyright © 2003, National Fire Protection Association, Quincy, MA 02169)*

are the typical racks that you would find in the backs of grocery stores and small storage rooms. Double-row racks are two single-row racks placed back to back, separated by a longitudinal flue space, meaning that there is typically a 6- to 12-inch space behind the racks (fig. 17–8).

***Fig. 17–8.* Example of double-row racks** *(Courtesy of Safeway inc.)*

Multirow racks are double-row racks placed back to back or a racking system that is more than 12 ft in depth (fig. 17–9). Solid pile storage is basically storage on the ground, without pallets. This can be storage of large commodities in boxes or goods and merchandise that is not packaged. Shelf storage is storage on a shelf or other small rack where the shelves are less than 30 inches wide (fig. 17–10).

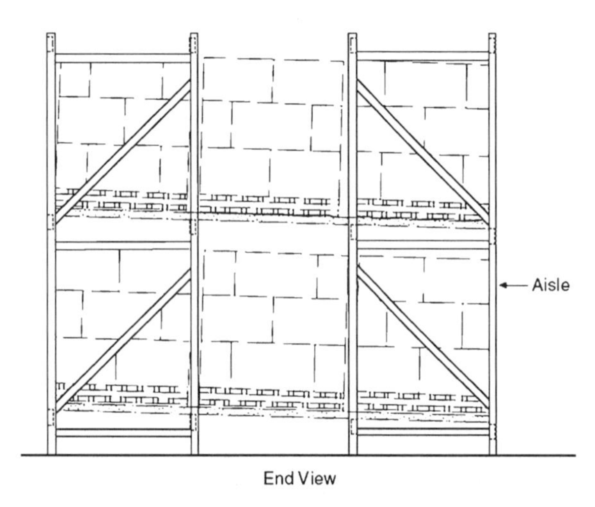

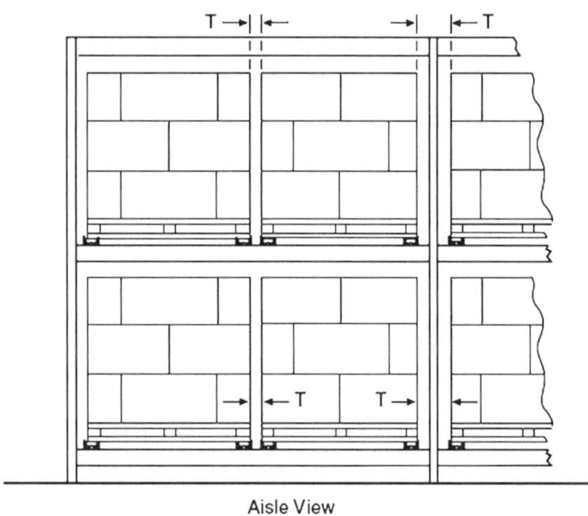

***Fig. 17–9.* Example of multiple-row racks** *(Reprinted with permission from the* **Fire Protection Handbook**, *19th Edition, Copyright © 2003, National Fire Protection Association, Quincy, MA 02169)*

***Fig. 17–10.* Example of shelf storage racks** *(Reprinted with permission from the* **Fire Protection Handbook**, *19th Edition, Copyright © 2003, National Fire Protection Association, Quincy, MA 02169)*

BUILDING CONSTRUCTION

As mentioned earlier, building construction can range from heavy timber in older cities to precast, tilt-up concrete structures commonly used today. Many new warehouses being built are tilt-up concrete structures with a lap-seam roof over C- or Z-purlins and are installed with an early-suppression fast-response (ESFR) sprinkler and fire detection systems. Warehouses can have a footprint of more than 1,000,000 ft^2, and there could be several warehouses in series. The building heights can exceed 100 ft, as is common in fully automated and computerized rack systems.

COMMON HAZARDS

Commodities stored in warehouses, because of their nature, can vary from week to week, month to month, or year to year. A monthly inspection of a warehouse could reveal that a Class I commodity stored last month has now changed to a Group A Plastic. This is extremely common in the warehouse industry. It is important that prefire plans or surveys are updated on a regular basis to reflect these changes in commodity storage. Because these commodity changes in warehouses occur so often, many new warehouses are constructed to a higher hazard rating with improved construction, fire protection, and enhanced storage racking system.

Storage occupancies are different from manufacturing or process facilities, in that they are static environments, meaning that there are no chemical reactions taking place or heavy machinery operating. Therefore, many of the operational hazards found in those types of facilities are nonexistent in a warehouse facility. Since many warehouses contain large floor spaces and have many different floors typically occupied by very few personnel, fires often go unnoticed or undetected for long periods of time.

Some common fire hazards that are generally associated with warehouses include incendiary, hot-work-related incidents, storage clearance from heat sources, smoking, electrical/lighting, and forklift trucks.

FIREFIGHTING TACTICS

Caution should be exercised when responding to a reported warehouse fire. Verification of contents and other pertinent information should be obtained before making an initial interior attack. Many newer warehouses have ESFR or other water-based fire suppression systems installed, and the first-responding trucks should immediately begin to support the fixed systems operations. During this time, information such as length of time burning, location of the fire, nearest door or overhead garage for entry, aisle markings, condition of the roof and roof vents, and electrical isolation to shut down any automated racking or delivery systems should be obtained by the incident commander (IC).

Water supply required for control and or suppression of a warehouse fire should have been determined during the preincident response planning phase. Using one of the recognized methods for determining water supply, criteria for manual suppression or firefighting supply for nonsprinkered buildings should be established prior to making entry into the building.

The use of high-rise or hotel packs will be beneficial to use during a warehouse fire. Connecting into a Class I or Class III standpipe (if provided) will help reduce extremely long hose lays and expedite manual firefighting efforts. Unless your department owns a truck-mounted positive-pressure fan (those that exceed about 80,000 cubic feet per minute CFM) roof operations should begin after the initial support of the fixed systems. Because of the large, open-area of a warehouse, venting is important so that smoke, heat, and other products of combustion can be removed for interior firefighters. Many warehouses, because of code-related issues, have automatic or manual smoke vents. These are ideal for rapid ventilation of a building. Many roof vents simply require the push of a button to activate several large vents for easy ventilation. If manual vents are present, simply removing the lock and pulling the lever will operate the spring-loading vent. These methods are preferred over cutting holes in the roof because they are easier to use and do not cause unnecessary damage to the building. Also, the use of thermal imaging cameras can aid in reducing the amount of time to find the seat of the fire, as well as providing the necessary visual information needed to find your way through the smoke.

Caution should be used during and after firefighting activities when operating in or around the storage racks. These racks often become unstable from the prolonged exposure to heat. Remember, these racks are not as strong as typical metal building components, so failure could occur sooner and at a lower temperature. In buildings that have automatic storage and retrieval systems and or robotic lift trucks, caution should also be used when operating in an emergency. The IC should verify that the power to the system or robots has been disabled to prevent accidental movement and injuries to firefighters.

Many warehouses consist of a single building, but sections are segregated by fire walls and doors. Additionally, fire walls may have been installed as a result of insurance recommendations in an attempt to isolate and minimize losses. Fire doors will typically not reopen after activation of the link or solenoid. Further, if an interior attack requires movement through a fire door, caution should be exercised. A rapidly closing fire door could sever the hose or injure a firefighter.

CONCLUSION

Responses to warehouse fires create a very challenging environment, ranging from the large amount of fuel loading to large, wide-open spaces. It is important that your department's planning includes identifying the commodities, types of storage, and locations of fixed system, including the use of thermal imaging cameras. Periodic inspections will aid in identifying changes in commodity or storage arrangements. Roof operations should take advantage of roof vents for ventilation of smoke and heat. Because of the large amount of combustibles, an adequate and reliable water supply is very important.

CASE STUDY

In July 2002 an explosion and fire occurred at a large printing plant complex in rural Wisconsin. The facility was a sprawling 2-million-square-feet complex consisting of 14 buildings, with the majority adjoining each other or interconnected. One of the structures was a 10-story building housing an automated storage and retrieval system (ASRS). This ASRS measured 80 ft wide by 770 ft long and used computers to control the transportation and storage of unbound magazines or catalogues contained on pallets. At this particular facility, 31,000 pallets of printed material can be stored.

Upon arrival at the scene, the fire department found that the ASRS area was completely involved in fire, and a large portion of the building had collapsed. The initial response vehicle had a 250 gpm pumping capacity and a 300-gallon water tank. Realizing the volume of water necessary, the department began a mutual aid response and set up additional water supplies to augment the on-site hydrant system. The on-site fire water system was supplied by fire pumps connected to an 800,000 gallon storage tank. Defensive positions were established, and a 3,000 gpm water delivery was maintained. Drafting operations from nearby ponds as well as tanker shuttles were established. One of the issues confronting the firefighters was the chance of fire spread to other areas of the plant through a damaged fire wall. Fire streams were established from the uninvolved side of the wall to prevent fire extension. Flaming debris carried by thermal updrafts and wind conditions also threatened additional buildings in the complex. The fire was completely extinguished 10 days after the initial response.[9]

Random thoughts

- Preincident response planning is essential at industrial and commercial occupancies. It was noted that at this incident no preincident response plan was prepared. Don't forget to include water supplies required and available in your preincident response plan.
- For large-scale fires where flaming debris is being carried by thermal updrafts or wind conditions, it is essential that brand patrols are established and surrounding building roofs are monitored for fire extension. Aerial platform apparatus or police helicopters can be used to assist with this function.
- Collapse zones must be established and enforced during defensive operations. It is always noticed that firefighters want to "get a little closer" at these incidents. Several safety officers should be appointed to cover all sides of the operation.
- At large scale or long duration operations relief of firefighting and support personnel is a must. During the incident action

planning, ensure that this is taken into consideration. Set a standard time limit for crew rotation, and plan to have the relief crews on scene in advance so that briefings can be conducted.

- Rehabilitation for these campaign-length operations should include more than fluids and snack bars. At this operation, dry socks and sunscreen were provided.
- Construction equipment may be necessary to pull debris from the collapsed structure to apply water to the seat of the fire and during overhaul. Plan early and have contingency plans available for contacting the companies that operate such equipment. It may be possible to use your town's department of public works to supply such equipment.

NOTES

[1] FM Global. 2000. Property Loss Prevention Data Sheet *8–1, Commodity Classification.* Norwood, MA: FM Global.

[2] FM Global, Property Loss Prevention Data Sheet *8–1.*

[3] FM Global, Property Loss Prevention Data Sheet *8–1.*

[4] FM Global, Property Loss Prevention Data Sheet *8–1.*

[5] FM Global, Property Loss Prevention Data Sheet *8–1.*

[6] FM Global, Property Loss Prevention Data Sheet *8–1.*

[7] FM Global, Property Loss Prevention Data Sheet *8–1.*

[8] FM Global, Property Loss Prevention Data Sheet *8–1.*

[9] White, D. (Ed.). 2003 (January/February). Racked Up. *Industrial Fire World,* pp. 6–9, 24.

18

Recycling Facilities

INTRODUCTION

Recycling in the United States has become a huge phenomenon for several reasons. One reason is the movement to reduce the waste stream and protect the environment. Another reason recycling is popular is for economic purposes. Because the costs of raw materials are increasing, the value of previously discarded materials has risen in value to a point that a profit can be made through recycling operations. The science of recycling has also matured, and now these products can be recovered and reused. In some cases, these recycled materials are even collected and sent overseas for processing.

Typical Contents of Solid Waste (Percent by Weight)			
	Domestic	Commercial	Industrial
Paper products	51.5	69.0	17.4
Glass	15.0	7.0	3.4
Cans and metals	7.0	10.0	8.0
Plastics	2.0	10.0	
Cloth, leather, rags	4.0		0.4
Food	10.0	4.0	8.4
Wood	2.0		17.9
Yard waste	8.5		
Minerals			36.7
Chemical waste			3.3
Rubber			1.6
Other			3.1

Fig. 18–1. *(Reprinted with permission from the* **Fire Protection Handbook**, *19th Edition, Copyright © 2003, National Fire Protection Association, Quincy, MA 02169)*

Examples of materials that once filled up the landfills and that are now recycled include paper; wood; oils; acids; various types of batteries; antifreeze; clothes and textiles; and metals such as aluminum, magnesium, and steel. Junkyards collect old cars, and the parts are sold at a discounted price. Some materials are kept in their original basic composition, such as glass, and other materials are burned and converted to energy (fig. 18–1). Recycling efforts greatly reduce the waste stream, save energy, and produce revenue.

What does this mean to the firefighter? The firefighter may be asked to respond to these recycling centers. Massive fires have occurred in these facilities that are beyond the capability of the local fire department to control. Response efforts have to be planned and organized to combat these fires. In some cases, recycling center fires emit toxic fumes that require large evacuations or produce hazardous runoffs that require EPA Superfund efforts to clean up. Firefighters must know what is in the junkyards and recycling centers, just as they need to what hazards are present in any other industrial facility.

DESCRIPTION OF THE PROCESS

This chapter examines some of these recycling facilities and addresses the hazards and firefighting techniques. As has been mentioned in the other chapters, firefighters must spend time preplanning for fires in these facilities and fully understand how these facilities operate. Recycling is too broad to cover in detail, but the information here will get you started in the right direction. Waste may be separated and collected by its particular type (fig. 18–2), or separated and certain components burned to reduce the bulk and ease disposal (fig. 18–3).

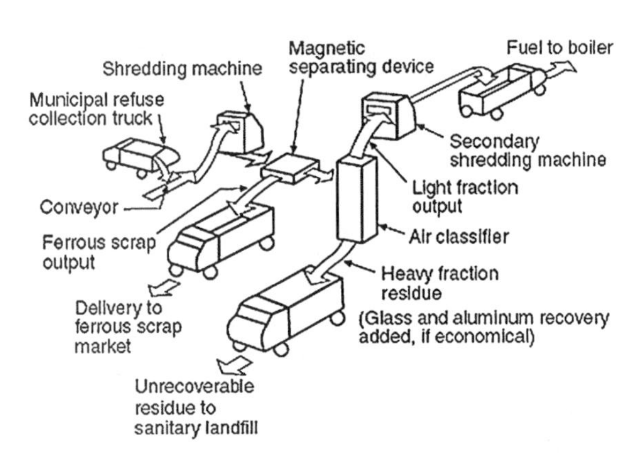

***Fig. 18–2.* Typical recycling of scrap and waste materials by various separation means** *(Reprinted with permission from the* **Fire Protection Handbook**, *19th Edition, Copyright © 2003, National Fire Protection Association, Quincy, MA 02169)*

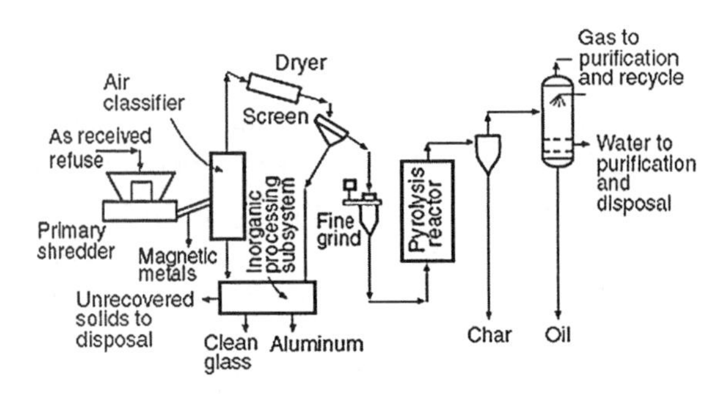

***Fig. 18–3.* Recycling of scrap and waste materials aided by a pyrolysis reactor** *(Reprinted with permission from the* **Fire Protection Handbook**, *19th Edition, Copyright © 2003, National Fire Protection Association, Quincy, MA 02169)*

Tire recycling

Tire recycling is a huge potential problem for firefighters. Tires are brought to an outside collection area to await sorting. Usable tires may be sold as is, or retreaded and sold. Unusable tires may be shredded or melted. The rubber compounds and steel from the tires can be separated and used for many purposes. Some tires may also be burned to produce energy or melted to recover the oil.

When tires arrive at a collection facility, they are stored for processing. Tire storage sites can cover hundreds of acres and contain millions of tires. Ideally, the tires will be stacked in piles that allow access for personnel and prevent the spread of fire. Unfortunately, many tire collection fires indicate that this is not always the case. Also, tires may be collected with other hazardous materials and present a significant fuel package.

Glass recycling

Glass is one of the most commonly recycled materials. The collection of glass and its storage presents little danger to firefighters. Recyclable glass includes mainly bottles and jars of various colors. The glass is melted down and most often made into new containers. Glass recycling includes the stockpiling of large quantities of consumer glass. This glass is typically shattered into small pieces to await melting. Use of this recycled glass significantly reduces the amount of energy and raw materials to make new glass.

Automobile recycling

Automobile recycling is big business and one of the oldest ways to recycle. Cars and their components are sold to consumers who need a part that is no longer manufactured or want to buy it at a discounted price. Like other recycled materials, cars will typically be sent to a central collection location. Unlike some other recycled materials, cars contain many types of plastics, rubber, metals, and fluids.

Typically, the first stage is to collect the cars at a junkyard or parts reclamation center. The gasoline, oils, various other fluids, heavy metals, and air-conditioning refrigerant are drained out of the engine and recovered. Depending on the type and sophistication of the automobile recycling center, engine and body parts may be removed from the vehicle before sending the vehicle to the yard. Next, the vehicle is sent to the yard, and other usable parts such as alternators are removed by either the recycling center staff or customers. The last stage for a truck or automobile will be to send it to be melted down for the metal, and maybe also the plastic and glass. Of course, tires, fluids, batteries, and other automobile parts will go to other recycling centers.

Metals

As mentioned, metals from old trucks and automobiles are frequently recycled. These metals may include steel, aluminum, lead, magnesium, and chrome. Other sources of metals could come from home recycling of materials such as steel and aluminum cans. Scrap metals from manufacturing wastes are also collected and sent to metal recycling centers (fig. 18–4). There are many other sources of scrap metal, for example, precious metals such as gold and silver or semiprecious metals such as brass and brass alloys. There could also be hazardous metals such as lead and mercury. Sources could be anything from an old dryer to an antiquated computer.

These metals are collected for processing. Once the metals are placed in a collection area, they are further separated by type, and any contaminates are eliminated if possible. Metals will typically be melted down and made into pellets, rolls, or bars, then sent to manufacturing facilities that need these materials.

Fig. 18–4. **Large metal recycling plant showing large scrap-moving equipment**

Oils and other fluids

Oils and other fluids are often captured and recycled. These oils and fluids include various hydrocarbons, alcohols, cleaning fluids, acids, and caustics. Even cooking oils are captured these days to be made into automobile fuel. Very few fluids can be put into the sanitary sewer waste stream. Fluids are collected in proper containers for their type, then sent to centers specifically designed to handle the waste. If the fluids are to be recycled, they are reprocessed and formulated to the required specification. Oil- and fluid-recycling centers have bulk storage tanks to hold both contaminated waste and potentially recycled materials.

Paper

The American Pulp and Paper Association reported that in 2005 each American used approximately 346 pounds of paper. They further reported that more than 50% of that paper was recycled. This means approximately that 51 million tons of paper is recycled in the United States each year. Typically, this paper is brought into industrial recycling centers from regional collection sites. Paper in the form of magazines, business paper, cardboard, and newsprint is collected.[1]

The basic process of recycling paper is fairly simple. Water is added to the bulk paper, and it is made into slurry. Surfactants are added to remove the inks, and the slurry is used to produce paper.

Plastics

Plastics are another major material recycled in the United States. All consumers are familiar with plastics used for consumer products like furniture, toys, food containers, and piping. Most of these plastic products are manufactured with recycling in mind. Recycled plastics are first collected by the type (many times identified by a number on the container to readily identify the type) and stored in bulk, awaiting processing. The number on the plastic product is called the plastic recycling code. The next step for plastics is to be ground, melted, and extruded to another form. Plastics are *polymers* usually made of carbon, hydrogen, oxygen, and silicon.

Power production

In some cases, trash is collected to be burned to produce some type of power. Trash is collected, and burnable material stockpiled in large warehouses (fig. 18–5). This keeps the materials dry. These materials vary by site, but common categories such as paper, rubber, cardboard, wood, and plastics may be used. Combustible materials are shredded and moved by conveyor belt and large cranes into the furnaces (fig. 18–6). In another form of recycling, trash is put into a landfill, and methane gas is collected from the decomposition of the trash. The trash must be carefully separated (first by hand, then by a mechanical tumbler) and made into a slurry. This gas could be used to fuel electric generators.

Fig. 18–5. **Large piles of assorted materials must sorted to be used either recycled, burned, or disposed of.** *(Courtesy of C. Guillemette)*

Fig. 18–6. **A crane is used to feed this material into a shredder before being sorted.** *(Courtesy of C. Guillemette)*

Textiles

As with other recycled materials, textiles are collected and stockpiled in large warehouses. Common initial collection points for textiles are charitable organizations such as the Salvation Army, which collects clothing for the needy. Donations of all types are dropped off at these organizations to be redistributed to people who need it. Cloth from donated clothing that is not suitable to for clothing is frequently made into rags or recycled for its fibers. These charitable collection facilities and regional collection centers can store large amounts of donated clothing.

Another type of textile recycling involves materials that are leftovers from manufacturing. Scrap materials may be sent from the manufacturing facilities, such as upholstery shops, back to a factory where they will be processed as fiber back into fabric. Materials arriving or waiting to be shipped are often baled to reduce the bulk and facilitate transportation. Scrap textile fibers are added back in cloth manufacturing process.

COMMON HAZARDS

The hazards to firefighters associated with recycling operations are numerous. The specific hazard of the material and the process of recovering have to be considered. Not only must you consider these hazards, but you will also have to consider what else may have come to be recycled by accident. Some materials may not be pure or could contain other hazards such as gases, liquids, or heavy metals. Sometimes exotic materials such as explosives and radioactive materials will end up at recycling facilities. The hazards of recycled materials vary by type or material and physical properties such as flammability and volatility. An example is used engine oil, which may be mixed with other contaminates (such as gasoline) and ignite readily. In addition to the properties of the material, the means in which the materials are stored increases or decreases the hazards, as discussed later.

Liquids and gases are collected in tanks and other containers. The hazards of these materials are typical to similar hazardous materials. These hazards should be identified during preincident response planning, and firefighters should observe any placards or signage, such an NFPA 704 symbol. Substances such as fluids and gases could also be collected in truck trailers, which may be equipped with Department of Transportation placards. Extra caution should be exercised on aboveground storage tanks. During size-up, the type of tank construction and support structure must be considered.[2] Any time liquids are collected in closed vessels of any kind, the potential for a boiling liquid expanding vapor explosion (BLEVE) must be considered (fig. 18–7). Just because the liquefied gas or other liquid is not flammable does not mean the vessel may not BLEVE from being exposed to heat from an associated fire. Any liquid in a vessel has the potential to BLEVE when heat input exceeds the vessel's ability to relieve the excess pressure. Also remember that vessels in recycling facilities may not be properly designed or be in the best condition.

Fig. 18–7. **Metal scrap collection and sorting (including various types of vessel) facility is located in the middle of a residential neighborhood in Pennsylvania.**

Most of the materials collected for recycling are not hazardous in and of themselves. These materials become hazardous because they are not yet in their end form. For example, finished magnesium pellets present little danger. However, scrap aircraft components containing magnesium could contain many associated hazardous materials. These hazards could range from hydraulic fluids to radioactive materials, which could be found in sources such as the instrumentation. Other hazards could be combustible metals. Most firefighters know that if magnesium ignites, it produces hydrogen gas. If large stockpiles of magnesium catch fire, there could also be small explosions that are made worse by the application of water. Recent fires illustrate the need to ensure that proper extinguishing agents are used by firefighters and used in fixed suppression systems.

Just the shear bulk of the recycled materials could present a significant hazard. Scrap materials could be very heavy and the piles unstable. If a fire should occur in one of the collection piles, firefighters may have to climb onto these piles to reach the burning materials or to separate materials to gain access. During firefighting operations, baled materials such as paper and cloth may become waterlogged and extraordinary heavy. Buildings can collapse from the sheer weight of the saturated materials or, in the case of paper pulp, from expansion.[3]

Paper pulp is stored in the dry state. If water is added, the pulp can expand and actually push the walls outward and collapse the building. An additional hazard to firefighters is the possibility that tall piles of collected materials may collapse onto them. The NFPA cites a study of a warehouse file that occurred in Idaho where piles collapsed onto firefighters, workers, and would-be rescuers. Five people died of suffocation at the scene of this relatively small, sprinklered warehouse fire containing high-piled, baled paper.[4] Similar hazards could come from cloth and plastic materials.

There are some specific hazards associated with plastics recycling. Plastics such as polyvinyl chloride (PVC) produce hydrochloric acid and otherwise toxic smoke when burned. If water is used to extinguish these fires, it mixes with the smoke produced from the burning plastic. Smoke and water runoff from these fires can contain hydrochloric acid, which presents a respiratory and burn hazard.

Since plastics are composed of materials like carbon, hydrogen, chlorine, and sulfur, the smoke should always be considered toxic. Large fires involving plastics can also release oils that were used to initially manufacture the materials.

Consider the hazards associated with automobile recycling. These operations could expose firefighters to flammable liquids and gases, acids from the batteries, heavy metals such as lead, as well as explosions of gasoline tanks and hydraulic cylinders. Because automobile collection sites are an eyesore and a potential hazard to the public, they are frequently located in rural settings. By the time firefighters arrive at the rural facilities, the fire may be well advanced. Some of these storage sights are massive, and vehicles are piled several high and often unstable. Firefighting water and foam streams could cause piles to topple over and injure or trap firefighters. Access for equipment will also be a challenge because of the lack of separation and fire lanes. Additional hazards could come from oxygen and acetylene tanks used to supply cutting torches. Intense fires, toxic smoke, and poor visibility will definitely pose hazards to firefighters.

When tires burn, they decompose into oil and other toxic by-products. So much oil is liberated from major tire fires that it begins to flow, and large catch basins must be constructed. The oil actually runs out of the fire area and can cause additional fires. The oil and other by-products pollute ground water and soil. These sites require extensive cleanup after the fire is extinguished.

In addition to the hazards associated with the materials waiting recycling, the equipment used to convert these materials to a usable state could also be a hazard. Typically, the machinery involves hydraulic crushers, front end loaders, cranes, material separators, pulverizers, conveyors, lifters, and furnaces. The conveyance and loading machines can also be a source of fire (fig. 18–8). There are also high-voltage power supplies and possible fuel storage tanks associated with these facilities. Power to these machines should be isolated before firefighters are allowed in the area.

Fig. 18–8. **A front-end loader fire at a recycling center**
(Courtesy of C. Guillemette)

FIREFIGHTING TACTICS

Fires at recycling facilities will probably start in one of two locations. The first location is in the mass storage location. The other likely place for a fire to start is where the material is being processed and refined into the end material or materials. The grinding, separating, and melting process sometimes leads to fires. In most cases, the storage location and processing location are likely close to each other.

As with other fire scenarios, you should first consider fixed or semifixed suppression or containment systems. Use of these systems should be the best method of extinguishing these fires. We say "should" because of the case in 2005 of a fire in a magnesium recycling facility. The facility was not designed to store magnesium, and the water sprinkler system had been left in place. When the water sprinkler system activated, it simply spread the fire out of control.[5] As most firefighters know, water is not the appropriate agent for a magnesium fire. So during your prefire planning efforts, ensure you evaluate any fixed systems that may have been installed.

As mentioned, fires in mass collection areas, like an automobile junkyard or tire collection area, can be huge. These fires have the fuel to burn for days or weeks. Ideally, the recycling center has considered the possibility of a fire in the collection area. Without adequate separation and access for firefighters, it may be impossible for firefighters to extinguish fires in these stacks of material. If a fire should occur in the collection stacks, you should consider separating the materials to remove as much potential fuel as possible. It may be safe enough for bulldozer and crane operators to perform this separation even as a fire burns. The safety of these operators must be considered to include adequate respiratory protection, cooling lines, and egress routes. Communications must be established for these operators, as well as for personnel directed to observe and report the progress of the fire to ensure the operator is not endangered.

External firefighting actions will commonly be required. These fires can be quite large, so carefully consider your water source. Thousands of gallons of fire extinguishment and cooling water may be required. In addition, depending on the material involved, special fire extinguishing media may be required. Class A and B foams may be effective on baled or other materials that you need to penetrate. These foams break down the surface tension of the water and allow water to penetrate the outer layers. Class B foams can be used at low percentages like Class A foams for this purpose. Containment and collection of water and foam runoff must be considered.

Flammable liquids such as hydrocarbons and alcohols require Class B foams for extinguishment and vapor mitigation. Consider other specialized foams for vapor mitigation, alcohols, or acid fires. Knowing the type of hazard you face is critical because you may need these specialized agents to combat these fires.

Of course, with metals fires, you will normally need a Class D agent. Most municipal fire departments probably will not have large quantities of Class D extinguishing agents. Fires such as magnesium may also be extinguished by clean, dry sand. Water is sometimes mentioned as an extinguishing medium; however, remember that water has in some cases spread magnesium fires. A combination of separation and the use of sand or Class D powder will be most effective.

Fires in associated machinery should be fought in the same manner as other machinery fires. Fire personnel must seek the expertise of plant personnel to close valves and shut down equipment. Some process equipment may have multiple power supplies and remain hot for some time after it is shut down. Extensive preincident response planning must be conducted for these facilities.

CONCLUSION

Because of the economics of the recycling industry and the concern for the environment, recycling will continue to be big business in the United States. As previously discussed, recycling of usable materials will be a challenge for firefighters. Some of these recycling facilities have been allowed to operate in an unsafe way for many years. Urban development may have moved to previously remote areas, and trash fires that would have been allowed to burn years ago may now have to be extinguished. These recycling sites and centers should be carefully preplanned in the same manner as other industrial facilities.[6]

CASE STUDY

A huge fire at an outside tire storage area on the Rhinehart Farm near Winchester, Virginia, occurred in 1983. Approximately 25,000,000 tires had been collected at the site, which operated as a tire reuse and disposal operation. The fire burned for nine months before being extinguished. The fire covered approximately 5 acres and consumed approximately 6 million tires. During this fire, great plumes of toxic smoke were produced, and approximately 800,000 gallons of oily waste were collected from the runoff. The runoff polluted nearby ponds and streams with oily waste and heavy metals, requiring the treatment of 75,000,000 gallons of water. The site was declared a Superfund project, and remediation took 11 years.[7]

Random thoughts

- Tire and waste recycling facilities pose an extreme hazard to communities. Hazards include toxic smoke and fumes, as well as the potential for catastrophic environmental damage.
- The challenges of large fires in tire and other waste collection and recycling facilities are easily beyond the capabilities of a single department to handle. Preincident response plans and mutual aid agreements must be done before an event.
- The Winchester, Virginia, tire fire could not be extinguished with either local or federal resources. The fire was grossly expensive and will negatively impact the environment for hundreds of years.
- Prevention is the best way to deal with fires in these facilities, and if they do occur, they will be manageable. Municipal fire departments must work with business owners, as well as local, state, and federal officials to ensure that waste and recycling facilities do not endanger their communities.

NOTES

1 American Forest and Paper Association, Environmental and Recycling Retrieved September 23, 2006, from www.afandpa.org

2 Noll G. G., and Hildebrand, M. S. 1997. *Storage Tank Emergencies, Guidelines and Procedures.* Annapolis, MD: Red Hat, pp. 27–41.

3 National Fire Protection Association (NFPA). 2003. *Fire Protection Handbook,* 19th ed., vol. I. Quincy, MA: NFPA, pp. 13–190.

4 NFPA, *Fire Protection Handbook,* pp. 13–190.

5 Associated Press. 2005 (January 16). Sprinkler System Caused Huge Fire at Magnesium Plan., Andersen, Indiana: ABC 7 Chicago. Retrieved September 23, 2006, from http://abclocal.go.com/wls/story?section=News&id=2623415&ft=lg

6 NFPA, *Fire Protection Handbook,* pp. 2–135.

7 United States Environmental Protection Agency (EPA). Rhinehart Tire Fire. Retrieved September 23, 2006, from http://epa.gov/reg3hwmd/npl/VAD980831796.htm

19

Pulp and Paper Manufacturing

INTRODUCTION

Paper is a thin, flat material produced by the compression of wood fibers. Wood is made up of small cellulose fibers, bound together by a gluelike substance called *lignin.* In the pulping process, these fibers are separated by cooking the wood with chemicals to dissolve the lignin. The fibers used is usually natural and composed of cellulose. The most common source of these kinds of fibers is wood pulp from pulpwood trees, (largely softwoods) such as spruce. Pulp and paper manufacturing is one of the oldest and more involved industrial facilities in operation today. The first machine ever built and operated was started in England in 1803. In 1804, the Fourdrinier Brothers took over the development of the machine and by 1807 had obtained all the patent rights and put into operation the Fourdrinier paper machine. These machines were the source of all paper products up to the 1950s.

Pulp and paper manufacturing facilities can vary in size but typically have large plot plans and sometimes a small city of their own. They are extremely complex and hazardous (fig. 19–1), with wood and paper dust hazards, flammable liquids, and recovery boilers, to name a few. This chapter will address the pulp and paper process in a general rather than describe various specific technologies.

Fig. 19–1. **Paper mill** *(Courtesy of David L. Green, Nov. 11, 2003)*

DESCRIPTION OF PROCESS

There are three basic processes for making pulp for the manufacture of paper: mechanical, chemical, and semichemical. Mechanical pulp production often is referred to as the "ground-wood process," because wood is ground via a grinding stone in an attempt to collect the wood fibers necessary for pulp production. Pulp production using chemicals is called chemical pulping. This is the most common method for obtaining wood fibers for the production of pulp. In this process, chemicals are used to degrade and dissolve the lignin (chemical that holds wood fibers together).

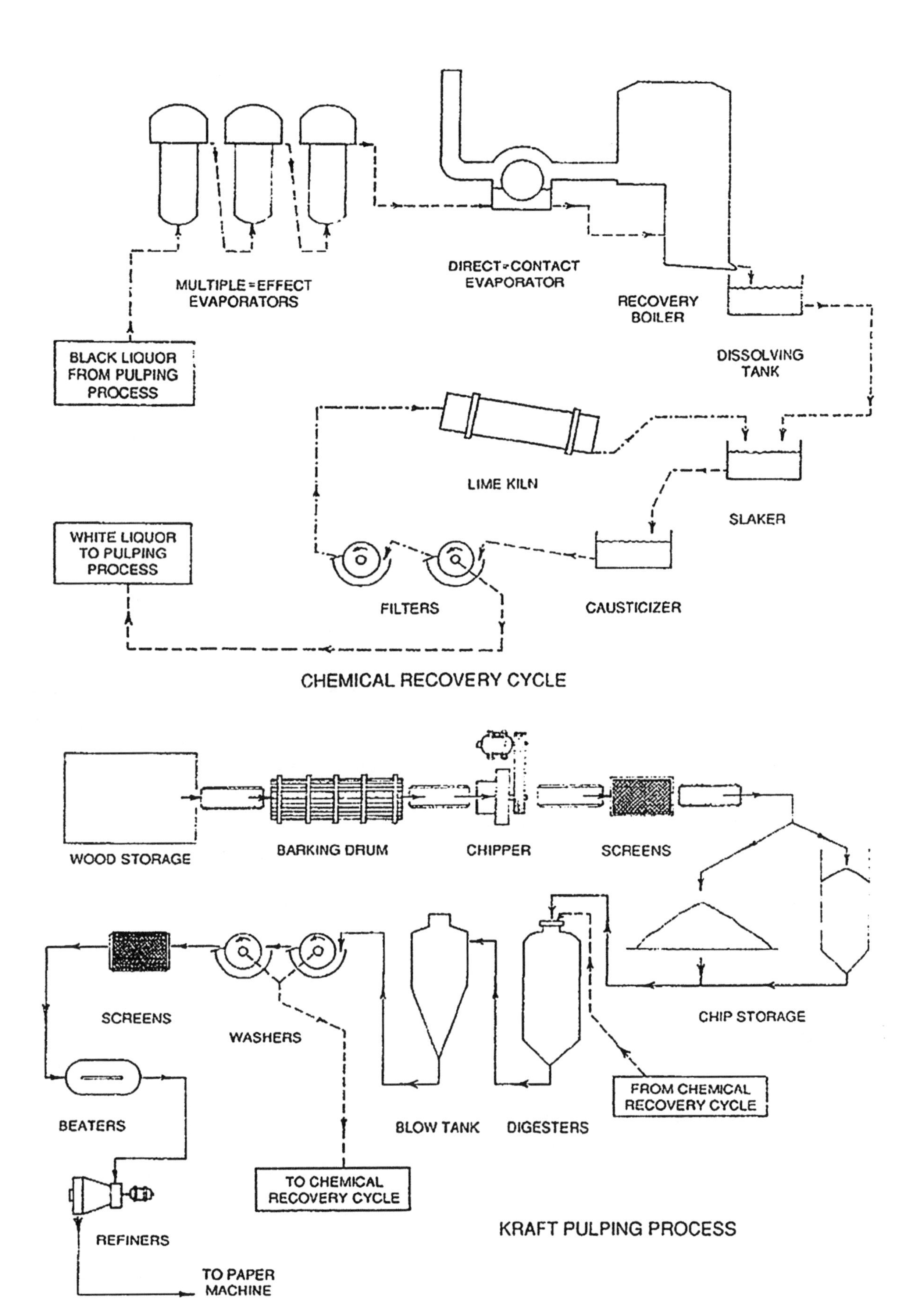

Fig. 19–2. **Kraft process overview** *(Reprinted with permission from the* **Fire Protection Handbook***, 19th Edition, Copyright © 2003, National Fire Protection Association, Quincy, MA 02169)*

This is done by chemical reagents at elevated temperatures and pressures in a process vessel called a digester (described later). The two main chemical pulping processes are Kraft and sulfite. The Kraft process, the most common of all, uses caustic soda and sodium sulfide, and the sulfite process uses sodium sulfite and sulfur dioxide (fig. 19–2). In this chapter, the Kraft process is described in detail as this is the most common type of pulp and paper process.[1]

The basic process layout of a paper mill consists of the following processes:

- Woodyard process
- Digesters
- Washers
- Cleaning and bleaching operations
- Refiners
- Paper machines

There are also two component processes that are subprocesses of the digester and washer processes:

- Chemical recovery
- Steam generation, or black liquor recovery boiler (BLRB)

Woodyard process

The process of paper manufacturing begins in the forest where wood, the basic raw material, is cut and shipped to the sawmill. To begin the process, pulpwood logs must be reduced to chip form. Prior to chipping, logs are passed through a debarking drum (large, open-ended cylinder). Within the drum, logs collide with one another and rub together, removing the bark. The bark falls through slots in the cylinder walls and is collected and burned as fuel in the power boilers. The debarked logs are conveyed to a chipper, which reduces them to small squares. Some facilities do not have this capability and therefore receive their raw material already in the form of chips. (See chapter 21 for details.)

Digesters

As previously mentioned, paper is made up of tiny wood fibers from softwood trees. The wood of these trees is made up of small cellulose fibers, bound together by a gluelike substance called lignin. In the pulping process, these fibers are separated by cooking the wood with chemicals such as white liquor (from the chemical recovery process), which dissolves the lignin. To accomplish this, the chips are loaded into large vessels called digesters on either a batch or continuous basis. Digesters are designed on the same principle as kitchen pressure cookers. The chips and chemicals are steamed under pressure for 1.5 to 4 hours until the mixture is reduced to a wet, oatmeal-like mass. The cooking frees the fibers so they can be suspended in water.

Washers

The pulp is blown from the digesters under pressure to separate the fibers. These machines wash out and remove the cooking chemicals (the white liquor is now black liquor) and dissolved lignin. These chemicals are sent to the chemical recovery process units to be regenerated and reused. At this stage, the pulp, which is naturally brown, can be used to make the famous brown paper bags that are popular at many grocery stores. (Though, over the last decade, plastic has increasingly become the replacement for paper bags.) This brown pulp can also be used to manufacture boxes and other brown-colored and rough-textured products. Sometimes this process is called the Kraft process (also known as Kraft pulping or sulfate process) and is used in production of paper pulp and involves the use of caustic sodium hydroxide and sodium sulfide to extract the lignin from the wood fiber in large pressure vessels called digesters. The spent pulping liquor, called black liquor, is concentrated by evaporation and burned to generate high-pressure steam for the mill processes. The inorganic portion of the black liquor is then used to regenerate the sodium hydroxide and sodium sulfide needed for pulping. In the case of softwood pulping, a soaplike substance

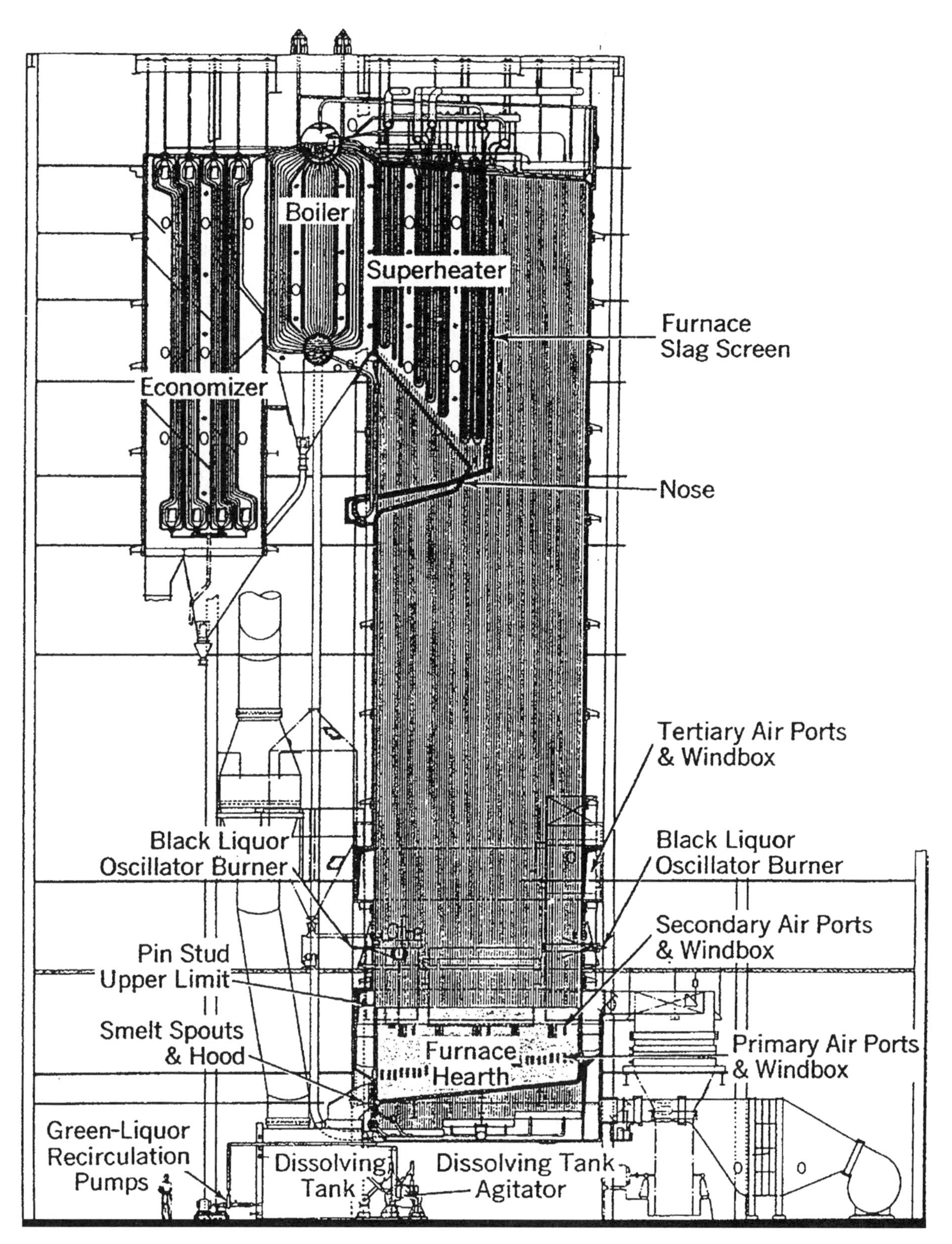

Fig. 19–3. Black liquor recovery boiler (BLRB)

is collected from the liquor during evaporation. The soap is acidified to produce tall oil, a source of resin acids, fatty acids, and other chemicals. Other by-products are the lignosulfonates, important deflocculants in drilling muds for oil industry.[2] (see chapter 32)

Chemical recovery and steam generation

In this process, the inorganic chemicals used in pulping are recovered and regenerated for reuse in the digesters. This is an extremely complicated process and is the location of many of the common hazards associated with pulp and paper manufacturing. The white liquor, consisting of sodium hydroxide (caustic soda) and sodium sulfide are used in the digester process to separate the cellulose fibers that make up the wood chips. Black liquor is the weakened liquor that results after the wood chips have been chemically separated in the digester. The washer, as described above, washes the black and white liquor (a waste product) from the pulp.

The first step in the chemical recovery process is the evaporator. The evaporator receives the white and black liquor from the pulp washers. The purpose of this equipment is to concentrate the black liquor by evaporating a portion of the wash water. This will raise the solid content of the liquor to almost 50% in preparation for burning in the BLRB (fig. 19–3).

The mix tank is where the black liquor is further concentrated by adding chemical ash (*salt cake*) collected from the bottom boiler hoppers. After being heated, the black liquor is sprayed into the BLRB, where the remaining water evaporates and creates steam. In addition, the organic compounds break down into volatile combustibles and char. The BLRB is a large boiler and furnace more than four stories tall, occupies an entire building, and is fueled using black liquor from the digester process. The BLRB is also an important part of the overall plant's production of steam and other plant utilities. It is a very dangerous operation, requiring careful consideration and preplanning. It is important to understand that the BLRB is both a furnace and boiler.

In the char bed at the bottom of the BLRB, salt cake in the black liquor is converted to sodium sulfide through chemical reduction (oxygen removal). The continued char burning coverts the residual caustic soda in the black liquor to sodium carbonate. These two compounds exit the BLRB as molten smelt, commonly called inorganic salts.

The molten smelt runs from the water-cooled smelt spouts into the dissolving tank where it is mixed with weak wash from the washers and becomes green liquor. The green color is caused by leftover iron compounds. From the dissolving tank, the green liquor is sent to the *recausticizing plant,* where the green liquor is clarified to remove insoluble impurities and then causticized in the lime slaker to convert the sodium carbonate to sodium hydroxide. The calcium carbonate (lime mud) is separated from the causticized liquor, and the clear (white) liquor is sent to the digesters, where the cycle begins again. The lime mud is washed, filtered, then burned in a kiln or fluidized bed to dry and calcinate (eliminate carbon dioxide) the lime. After the calcining operation, the lime can be reintroduced in the lime slaker to convert the green liquor to white liquor.

Cleaning and bleaching

Clean stock (pulp) from the washers is sent to a screen and onto the cleaners. At the cleaners, the clean stock is removed from the top, and stock containing dense contaminates is removed from the bottom. Centrifugal force causes the dense materials to lose their momentum on the inside walls of the cleaners. This allows the dense material to settle much more quickly than the fibers.

Pulp to be used for printing papers cannot be used in its current state due to its brown color; therefore, it needs to be bleached. This is done by chlorination and hypochlorite bleaching. The pulp is sent to the bleaching, unit where the chlorination process begins. The next step is to remove the remaining caustic via extraction and soaking the pulp. The bleaching process continues based on the desired results for customer demands. Additionally, specialty chemicals and further treatment of paper continues, which can include dying or kaolin (a layered clay silicate mineral) or calcium carbonate to make glossy paper.

Refiners

From bleaching operations, the pulp is passed through refiners. These refiners roughen the surface of the individual pulp fibers by loosening the threadlike elements from the fiber wall, so they cling together when formed into a sheet. The fibers of the bleached pulp are cut and brushed. This improves their bonding properties and strength.

Paper machines

The *Fourdrinier process* involves putting a single ply of paper on top of a single-layer, moving wire mesh. Paper is formed and pressed by the machine, then air-dried. The process itself is so efficient that the basic method is still used today. However, the capacity and speed of modern machines far exceed the original design. Most mass-produced paper is made using the continuous Fourdrinier process to form a reel or web of fibers in a thin sheet.

The pulp from the refiners is now ready to be formed into paper. The pulp arrives at the headbox, or sometimes called the "wet end" of the paper machine. This is where the pulp mixture is further diluted with water, resulting in very thin slurry. This *furnish*, or stock, as it is called, is then run onto the forming fabric or wire of the paper machine under pressure in the headbox. It is now sometimes referred to as the *slice*. The forming fabric is an endless mesh screen that circulates at the wet end of the paper machine. Traveling at speeds of more than 3,000 feet per minute, the paper is pressed between water-absorbing fabrics. As the screen moves away from the headbox, various suction devices drain water from the pulp, leaving a sheet of matted pulp that is still very wet. A wire-covered roll holding a wire design, called the dandy roll, may travel over the surface to impress a watermark. The sheet then moves onto a woolen felt screen, which takes it through a series of presses, where more water is removed, and the paper gets a bit drier. Finally, the sheet of paper passes over a number of heated drums that evaporate the remaining water. Many new papermaking machines incorporate two or three moving wire-mesh screens (called the Triple Fourdrinier), between which the pulp is pumped, and water is extracted from both sides. The twin-wire machine produces a paper practically identical on both sides, an important property in printing.

The cylinder machine differs from the Fourdrinier principally in the wet end, or forming operation. Instead of the moving wire screen, a screen-covered rotary cylinder is half-submerged in the pulp vat. As the cylinder rotates, a sheet of matted pulp is formed on its exterior surface and is then picked up by a moving belt, where it is treated to remove the remaining water, as in the Fourdrinier process. A series of cylinders may be used, each one depositing an additional layer of pulp on the belt, so that multilayer sheets are built up. Cylinder machines are used for making thicker papers and paperboard.

As it leaves the paper-forming machine (now referred to as the dry end), it travels through a series of steam-heated cylinders called dryers, where the last of the water in the sheet is removed. The dried paper is wound onto large reels. The rolled paper may be slit to the widths required, cut into sheets, trimmed, and packaged. Finishing operations include calendering, or passing the paper through a series of steel rolls that impart one of a number of finishes; and coating, where one or both sides of the paper are glazed with a mixture of pigment, dispersant, and adhesive to produce a glossy finish or of to improve the smoothness or opacity of the paper. Operations that convert the paper roll into bags, boxes, corrugated shipping paper, and other products are done at this stage of the papermaking process.

COMMON HAZARDS

As with most large industrial facilities, pulp and paper manufacturing have some common hazards that should be considered either during the preincident response planning or emergency response. Some of the more common hazards include those associated with dust (paper machines), flammable and combustible liquids (paper machine and chemical recovery), explosion hazard (BLRB), hazardous chemicals, roll paper storage (warehouse), and electrical cable trays. Another hazard that exists is the large amount of outdoor storage of wood (see chapter 15).

Dust hazards

Dust is certainly one of the more common hazards in the pulp and paper industry and requires close attention. As previous mentioned, in processing or storage of grains, an explosion may originate in the process equipment (equipment explosion hazard) or in the ambient environment in the room containing the process equipment (room explosion hazard). In the case of pulp and paper facilities, dust hazards are more closely associated with the spread of fires due to the continuity of combustibles. Dust accumulation on the joist and beams above the paper machine and dryers are the more common locations. A program of strict housekeeping should be followed in these areas to prevent the spread of fire.

Flammable and combustible liquids

As with most industrial facilities, the presence of large amounts of flammable and combustible liquids is a common hazard at pulp and paper facilities. Hydraulic and lube oil systems are used in most of the processes in the production of paper. These systems can be found in the wood preparation area and other areas where large machinery is in service. The largest concentrations of these systems are in the area of the paper machine. This is due to the large amount of rotating equipment such as the calendar rolls and the wire-fabric. Housekeeping is also a must in this area to limit the amount of flammable and combustible liquids stored or leaks from pipes and pipe joints.

Explosion hazard

The BLRB in the chemical recovery process is one of the greatest hazards at a pulp and paper facility. The accumulated hot smelt or inorganic salts at the bottom of the BLRB could cause a high-pressure gas expansion explosion if it comes in contact with water. This explosion could cause a great deal of damage to the BLRB building and surrounding equipment. It could even cause damage to other power or gas-fired boilers and cause a larger gas leak and secondary explosions. This could be a result of water accidentally reaching the smelt bed of the BLRB from leaks in the pressure components or boiler tubes or from the use of weak black liquor.

Hazardous chemicals

A variety of chemicals is used in the production and manufacture of pulp and paper. These chemicals and their associated processes are present throughout the process flow, and they are typically found in large quantities in the process area and in bulk storage. Some of the more common chemicals include sodium hydroxide, sodium carbonate, sodium sulfide, calcium carbonate, and sulfur dioxide. These pulping chemicals are found in the chemical recovery and digester areas. Other chemicals such as chlorine, sodium chlorate, chloride, chlorine dioxide, sulfuric acid, and caustic soda are commonly found in the bleaching area and in bulk storage areas. Although most of these chemicals are noncombustible, they will promote the burning of other nonrelated combustible materials. Additional details pertaining to hazardous materials can be found in chapter 11.[3]

Roll paper storage

At the end of the pulp and paper process, paper is normally wound onto rolls for storage and shipping. Roll weight and sizes vary based on customer demands and can exceed 2 or more tons and over 6 feet in diameter. There are different types of rolled paper, and each has its own burning characteristics. Some of the more common categories include linerboard, light, medium, and Krafts, and tissue. The heavier paper types such as linerboard or Kraft are wound and held in place with steel wire bands at the top and bottom of the roll. Other papers like light and medium weight are wrapped with heavy plain paper such as Kraft paper. Tissue paper is normally not bound with anything. These rolls are usually stored in the vertical position or on-side, up to three rolls high. To maximize storage space, rolls are generally tightly packed with only a narrow 12- to 24-inch wide space between stacks, with an 8- to 12-foot aisle width.[4]

As mentioned, each type of paper has different burning characteristics. One of the main characteristics is called peeling, delaminating, or exfoliation. The severity of the roll paper fire is dependent on the peeling rate, which is associated with the rate of unwinding of the paper as the fire burns through and exposes new layers, thus fueling the fire. Therefore, the type of paper is a significant fire hazard. Lighter papers such as tissue will burn faster than heavy papers such as linerboard or Kraft. It is common to use Kraft paper to totally wrap lighter-weight papers such as newsprint to reduce the fire hazard, as well as a form of banding.[5]

Electrical cable trays

Due to the size and amount of equipment, most pulp and paper mills have several process and control rooms. Typically, these rooms control a particular section of the process such as the bleaching process, chemical recovery process, or the paper machine process. All these control rooms require a large number of electrical cables, which are usually grouped into large cable trays and located in raised floors, vertically along walls, overhead at the ceiling, or in horizontal underground tunnels (cable tunnels). These cable trays are exposed to corrosive atmospheres and accumulation of dust and combustibles, adding to the fire hazard. Because of the nature of the cable tunnels, which may pass through other building(s) or process compartments, many times they contribute to the spread of fires.

FIREFIGHTING TACTICS

A well-planned and tested emergency response plan is needed. Preincident response planning should be carried out for this type of occupancy because any fire in a paper plant has the potential to develop into a major incident. The preincident response plan and its development should include the plant personnel, as well was other experts in the field. As mentioned in chapters 4 and 5, the plan should include details about command structure, resources, and details of response activities.

As with all responses, firefighting strategies for pulp and paper facilities must consider the availability of resources and the ability of responders to deploy these resources. One of the most common and useful resources in industrial response is water streams. So, establishment of a water supply is primary. Many of these facilities are located in rural areas or small towns, and the only water supply source is the local plant system. Water supplies might come from ponds or reservoirs, and long hose lays may be necessary.

For fires involving the wood preparation areas, caution should be used around log piles due to instability that may result from shifting due to burning. Because of the large amount of combustibles, expect these fires to be extremely hot. Some of these fires may involve the process equipment. Emergency stops or power disconnects should be located prior to fighting fires in these areas. Chapter 21 has additional details regarding log storage fires.

When responding to fires at pulp and paper facilities, caution should be used when an explosion has taken place. If an explosion has taken place, all operations should cease, and the location of the explosion should be surveyed. As mentioned already, the likely area would be the BLRB building area. If this is the case, the BLRB should be immediately shut down via the emergency shutdown device (ESD; see chapter 9 for details). The building integrity should be evaluated prior to the commitment of personnel to interior firefighting. This is particularly critical considering the age of many of these types of occupancies and the type of construction.

These facilities present a significant risk for firefighters to be exposed to hazardous materials during responses to fires and other emergencies. The presence of hazardous materials may vary significantly in toxicity and volume from what the typical firefighter may be familiar with. When arriving on the scene of a hazardous materials incident at a pulp and paper facility, you should first try to determine the material in question and whether the leak in still active. Most common areas are the digester, chemical recovery, and bleaching areas. Details regarding hazardous chemicals are in chapter 11.

Fires involving roll paper storage are also very challenging. There are different types of rolled paper, and each has its own burning characteristics. These fires can spread extremely fast and quickly overtax the sprinkler system (if provided) and the local water supply. The use of high-rise or hotel packs will be beneficial to use during a pulp or paper fire. Connecting into a Class I or Class III standpipe (if provided) will help reduce extremely long hose lays and expedite manual firefighting efforts. Unless your department owns truck-mounted positive-pressure fan (those that exceed about 80,000 cubic feet per minute [CFM]), roof operations should begin after the initial support of the fixed systems. Because of the large, open area of a pulp or paper facility, venting is important, so that smoke, heat, and other products of combustion can be removed to reduce the exposure of interior firefighters. Additional details are in chapter 17.

Fires involving the cable trays, whether overhead or in tunnels, can spread to other areas of the facility. If a fire occurs in the cable trays, fires can spread to the control room, causing the loss of plant operations and control. If this happens, it would create a very dangerous situation. Therefore, all efforts should be made to protect, isolate, and contain cable tray fires to prevent involving control rooms or motor control centers in the fire. Additionally, cable tunnel fires are extremely dangerous. Similar in nature to basement fires, these cable tunnels can be a few feet underground, collect heat and other products of combustion, and cause a loss of communications because of the thick concrete walls. This is especially true if your department uses a newer digital radio system. Different forms of communications should be evaluated prior to fighting a fire in a cable tunnel. Cable tunnel fires also create a challenge with respect to supplying water to handlines. These cable tunnels can be rather long (several hundred feet), and it may become necessary to set up attack lines from the truck gated down to smaller handlines to extend the effective range. If Class I or III standpipes are provided, using high-rise or hotel packs will eliminate the need for several hundred feet of fire hose and expedite the attack.

CONCLUSION

Fires in pulp and paper mills are challenging, as are fires in most continuous process facilities. Knowing the general process hazards, preincident planning, and proper identification of recourses will aid in the success of response to these plants. Because of the large area these plants cover, familiarity with building names and processes are important. Extreme caution should be exercised for events involving the BLRB or cable tunnels. Additionally, fires in roll paper warehouse areas also create dangerous environments because of the large quantities of combustibles.

NOTES

[1] National Fire Protection Association (NFPA). 1990. *Industrial Fire Hazards Handbook,* 3rd ed. Quincy, MA: NFPA, p. 341.

[2] NFPA, *Industrial Fire Hazards Handbook,* p.341.

[3] NFPA, *Industrial Fire Hazards Handbook,* p.339.

[4] NFPA, *Industrial Fire Hazards Handbook,* p.347.

[5] NFPA, *Industrial Fire Hazards Handbook,* p. 353 .

20

Plastics Manufacturing

INTRODUCTION

Plastics are found in everyday things from cups, bowls, and toys to automotive parts, cameras, and computers. Without plastics, life would be very different. Plastic has become a very important part of how we live and work. Plastics are now found in places we never imagined. Because of the high demand for plastics, more and more facilities are being built to accommodate this increase. This has increased both the production of raw materials used to manufacture plastics and the manufacturing plants that turn the raw resin and ethylene into our favorite coffee cup. With no end in sight, plastics and their associated hazards are a growing concern.

First, we will define plastics for our use in this chapter. *Plastics* consist of combinations of carbon with oxygen, hydrogen, nitrogen, and other organic and inorganic elements. In most cases, plastics are produced from a single chemical unit known as a *monomer.* These monomer units are low molecular weight formations of organic elements. By chemical reaction or *polymerization,* these single monomer units are connected to each other to form long polymer chains. *Polymer* simply means long chain.[1]

In the finished state, plastics are solid, but at some stage during manufacture, a plastic is liquid, molten, or softened and is capable of being formed into various shapes, mostly through a process and application that involves either heat or pressure or a combination of both. Plastics are categorized into one of two groups: thermoplastics or thermosets.

Thermoplastic resins can be softened and hardened over and over again by heating and cooling without a chemical change in the plastic. *Thermoset* resins, when heat-treated, undergo a chemical reaction, and once a solid, they cannot be softened or heated without causing a significant damage.

DESCRIPTION OF PROCESS

The manufacturing of plastics involves the polymerization of the product in which the monomer is reacted in a reactor to form the basic polymeric material. During this step, the basic material is formed into powder, granules, or viscous liquid. This chapter will discuss the process of converting the plastics into everyday items, and some hazards associated with the industry. (For more details on the hazards of the polymerization plant and other similar facilities, see chapter 13.)

Conversion of plastics into useful articles such as cups, toys, car accessories, and the like, is typically done by molding, extrusion, foaming, or casting involving heating the plastic so it will flow into a predetermined shape. Although chemical reactions are not often considered a significant part of these operations, as with other chemical and petrochemical processes, this process is typically referred to as *converting.*

Plastics are generally a combustible organic compound that can burn or melt under various conditions. Aside from the inherent combustibility of plastic formulations influenced by the basic polymers used, the nature of plastic additives or raw material and the conversion of plastics into finished articles have associated hazards. These include combustible dusts, flammable solvents, electrical faults, hydraulic fluids, the storage and handling of large quantities of combustible raw materials and finished products, and housekeeping within the processing area. Many of these issues are also discussed in other chapters of this book.

Basic processes of manufacturing systems

The following descriptions of processes cover the basics of the major and more common manufacturing systems such as blow molding, casting, compression molding, extrusion, and injection molding. All of these systems have a unique way of turning plastics resins and low-linear or high-density polyethylene into a desired commodity.

Blow molding is generally used only with thermoplastics and is one of the more common manufacturing processes used. It is the primary technology relating to the production of hollow plastics products, such as water bottles and tanks used in automotive assembly such as coolant and fuel tanks. Blow molding involves heating and melting the thermoplastics resin, then forming the molten, highly viscous polymer into a tubelike shape. The ends of the tube are sealed and air is injected into the tube. The tube still hot and in a softened state, then is inflated inside the mold and forced against the walls of the mold, where it cools, solidifies, and is ejected from the mold. The plastic tube is extruded from the mold, then rotates to the second station, where air is injected into the still hot tube to blow it out to the shape of the inside mold cavity. At the third station, the blown part is allowed to cool and set, then is ejected from the mold.[2]

Casting can be used either for thermoplastics or thermosets to make products, shapes, rods, and tubes by pouring a liquid monomer solution into a mold, where it finishes polymerizing into a solid. Pressure need is not used with the casting process, unlike the blow molding process. This process is similar to making a cake or gelatin, using the mix and a mold.[3]

Compression molding is another one of the more common methods of forming thermoset materials. Compression molding consists of squeezing of material into a desired shape by applying heat and pressure to the material in a mold (fig. 20–1). In the case of thermoset resins, the material is heated until curing of the material takes place and the part is ejected from the mold hot.[4]

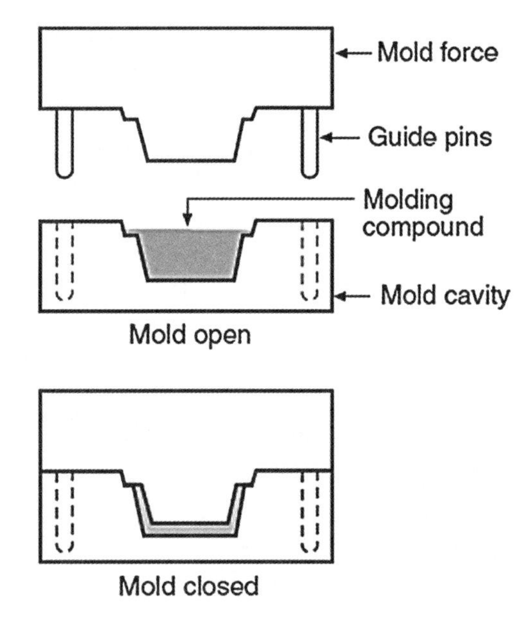

Fig. 20–1.* Two-piece compression mold** *(Reprinted with permission from the* **Fire Protection Handbook, 19th Edition, Copyright © 2003, National Fire Protection Association, Quincy, MA 02169)*

Extrusion is employed to form thermoplastic materials into continuous sheeting, film, tubes, or rods. The most common of devices is the single-screw conveyor, in which plastic pellets are fed into the hopper and then through a screw that rotates in a heated barrel. The rotation of the screw conveys the plastic forward for melting and delivery through the breaker plate through the adapter, then into the die, which dictates the shape of the plastic.[5]

Injection molding is the method of forming objects from pelletized, granular, or powdered plastics, most often of the thermoplastic type, in which the material (usually plastic pellets) is fed from a hopper to a heated chamber in which it is softened, after which a ram or screw forces the material into a mold (fig. 20–2). Pressure is maintained until the mass is hardened sufficiently for removal.[6] Figure 20–3 shows half of a die producing a decorative paper clip mounted to the movable platen in the work area of an injection molder. The injector nozzle is shown retracted at the right side of the photo. The limit switch to control the hydraulics (which opens and closes the die) is visible in the foreground. A smaller version of an injection molder (fig. 20–4) shows movable guard, hydraulic ram, die and the hopper for plastic pellets.

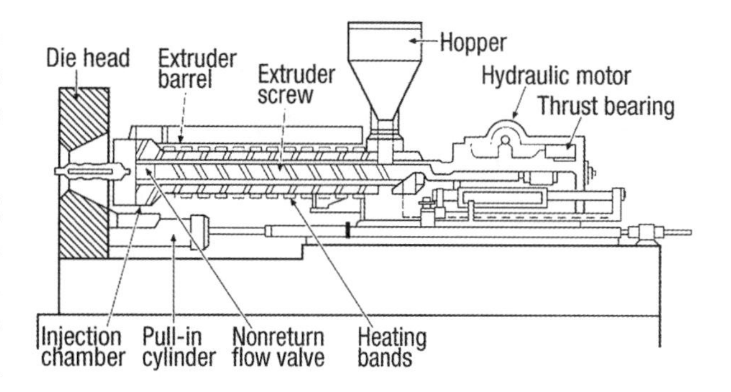

Fig. 20–2. **Diagram of an injection mold machine** *(Reprinted with permission from the* **Fire Protection Handbook,** *19th Edition, Copyright © 2003, National Fire Protection Association, Quincy, MA 02169)*

Fig. 20–3. **Injection mold machine** *(Courtesy of Glenn McKechnie, 2005)*

Fig. 20–4. **Small Injection Mold Machine** *(Courtesy of Glenn McKechnie, 2005)*

COMMON HAZARDS

Some of the more common hazards that exist in these types of facilities include dust, flammable and combustible liquids, and high hazard storage.

Plastic dusts

Plastic dusts and the potential for explosions should be considered possible when manufacturing operations include the use of pulverized plastic pellets, convey plastic granules and resin through pneumatic conveying systems, or produce dust by machining or sanding in finishing work. A high ratio of surface area to unit volume or unit mass will significantly increase the possibility of ignition leading to an explosion.[7] Specific explosion hazards are discussed in other sections of this book.

Plastic pellets for injection or extrusion molding are commonly known as molding powder. Typically these come from the manufacturers in hoppers via rail or truck and are stored in large vertical vessels at atmospheric pressure. They are then conveyed via pneumatic conveyor systems via a piping network throughout a facility. Although the pellets themselves in a static state do not generate dust, some dust can be generated by abrasion and collisions of these plastics particles when conveyed in a long pneumatic conveying systems or where the velocity is very high.

Some of the preceding manufacturing techniques generate excess pieces or trimmings from such processes as compression or injection molding. In an attempt to reduce waste, increase production, and decrease the use of fresh plastic pellets, many facilities take the trimmings, which are too small for reuse alone, and mix them in a blending process along with new pellets. This regrinding process usually generates some fine powder, thus creating a dust hazard.

If a facility has a dust-related problem, care and consideration should be used to help reduce or eliminate the hazard. Dust collection point, improved housekeeping, or better production methods may need to be considered. In any situation, it is important that the dispersion of plastic dust into the atmosphere of a facility be kept to a minimum. It is important that precautions are made to limit the possibility of ignition sources, relieve explosion pressure, and confine and control fire.

Flammable and combustible liquids

Flammable and combustible liquids are found in nearly every plastics manufacturing facility. Sometimes they are used in small quantities to apply adhesives or paints to molded or fabricated items or to clean tools and surfaces of equipment. Larger quantities can be found in and around the manufacturing equipment in the form of hydraulic oil. Much of the equipment used in the manufacturing process described earlier use fluid power systems to operate the screws, rams, and other essential equipment. These fluid power systems use hydraulic oil as the medium to convey work. Some hazards associated with this include spray type fires and loss of containment fires leading to a pool fire.

Spray fires

Spray fires are very common where fluid power systems are used. These types of incidents usually generate from the failure of a flange, gasket, pipe, or some piece of machinery. If the equipment is not designed to shut down on loss of system pressure, overpressure, loss of fluid in the tank, or other off-normal conditions, this could lead to a large fire hazard. Other issues are then considered, such as spacing between machines or the probability of involving other equipment.

Loss of hydraulic fluid containment is another hazard. This can result from external damage to the tank or overfilling. Although hydraulic fluid is generally not as flammable as other similar fluids, a pool fire can develop if a suitable ignition source is found. This type of incident, much like the spray fire, can spread and involve other process or manufacturing equipment in the immediate vicinity.

In both cases, either spray fire or loss of hydraulic fluid with ignition, housekeeping is important. In the event of a fire, if the continuity of combustibles from poor housekeeping is present, not only can this lead to increasing the fuel loading but also to allow for faster or farther flame spread. This is further complicated by limited spacing sometimes found between machinery.

Storage

Storage is another location of hazards. The hazards can be associated with storage of plastic pellets and resin or related to the storage of finished product.

Storage of plastic pellets or resin typically consists of large vertical vessels located outside along the facilities wall or in 50-lb bags, similar to how concrete mix or flour is packaged. If large vessels are used, they are stored at atmospheric pressure and are used in conjunction with pneumatic conveying system to deliver the pellets to the machines. Even more hazardous is the storage of pellets in bags on a rack in a warehouse. This large amount of fuel loading creates a very dangerous situation for rapid fire spread and building failure. Aisle spacing, types of racks, and the presence of fire sprinklers offer the best choices for protection. (for details regarding storage arrangements and protection, see chapter 17.)

FIREFIGHTING TACTICS

As with all firefighting tactics, and discussed many times in this book, a well-planned and tested preincident response plan is needed. Preincident response planning is carried out in cases where a building is identified as having the potential to develop into a major incident, and a plastic manufacturing plant is no exception. In the event of a fire at this type of facility, it is advantageous to have information readily available that relates to the structure, fire safety provisions, location of water sources, presence of hazardous materials, and any other information that may be relevant. Good loss prevention starts with the fire safe design of the plant. The use of sprinklers, noncombustible construction, and proper housekeeping is a preferred choice for buildings occupied for storage, processing, and manufacturing of plastics.

When responding to fires at plastic facilities, caution should be used when approaching the area. An adequate water supply should first be identified and secured before committing resources. Fires in these types of facilities require large amounts of water, sometimes 10 or 12 times that required for a simple house fire. Additionally, the incident commander in charge might want to consider the use of multiple water sources to ensure that the local or private water supply system is not overtaxed.

After a water supply has been established, supporting the fixed systems, if present, should be the next step. Many times, fixed systems can keep a fire in check and, more often that not, actually suppress a fire.

Details of the plant such as the location of the fire, process equipment, and utility connections need to be identified. Because of the large amount of process equipment inside and the presence of other related equipment, it is highly advisable to disconnect and isolate all power to the building or, at a minimum, to the machines in the area of the fire.

A firefighter should use extreme caution when maneuvering around inside a building. A thorough preplan and walk-through will help the firefighter understand the layout of the equipment. Because of the nature of plastic fires and the expected amount of black accumulating smoke, the use of a thermal-imaging camera is highly recommended. The camera will also help to locate process equipment and conveyor systems in the area.

A fire involving the storage area will be a very hot fire. Caution should be used when operating within or near storage racks. Under extreme heating, these racks may fail and fall over under the large heavy load. If roof vents are present, opening these vents can provide needed relief of the accumulated heat and smoke; but if sprinkler systems are present, this should be accomplished only after an interior attack is progressing to prevent overtaxing of the sprinkler system.

In either situation, whether involving a piece of equipment or storage, the firefighter can expect the need for a reliable water supply with ample volume, long interior hose lays, and large amounts of smoke and heat. Thermal-imaging cameras and a detailed incident preplan will aid in the success of fire extinguishment.

CONCLUSION

Because of the high demand for plastics, more and more facilities are being built to accommodate this demand. Early planning conducted by your department can help reduce unknowns regarding these types of occupancies. At facilities such as the one mentioned in this chapter that contain more than one or two common hazards, it is important to identify, upon initial size-up, what exactly is the problem. Because fast-moving fires are common in these facilities, water supply and fixed system support is important. In addition, isolating the power to equipment will also prevent exposure of firefighters to energized equipment. Only training and site familiarization will help address some of these issues.

NOTES

[1] National Fire Protection Association (NFPA). 2003. *Fire Protection Handbook,* 19th ed., vol. I. Quincy, MA: NFPA, p. 6-261.

[2] NFPA, *Fire Protection Handbook,* p. 6-264.

[3] NFPA, *Fire Protection Handbook*, pp. 6-264-265.

[4] NFPA, *Fire Protection Handbook,* p. 6-265.

[5] NFPA, *Fire Protection Handbook,* p. 6-265.

[6] NFPA, *Fire Protection Handbook*, p. 6-266.

[7] NFPA, *Fire Protection Handbook,* p. 6-268.

21

Woodworking Facilities

INTRODUCTION

Woodworking facilities are a prominent site around many sections of the United States. These facilities can range from a single-person shop to large commercial occupancies with more than 300 employees. The Southeast (North and South Carolina, Georgia, Alabama, and Mississippi) and the Northwest (Washington and Oregon) have a higher concentration of these large plants, due to the abundance of softwood trees, the major raw material for the industry. [1] This chapter will discuss some of the more common facilities relating to woodworking, such as sawmills, manufacture of composite panels and plywood, and the manufacture of household furniture. Other industry-related occupancies such as pulp and paper processing and paper manufacturing have been addressed in other chapters of this book.

The two main hazards associated with woodworking facilities are fire and explosion, similar to grain and milling occupancies. Fire hazards include the storage of logs in large piles and storage of finished wood products (plywood and particleboards), which if ignited, make fire control and extinguishment very difficult for firefighting crews. The explosion hazard derives from the production of fine wood dust during both the sawmilling activities and furniture manufacturing operations, plus other raw-material-reducing processes that also generate large amounts of wood dust.

DESCRIPTION OF PROCESS

Sawmills

A sawmill's main function is to transform logs (raw material) into a product called finished lumber, which is the raw material used by all other woodworking facilities to produce anything from general lumber such as building materials (two by fours) to materials used for the manufacture of furniture. Sawmills consist of four basic areas:

- Log storage
- Sawmill processing building
- Sorting
- Drying, finishing, and storing

Raw materials for sawmills include a cross section of various softwoods from local forests. The supply is typically focused on areas where trees can be logged and shipped to the sawmill within a reasonable shipping distance. Trees are cut and shipped to the sawmills via truck or railcar. Once at the sawmill, logs are arranged by length or diameter, or ranked by other parameters.

Storage of logs creates a large fire hazard and so should be kept away from main processing or other critical buildings. Log storage yards are usually large open areas of unpaved roads. They are often the first access area to the initial logging operations. These areas are simply open yards used to store cut logs in an orderly fashion.

Sawmill processing buildings are typically very simple buildings, usually only one story with a basement. Basic operations begin with moving the logs from the storage yard via large front-end loaders fitted with clamps or dragged with a chain to a conveyor that will carry the logs to the sawmill building. On arrival to the sawmill building, the logs are worked and cut into a more manageable length, then the bark is removed by a piece of equipment called a *debarker*. After the log is cut and debarked, it is placed onto a conveyor and passed through a head rig. The *head rig* is either a band saw or circular saw and is used to cut the log. Band saws tend to be more common because they usually increase productivity. The purpose of this saw is to begin the process of producing lumber boards. After going through the head rig, the log is then passed through a series of gang saws. Gang saw processing, often called resawing, is used to further reduce the size of the log into the lumber boards. At this point, the log now has a diameter of 10 inches or larger. The process continues with equipment designed to reshape the log into a square profile, as well as remove any remaining bark. The remaining process involves running the boards through a trim saw and mill area to further reduce the log's profile and remove knots and other defects. Once the board has passed the trim area, it is sent to the milling area for further treatment, processing, and sorting in preparation for shipment.[2]

Sorting usually takes place in an open-sided but covered building near the sawmill. A mechanical sorter or other piece of mechanical equipment is used to sort the lumber boards by length, width, and thickness. These mechanical sorters are made up of a loading conveyor, storage racks or trays, and an unloading conveyor. As the boards are loaded into the conveyor, diverters are used to help guide the boards to designated racks or trays based on the desired size. Each rack or tray diverter is able to accept predesignated lumber sizes. After the lumber is sorted, the lumber is moved on rollers or tracks while on the tray, then moved toward the discharge end of the mechanical sorter. From there, a mechanical arm and conveyor is used to unload the trays onto the unloading conveyor, and the lumber is now shipped to the driers.[3]

Driers are used to prepare the lumber for storage and shipping. The intent of the driers is to remove excess moisture from the lumber to a predetermined level, based on customer requirements. This drying process is typically accomplished by use of large baking ovens or kilns. In some areas, depending on local climate, the lumber is stacked in piles in the storage yard for air drying. Kilns also present a fire or explosion hazard. Because of the basic operating principles of kilns, they can present a fire hazard due to the presence of fuel (wood in the kiln or supplied natural gas). Natural gas to fire the kiln can present a fire or explosion hazard. The condition and company inspection records should be evaluated to ensure that adequate preventive maintenance has been completed on these kilns.

Similar to the log storage yard, this large amount of fuel poses a great concern for firefighting operations. The arrangement of these piles should be noted on annual inspections and preincident planning. Consideration should be given to heights, aisle widths, access, and water supply availability. These items should be noted on the inspection report and on response plans.

Finishing and storing are the last steps in the sawmill process. Finishing includes the use of a planer, rough sander, or trimmer to finalize the lumber profile for shipping. The purpose of the finishing operation is to further remove knots, bends, and other imperfections. The shaved lumber boards are now cut to the desired size, commonly two by fours and two by sixes. The finished lumber is then classified, labeled, banded, and packaged for shipping and sale. The finished product is usually stored in open-air sheds or outdoor storage yards ready for customer pickup or shipped to a third-party warehouse or hardware store.

Manufacture of composite panels and plywood

As with most industries, nothing is left to waste in logging and wood processing. Waste from the sawmill process is sent to other plants for processing as composite panels and plywood. These products consist of particleboard, medium density fiberboard (MDF), and oriented strand board. Although these products differ in strength and use, they are produced in the same way, that is, wood chips or fibers bonded together using a resin and then pressed into boards.

Manufacture of composite panels

The manufacture of composite panels consists of receiving the raw materials, grinding and drying, forming and pressing, and finishing (fig. 21–1).

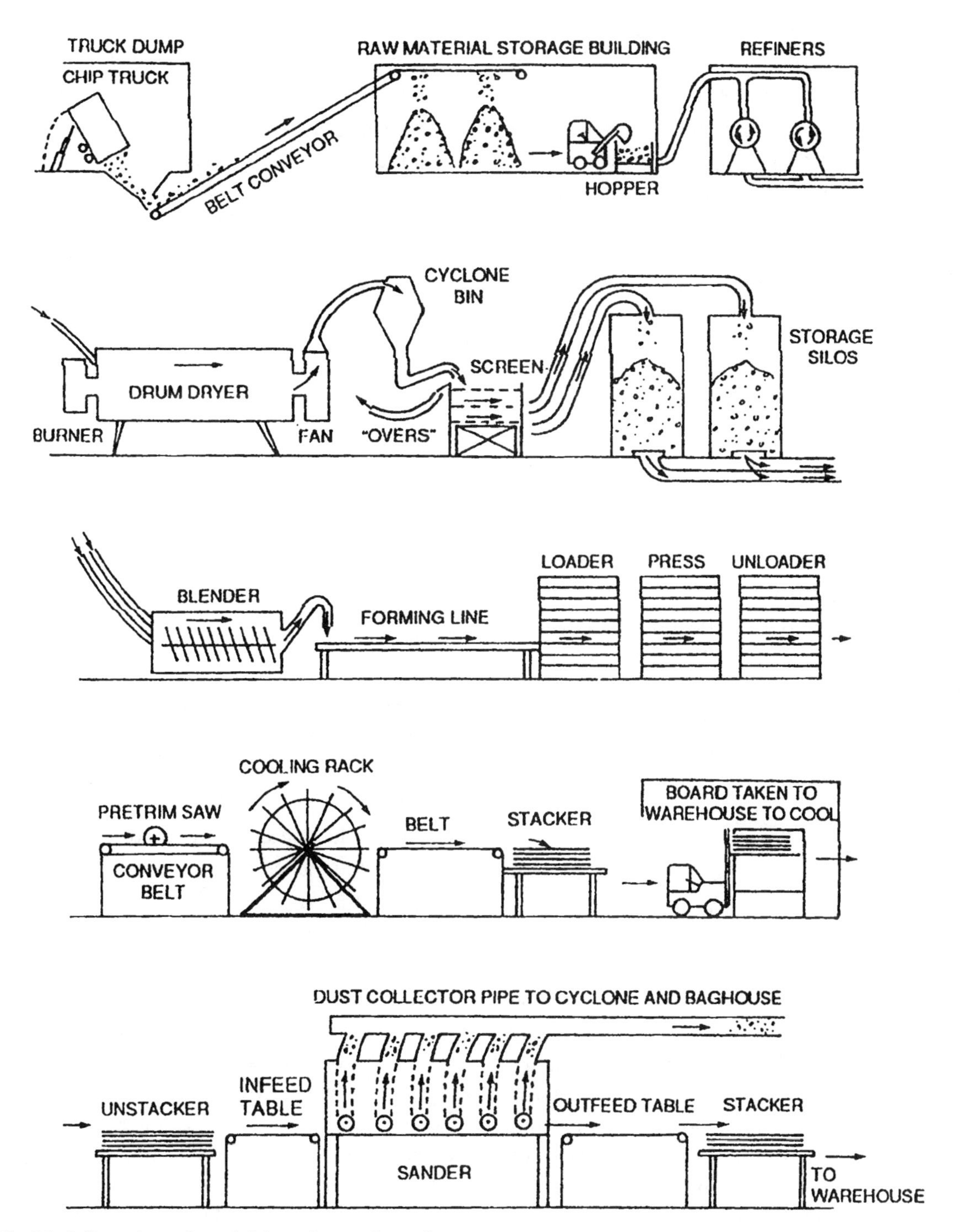

***Fig. 21–1.* Overview of particleboard manufacturing** *(Reprinted with permission from the* Fire Protection Handbook*, 19th Edition, Copyright © 2003, National Fire Protection Association, Quincy, MA 02169)*

Raw materials consisting of wood chips, flakes, and saw dust are delivered from various sawmill facilities and are typically stored in open-air, covered buildings. Front-end loaders are commonly used for transport and movement activities. In addition to the explosion hazards already present, these areas are at risk for explosions created by the suspension of wood dust in the air from the loading and unloading process of the raw materials.

Grinding and drying is the first step in the process of converting the raw materials into composite panels. The grinding process consists of taking the raw materials and transforming them into a desirable size by the use of hammer mills (see chapter 15) or other similar mechanical grinding and milling equipment. The first step in the grinding process passes the raw material through separating equipment, similar to that mentioned in chapter 15. The intent of the separators is to remove foreign objects such as stray pieces of metal or other objects that can be either ignition sources for fires or explosions or are not desirable in the finished product. After the grinding process, the now reduced wood chips and fibers (wood particles) are sent via pneumatic conveyor to dryers. The purpose of the dryers is to reduce the amount of moisture, as well as create a uniform moisture content level in the batch of wood chips or fibers. These dryers have fire and explosion hazards similar to the kilns discussed in the sawmill section of this chapter.

Forming and pressing is the next step in the process of manufacturing composite panels. Wood particles are fed from the grinder through storage silos into a blender on the forming line. The purpose of the forming machines is to mix the wood particles with the resin to form the mixture needed to form the boards. The resin is metered by weight or volume in the mixture in an open tray or forming box. With the use of a high-vacuum arrangement on the tray or forming box, the wood particle and resin mixture is formed and pressed into the shape of a board. The final properties of the board depend on the size and orientation of the wood particles and the specific type of processing used. From the forming process, the boards are now sent to high-temperature presses. The hot presses are used to further compress the boards. These presses, under high heat and pressure, cause the resin to liquefy and flow around the wood particles. The thermosetting resin solidifies under this process and forms the solid sheet of composite wood boards. Occasionally, the boards are sent to ovens and humidifiers for further drying and treatment, based on individual client needs.

Finishing, the final step in the process, involves trimming saws, cooling racks, and sanders, and is done prior to the product being sent to the warehouse for storage. The trimming saws cut the edges of the boards to help form a more square or rectangle profile, as well as remove imperfections caused by the production process. Cooling racks permit the board to cool in the open air, allowing the heat from the forming and pressing operations to dissipate. The boards are then sent to sanders as the final step in the production process. Large belt sanders are used to further fine-tune the board profile and make it ready for storage and shipping to the customer.

Manufacture of plywood

Plywood is one of the most widely used wood products in construction industry. Plywood offers an economical choice to composite boards and provides good strength and rigidity for a multitude of uses. Plywood is made from finely cut sheets or veneers from the debarking process of logs.

The production of plywood consists of wood *veneers* that are glued in a series to form a single sheet of finished product. This type of production is a simplified version of a composite board process and involves some of the same techniques.

The pieces of veneers are produced by rotating a log on a lathelike machine and using a sharp device to cut thin slices of wood, similar to using a cheese slicer on a block of cheese. The sliced pieces of veneer are then dried and graded and sent to the next step in the process. After the veneer is graded and sorted, it is stacked to form multiple layers and glued. After being glued, the now-transformed veneer pieces are formed to a single sheet of plywood and sent to the press to allow for setting the glue.[4]

Manufacture of furniture

The manufacture of furniture and other household goods such as kitchen cabinets and countertops uses the same processes as described in sawmill and composite board manufacturing. Therefore, many of the same fire and explosion hazards exist, with the addition of other fire hazards due to the use of various flammable materials such as plastics, foams, coatings, and paints.

Many of these facilities are very old and therefore do not have some of the fire prevention measures that you would normally expect in a newer facility. These older structures are often of heavy timber construction or another highly combustible construction technique. These facilities are filled with wood and wood dust, fabric, and finishes that have accumulated over time. High ceilings, old exhaust and dust collection systems, and poor housekeeping lend themselves to a dangerous situation. As with other woodworking occupancies, the two main problem areas in furniture manufacturing are dust collection and spray-finishing operations.

Furniture can be either solid, where the principle raw component is lumber, or veneered furniture, which uses the same lumber, but the logs are sliced and peeled for laminating to form the desired piece. Newer furniture is made with plastic components, which tend to be lighter and more cost-effective, but can also lead to more of a fire hazard with respect to its gases and other products of combustion that are released during burning.

The production process varies but basically consists of

- Lumberyard storage
- Roughing mill
- Veneer operations
- Finishing and sanding
- Assembly
- Packaging
- Storage and shipping

Furniture begins in the lumberyards and roughing mills as described previously. Many of the products that are produced in sawmills and composite and plywood board processes end up as the raw material in the manufacturing process of furniture.

The veneering process is similar to the manufacture of plywood. The pieces of veneers are produced by rotating a log on a lathelike machine and using a sharp device to cut thin slices of wood. After the veneer is graded and sorted, it is stacked to form multiple layers and glued.

The machining operation consists of finishing and sanding operations. This is where the individual pieces of raw lumber are cut to specific sizes to form legs of a table or chair, table top, arms of a couch, or other components of furniture. After the lumber is cut to design, it is then sanded to remove rough corners and splinters to form a finished furniture component. These operations usually consist of lathes, saws, routers, miter boxes, bore machines, and other typical woodworking equipment. These areas are prone to dust and waste accumulation and should be monitored for housekeeping.

Assembly is where the furniture changes from a stack of raw material into a piece of finished furniture product. Here, the pieces of lumber that have been machined and sanded are assembled using screws, nails, glue, dowels, staples, and other friction-type devices. Color is added along with fabric, cushions, and other finishing components. Inspection is conducted to ensure quality, and the finished product is sent for packing, storage, and shipping.

Packaging consists of a variety of ways from palletizing and encapsulating with plastic wrap to placing the finished product in a box with packing material.

Storage and shipping is the last stop in the manufacture of furniture. Large warehouses store the finished product in either racks or a solid pile.

COMMON HAZARDS

Fire hazards

The storage of logs creates a large fire hazard and should be kept away from main processing or other critical buildings. Log storage yards should be segregated into small piles of more manageable size to prevent an uncontrollable fuel load. Additional requirements for log storage should include ample aisle width and limited heights for permit ease of access and limit collapse potentials. Log storage yards should also maintain a clear-cut distance from surrounding brush and other types of vegetation to prevent grass and brush fires from igniting logs in the storage yard.

Machining and sanding consist of a wide range of waste material ranging from dust to slivers or blocks of cut wood. The physical presence of large amounts of wood waste products introduces additional combustible material to the area, and housekeeping should be strictly followed. Large amounts of combustible material can lead to a fast-moving fire.

Finishing operations' main hazard is the spray application use of flammable and combustible liquids and their vapors. It is important to ensure that a properly designed and maintained spray booth is installed. These booths should ensure adequate mechanical ventilation and removal of these vapors to a safe location to prevent the buildup of an explosive atmosphere or fire hazard. Also, the classification of electrical equipment in this area should conform to applicable National Fire Protection Association (NFPA) standards using a listed or approved device(s).

Storage of the finished product allows for a large amount of combustible material to be stored in geometrical arrangement that promotes fire growth. Refer to chapter 17 on storage and warehouses for more information. The storage of wood chips and sawdust can present fire problems due to spontaneous ignition. Such spontaneous heating may take place when heat from decomposition cannot be readily dissipated.

Explosion hazards

Any occupancy that handles dust or produces dust as part of the process, whether grain as described in chapter 15 or wood as described in this chapter, has the potential for an explosion. Housekeeping and the prevention of fugitive dust emissions is very important. Explosion hazards could be present in the logging, machining/sanding, or finishing areas of the production process. The removal of dust is usually done with a dust collection system (fig. 21–2). These systems are located around machines where a large amount of dust is generated, thus creating the accumulation of wood dust (figs. 21–3 and 21–4). More information about dust explosions is addressed in chapter 15 of this book.

Fig. 21–2. **Single-stage dust collection system**
(Reprinted with permission from the **Fire Protection Handbook,** *19th Edition, Copyright © 2003, National Fire Protection Association, Quincy, MA 02169)*

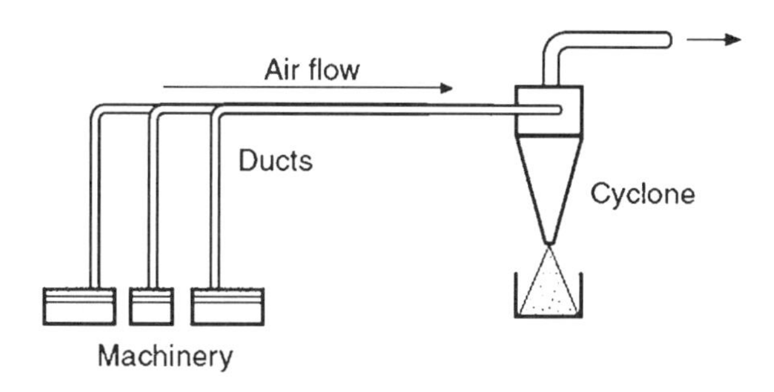

***Fig. 21–3.* Single-stage dust collection diagram**
(Reprinted with permission from the **Fire Protection Handbook,** *19th Edition, Copyright © 2003, National Fire Protection Association, Quincy, MA 02169)*

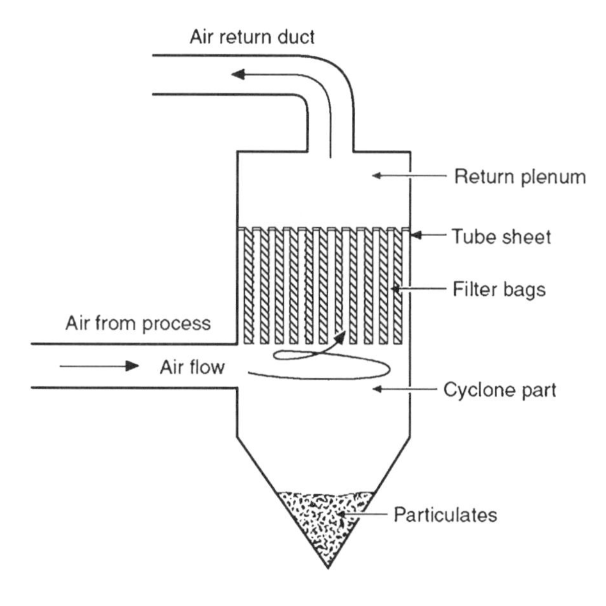

***Fig. 21–4.* Single-stage dust collection with baghouse diagram** *(Reprinted with permission from the* **Fire Protection Handbook,** *19th Edition, Copyright © 2003, National Fire Protection Association, Quincy, MA 02169)*

FIREFIGHTING TACTICS

As with all firefighting tactics, a well-thought-out and tested preincident response plan is needed. Preincident response planning should be carried out in cases where a building is an industrial occupancy or has been identified as a high-risk or target hazard occupancy. In the event of a fire at such a building, it is advantageous to have information readily available that relates to the structure, fire safety provisions, location of water sources, presence of hazardous materials, and any other information that may be relevant.

When responding to fires at woodworking facilities, caution should be used when approaching the area. It should first be determined what is burning and where. As a precaution, if an explosion has taken place, all operations should cease, but the dust collection system be permitted to operate under the control of a qualified operator and under the supervision of the fire department. It should be immediately shut down if its use will increase the possibility of further explosion or fire. This will aid in the removal of dust and, potentially, smoke and heat. Building integrity should be evaluated prior to the commitment of personnel to interior firefighting. This is particularly critical considering the age of many of these types of occupancies.

For fires in the logyard, large quantities of water may be needed. This is extremely challenging due to the lack of hydrants and/or an adequate water supply. Firefighting operations in logyards may require tanker-shuttle operations to be established. There are many resources that can be used to determine the amount of water supply needed for such an operation. Insurance Services Office (ISO), NFPA, and the International Fire Code (IFC) offer some good references to this regard. More water is better than not enough. Caution should be used around log piles due to instability as a result of shifting from burning.

Fires involving the manufacturing area could include long hose lays and large amounts of water due to the abundance of combustibles. Caution should be used before entering an area that may have operating conveyor belts, saws, sanders, and other commonly found woodworking devices. If able, power should be disconnected to these devices, but ensure that the dust and vapor collection system remain in service.

For fires involving wood chips or sawdust, the piles will have to be broken down because these fires will usually be burrowing and deep-seated. Penetrating nozzles can be used, but in many cases where the smoke is does not always indicate where the seat of the fire is located. In most cases, the piles are broken down with the product spread on open ground where the pockets of fire can be extinguished. Care must be exercised so that flying embers are not distributed by the wind, further complicating the problem. In addition, the pile may have to be dampened prior to its being spread to prevent dust generation with a resultant explosion.

CONCLUSION

Responding to incidents at woodworking facilities creates logistical issues such as their remoteness to a water supply, large amount of combustible wood, or explosion potentials. The main fire hazards include the large storage of logs in large piles and storage of finished wood products (plywood and particleboards), which if ignited, make fire control and extinguishment very difficult for fire fighting crews. The explosion hazard derives from the production of fine wood dust during both the saw milling activities and furniture manufacturing operations. In your preincident response plans, locating an adequate water supply and staging of resources is very critical. Drills with local facility personnel and periodic visits will help ensure your department is prepared.

NOTES

[1] National Fire Protection Association (NFPA). 1990. Industrial Fire Hazards Handbook, 3rd ed. Quincy, MA: NFPA, p. 297.

[2] NFPA, Industrial Fire Hazards Handbook, p. 300.

[3] NFPA, Industrial Fire Hazards Handbook, p. 302

[4] NFPA, Industrial Fire Hazards Handbook, p. 317

22

Food and Beverage Facilities

INTRODUCTION

Food and beverage facilities that manufacture and store all the items we love are common in North America. These facilities mass-produce products and send them to restaurants and retail establishments. Most goods must be supplied at the rate that the consumer will purchase them. Some items such as dairy and bakery have very short storage lives. Other products, such as canned and preserved food, can last a year or more. To ensure products are fresh to the consumer many of these factories operate around the clock.

The National Fire Protection Association's *Fire Protection Handbook* lists the primary cause of fires in food and beverage occupancies as process equipment, combustible dusts, flammable gases and liquids, conveyors, refrigeration systems, fuel-fired equipment, and storage fires.[1] A good example of a food-processing facility is the American Italian Pasta Company (AIPC) located in Tolleson, Arizona. This plant can produce 109 million pounds of pasta per year. The facility is almost 300,000 square feet and has its own storage silos, robotic packaging system, and pasta manufacturing machinery. Raw pasta is extruded at approximately 1,800 pounds per square inch. The pasta is dried with heat produced from boilers capable of generating 25 million British thermal units (Btus). The facility processes 83 pallets of pasta per hour. This facility was retrofitted, and an addition was built to house the new pasta-manufacturing machines. AIPC expects to further expand this facility to meet growing demand.[2] Expansion and change of occupancy are common in food- and beverage-manufacturing facilities.

DESCRIPTION OF PROCESS

Food and beverage facilities present many of the same challenges as other large industrial facilities. Raw materials are blended and manipulated through heating, extruding, fermenting, and freezing to produce the desired appearance and taste. These raw materials could be anything from fruit to cooking oil to plastics used to wrap the food. The massive furnaces that bake the bread and vats that ferment the beer all present their own hazards.

There are a wide variety of food and beverage facilities, and many hundreds of processes within the facilities. The raw materials that enter these facilities could be grains, powders (such as flour, sugar, starch, etc.), vegetables, plastics, and glass, to name a few. These materials are unloaded and may be temporarily stored or move directly into the mixers or some other process. In the food and beverage facilities, these materials are kept in sanitary conditions to meet the requirements of the specific material.

A common process is the production of fried and baked snack foods (fig. 22–1). Potato chips remain popular across the country. The production of potato chips involves the cleaning of the potatoes, then the peeling. Most often abrasion is used to remove the skin. The potatoes are then moved into a rotary slicer. The slices then move to a dryer, which removes some of the moisture. Now slices are ready for frying. There are two types of frying methods. The chips could be batch fried or fried with the continuous process. As the term indicates, the batch-frying process involves one batch of chips at a time. In the continuous process, chips continuously move though the fryer. Some of the common oils used in frying include cottonseed, corn, and peanut oils. After the chips are fried, they are allowed to cool, then packaged and shipped.[3]

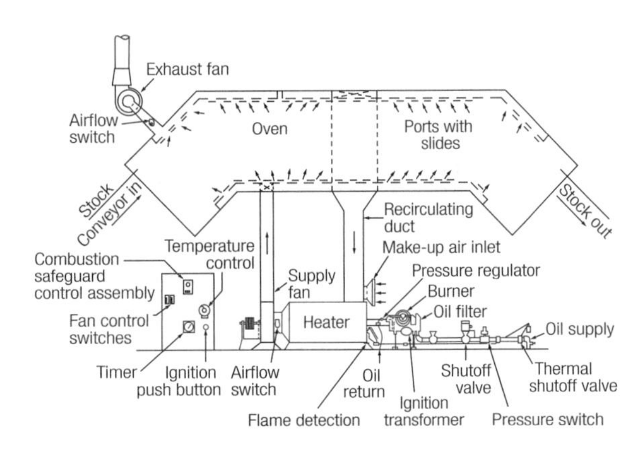

Fig. 22–1.* Typical large conveyor type baking oven** *(Reprinted with permission from the* **Fire Protection Handbook, 19th Edition, Copyright © 2003, National Fire Protection Association, Quincy, MA 02169)*

Beverages containing alcohol are also produced across the country. Some of these beverages include beer, wine, brandies, and distilled spirits such as whiskies and rum. The raw materials for these beverages include various grains, such as corn, wheat barley, and rye, and sugars. The raw materials are received in trucks or railcars. The raw materials are stored in silos until ready to use. Pneumatic conveyor systems, auger systems, or bucket elevators move the grain at the facility. Grains are milled to expose the starch. It is the starches and sugars that react and turn to ethyl alcohol. Some beverages such as whiskey are distilled to extract the alcohol. After distillation, the product may be stored in barrels and aged.[4] Bottling comes after the aging, and then the bottles are shipped from the plant.

Construction of food and beverage facilities are normally noncombustible. Some of these facilities may even be protected noncombustible. Many of the older facilities are a mixture of construction types. Newer construction may incorporate better construction materials and methods. Some factories and associated storage facilities may even be heavy timber. Fire walls may be in place and equipped with heavy fire doors. These doors present a hazard and could close on hoselines or disorient firefighters.

A conveyor belt may be used anywhere in the process such as bottles during container filling (fig. 22–2). In a baking facility, a conveyor belt may be used to move either the processed flours or grains or the product through the oven. Products and raw materials may also be moved in pipes, through automated robotic machines, in tubs, or in hoppers, to name a few common methods. Other processes are too numerous to describe, but may involve cattle, poultry, pork, cereals and pet food, sugar cane and sugar beet, vegetable oils, fish, fruits, nuts, and vegetables, among others.

***Fig. 22–2.* Conveyor belt in a bottling plant** *(Courtesy of Safeway Inc.)*

COMMON HAZARDS

One of the hazards in these facilities is refrigeration equipment for associated equipment and warehouses. These refrigeration units may use either ammonia or Freon as a cooling agent. Ammonia is a hazardous gas if released during a fire situation. Freon is less hazardous but may be released in high quantities depending on the size of the freezing unit. Freezers may be equipped with dry-pipe sprinkler systems. These systems should initiate on their own but may need to be supported after activation.

Machinery in the food- and beverage-processing facilities will likely be abundant (fig. 22–3). All types of processes are performed by these machines. Workers are generally only employed to monitor and to maintain the machines. The low number of workers in a facility can be a help or a hindrance. Fewer workers means fewer potential victims, but it also means fewer people to control an emergency situation that may require manual manipulation of controls and valves. Many plants will have at least one maintainer on duty to quickly repair any machine to keep production flowing. In the case of a fire, these maintainers should be consulted to determine what steps to take to power down machinery. Other hazards present could include heat and cold, flammable gases and liquids, pressurized vessels, and hazardous materials, to name a few.

Fig. 22–3. **Bottle-filling machinery** *(Courtesy of Safeway Inc.)*

Silos may be present in several types of food and beverage facilities such as bakeries, pasta plants, breweries, and distilleries (figs. 22–4 and 22–5). Silos present various types of hazards to firefighters. These hazards include their elevation, the potential for a contents fire or dust explosion, and risk of trapping firefighters. The elevation of some silos may require high-angle rescue techniques. There have been many cases of dust explosions inside of silos. Furthermore, there is the possibility of additional dust explosions after the first. As a result of the blast wave of the first explosion, dust is liberated and suspended in surrounding silos. This phenomenon is called cascade explosions. Secondary explosions are in many cases more severe.[5] This dust is then ignited, and further explosions occur. The initial explosion may be sparked by static electricity, lightning, or faulty equipment. It is essential that firefighters use only explosion-rated equipment at these incidents.

Fig. 22–4. **Large dry material indoor silos** *(Courtesy of Safeway Inc.)*

Fig. 22–5. **Exterior silos and vertical liquid vessels. Note the signage.** *(Courtesy of Safeway Inc.)*

Dusts and vapors can be present in many areas. Movement of flours and grains through hoppers and conveyors, or the mixing or blending process can liberate sufficient dust to create an explosive atmosphere. Dust and vapor management, and extrication systems must be used to ensure that explosive levels do not accumulate in these facilities. The potential for explosions from flammable vapors is well known. Perhaps less known is the potential of common food ingredients such as cocoa, corn, corn starch, rice, sugar, grains, and flours. All dusts must be considered for their potential to explode in air when disturbed and suspended. These explosions may take place in a piece of equipment, a room, or storage area. Sympathetic explosions may occur after the first explosion as more dust is liberated and becomes airborne.

Another hazard found in the food industry is ethylene gas (fig. 22–6). This gas is used to hasten the ripening of many fruits and vegetables. Ethylene gas is moderately toxic and very flammable. Ensure that you know where ethylene gas is used in any food storage warehouses you might respond to.

Fig. 22–6. **Fruit-ripening rooms** *(Courtesy of Safeway Inc.)*

As with other industrial facilities, there could be flammable liquids such as alcohol and alcohol-based additives. For these types of fires, you will need alcohol type foam. Flammable gases used to power heating equipment may be present. As with all flammable gas fires, you must shut off the source of the fuel before you extinguish a gas-fed fire. Attempt to contain the fire until the gas is shut off. Any unburned fugitive gas is an explosion hazard. Use appropriately rated ventilation equipment and fog streams to dissipate any gas or flammable liquid vapors. Evacuate if these vapors cannot be dispersed.

One specific hazard that may be present in food and beverage facilities is the propensity of certain materials to self-heat and rise to ignition temperature. Animal and vegetable oils can polymerize and self-heat, especially if they are in cloth.[6] Fires may start in a mop closet, or a rag or laundry collection area. These oils may also fuel hot, smoky fires.

Some process areas within food and beverage facilities handle flammable liquids and gases. For example, in the brewing industry, the process of fermentation takes place to produce alcohol. Depending on the end product, various purities of alcohol are produced. The purity of an alcohol could be measured in proof or percentage. The process areas designed to safely handle the flammable liquids should be explosion proof. It is important that firefighters ensure that they use only explosion-proof (intrinsically safe) equipment during search and rescue efforts in areas designated as potentially explosive.

Another hazard of silos and other mass storage is the potential for engulfment. Many rescuers have tried to free trapped victims and become trapped themselves. Mass stored materials may appear firm around the edges. As the rescuer moves toward the center, gravity pulls him or her inward and down. Rescues of this type must be carried out using proper confined space techniques.

No firefighter should enter, step on, or even touch machinery that has not been confirmed as deenergized. Machinery should be properly locked out and tagged out to ensure that firefighters are not injured or killed in case machinery is accidentally energized. Machinery could use several different types of energy. These sources could include various electrical voltages, flammable liquids and gases, steam, pneumatics, and hydraulics,. Equipment may be equipped with emergency shutdown systems that will isolate all sources of power. Preincident response plans must list the safe shutdown procedures for all major equipment. Remember that some systems store energy even after they are deenergized. Examples of stored energy could be hydraulic pressure and electricity for some machines.

Another source of potential hazards are large vats and tanks. Workers may accidentally fall into these vats, which could contain thousands of gallons of liquid. Another liquid hazard could be the drainage systems. Large-capacity drainage systems are designed to quickly move copious amounts of liquid to either the sanitary sewer system or a holding tank. The waste in these ponds could be hazardous or present a drowning hazard. Further, the tunnels, separation tanks, and lagoons could give off toxic fumes or vapors that could overwhelm firefighters or plant workers. In any case such as this, proper hazardous materials and confined space rescue procedures must be followed.

Heat exposure is another hazard. Sources of heat could be from electric or gas-fired ovens (fig. 22–7). Steam is sometimes used to heat liquids so that they can be processed, mixed, or fermented. The release of hot liquids such as molasses and oil can seriously injure or kill firefighters. Steam is also used to purge and sterilize containers and vessels. Heat is a common hazard in these facilities that must be considered.

Fig. 22–7. **A large baking oven** *(Courtesy of Safeway Inc.)*

Associated with food and beverage facilities will be packaging and storage facilities. Packaging could include bottling, bagging, boxing, or mass storage of products. In many cases, automated systems move products off the production line to be packaged. These bulk products could be placed on wooden or plastic pallets. These products could be loaded onto a truck or even a railcar. Hazards include exploding containers, high storage, plastics, and wood fires, as well as the associated transportation systems such as conveyor belts and forklifts. Storage facilities are mentioned briefly here and are covered in more detail in chapter 17.

FIREFIGHTING TACTICS

Firefighting tactics in food and beverage facilities are similar to those in some other large facilities. Because of the size and complexities of these facilities, a preincident response plan is essential. Size-up is crucial before forces are committed. Facilities will normally be equipped with fire detection and suppression systems. Water-based suppression systems will normally be equipped with fire department connections. Fire detection and suppression systems must be clearly identified on the prefire plan.

On arrival, the incident commander should make contact with a plant representative. Fire alarm and enunciator panels should be checked for system activation and detection. An addressable fire protection system could indicate the exact area of the fire or the specific piece of equipment. Once the specific location of a fire is determined, a plan can be established. This plan will most probably include shutdown of process equipment. As with other industrial equipment, there will be an established method and sequence to shut down equipment. If possible, plant personnel are the best people to perform this task since they are most familiar with the equipment.

In many cases, plant personnel have access to instrumentation that could give you insight into what is occurring. They could say the conveyor belt stopped moving for 10 minutes, and then there was a fire warning light in the tunnel. Before you start tearing open expensive plant machinery, you should identify fire access doors, fixed systems, and process shutdown devices. No need to try to fight a fire in a gas-fired oven while the gas is still on.

The use of an installed suppression system is the safest procedure for firefighters and normally the most effective. If automatic systems fail to operate, then firefighters should manually try to initiate them. These systems may include water deluge, water spray, gaseous, wet chemical, inerting, or powder systems. Suppression systems could be specific to a hazard such as a cooling system for a tank, wet chemical system for deep fat fryers, an internal misting system for a conveyor system, or a total building coverage water sprinkler system such as an early-suppression fast-response (ESFR). Foam suppression systems may also be present for flammable liquids. As described in chapter 8 on fixed fire suppression systems, firefighters must ensure that water streams do not dilute or negate the effects of firefighting foam.

Special water-based fire suppression systems may be utilized throughout these facilities. Such systems are normally designed to deliver much higher quantities of water (application density per square foot) than conventional sprinklers. The demand for a conventional sprinkler system may only be 150 gpm. The demand for these special cooling and suppression systems may be greater that 1,500 gpm. The capacities of the fire water system and the locations of the fire department connections must be known. This is why preincident response planning is essential. Support of the installed fire suppression systems is the highest priority. Do not rob the water supply from these systems to attempt manual firefighting efforts unless you know that the fixed or semifixed system is out of service.

Other useful systems to firefighters are dust and vapor extraction systems. Firefighters should attempt to use these systems where available. It is mush easier and more efficient to use installed vapor and dust extraction systems in these facilities than it is to set up high-flow positive or negative pressure ventilation fans. The atmospheres in tanks, vessels, and facilities must be properly monitored. This may be a case where it is safer to leave the power to facility on so that the facility's extraction system can be used. By shutting down the power, you may also shut down the facility's high flow ventilation system.

In the event of a fire in a large fryer, firefighters must take care never to plunge hose streams into the liquid. After the power is shutoff to these fryers, the oil will remain hot for some time. The best means of extinguishment is with dry- or wet-chemical extinguishers. Many times there will be covers that release and cover the fryer surface. Ordinary Class B foams will be extremely hazardous to apply and may just spread the fire. The water in the foam will continue to flash off to steam and react with the hot oil. This reaction will probably cause some of the still hot oil to splash out of the fryer and spread the fire. Try to contain the fire to the area of origin, let the oil cool with the lid applied if available and use dry- and wet-chemical extinguishers on oil fires.

Confinement of the fire in one portion of the structure may be possible. In many cases, these facilities are built with proper fire cutoffs and firewalls.[7] The fire may rage in one section of the facility, but automatic and manual efforts to confine the fire to the immediate area of origin may be the best tactic. Once confined using fire doors and dampers, all power sources to the affected area should be shut down. The fire suppression systems should be supported, and manual hose streams readied to ensure that the fire does not spread to the unaffected area. Proper fire walls may contain the fire for one or more hours. Once isolated, firefighters should resist the temptation to open fire doors and check the status of the fire.

CONCLUSION

Fires in food and beverage facilities may not seem to present the same level of hazards as other, more exotic industrial facilities. In reality, these facilities are more abundant and located close to populated areas. They present a full range of hazards from flammable liquids and gases fires to dust explosions. These facilities have tanks and tunnels and silos that can trap or drown firefighters. They may have been built in older structures with low fire resistance and may lack modern fire protection features. These facilities deserve your time and attention. Comprehensive preincident response plans and full-scale drills will ensure that your department is ready for the challenge.

CASE STUDY

A fire in a large, cold storage and fruit-packing plant resulted in a total loss of more than $4,000,000. The 111,000 square foot structure showed smoke from the third and fourth floors when firefighters arrived 8 minutes after receiving the call. The structure had no fire protection systems, and the preincident response plan indicated that 6,000 gpm were required to combat the fire if 50% of the structure was involved. Firefighters reverted to a defensive attack within an hour because of a limited water supply in the area. Partial structural collapse occurred approximately 2 hours into the fire. Additional firefighting forces were called and battled the blaze for approximately 17 hours. The lack of suppression system, the heavy timber construction, and the limited water supply most certainly contributed to the failure of firefighting forces to save the structure. Thankfully, a preincident plan was available, and no injuries were reported.[8]

Random thoughts

- The possibility for large-loss fires in food and beverage facilities exists throughout the country. Almost every municipal fire department has a food- or beverage-manufacturing facility within its response area. Preincident response planning was essential in preventing injuries at the cited case of a fire at the fruit-packing plant.
- The modification of older facilities and renovations may not fully address fire safety. Some older structures, and even some of the modern structures, may not be suitably designed for the process installed. Introduction of hazardous processes into the area of the building could result in catastrophic fires.

- The senior fire official must seriously consider defensive strategies for these facilities, especially once all persons are known to have left the building. These facilities may experience sudden collapse or have fire flow demands in excess of the capacity of the water system.
- As always, public and firefighter safety comes first. If the senior fire official does not have accurate knowledge of the building's contents, building construction, or the extent of the fire, the firefighter's safety could be seriously jeopardized.

NOTES

[1] National Fire Protection Association (NFPA). 2003. *Fire Protection Handbook,* 19th ed., vol. I. Quincy, MA: NFPA, pp. 13-229–235

[2] Food Processing Technology, American Italian Noodle Production Plant, Retrieved September 23, 2006, from http://www.foodprocessing-technology.com/projects/aipc.

[3] United States Environmental Protection Agency (EPA), *AP 42,* 5th ed., vol. I, chap. 9: *Food and Agricultural Industries.*

[4] EPA, AP 42.

[5] Kennedy, P. M., and Kennedy, J. 1990. *Explosion Investigation and Analysis, Kennedy on Explosions.* Chicago, IL: Investigations Institute, pp. 137–139.

[6] National Fire Protection Association (NFPA). 2004. NFPA 921, *Guide for Fire and Explosion Investigation.* Quincy, MA: NFPA, sect. 5.3.6.

[7] FM Global. 2000. Property Loss Prevention Data Sheet, *7–74, Distilleries.* Norwood, MA: FM Global.

[8] Tremblay, K. J. (2005, September/October). Fire Destroys Cold Storage Fruit Packing Buildings. *NFPA Journal.* Retrieved September 23, 2006, from http://www.nfpa.org/publicColumn.asp?categoryID=&itemID=25645&src=NFPAJournal.

23

Laboratories

INTRODUCTION

Laboratories are common in industrial facilities and serve a variety of functions. Some of these laboratories are for internal quality control of finished and raw materials. Other laboratories are dedicated to research and development. Laboratories may contain any type of material to include explosives, flammable liquids and gases, radioactive material, etiologic, and other exotics. In addition, the processes within the facilities could be quite dangerous. Frequently, laboratory work includes mixing of chemicals, melting and heating, freezing and cooling. This work could also include exposure to radiation. Furthermore, work could be on live animals that have been exposed to some type of disease. Live viable organisms could also be grown for medical, veterinary, agricultural, or other research use.

Laboratories are generally well designed and controlled. Typically, highly skilled technicians and scientists perform work in these facilities. Laboratory work and facilities are expensive. The high-precision instruments and equipment are designed with safeguards for the operators and those in the vicinity of the work. Facility size and complexity varies depending on the nature of the work. These laboratories could be part of the industrial facility or a stand-alone structure. Laboratories are allowed to be mixed with other industrial and even business offices depending on the type of testing being conducted.

Depending on the nature of the facility, laboratories could be designed with features such as noncombustible construction, isolation areas equipped with fire walls, and partitions. Exhaust fans and hoods may be installed to keep a desirable negative or positive pressure in the facility to prevent accumulation of unwanted gases, fumes, dust, or vapors. Airborne materials can be extracted from the work area and not allowed to collect in harmful concentrations. These ventilation and extraction systems are designed to operate even when the fire alarm system is activated. Failure of an exhaust system to function during a fire could result in the rapid spread of the fire or release of toxic fumes when the system is needed most. Hood and exhaust systems must be made of noncombustible material so that in the event of a fire, the heat and flames are contained within the system and exhausted. The exhaust stacks from these systems may be quite high above the roof elevation. They must exhaust toxic or flammable vapors at an elevation high enough above the roof so that they are not drawn back into the building's ventilation system or present a hazard to other occupancies in the area.

Many laboratories are equipped with some type of fire protection system. Depending on the nature of the hazard, they may have a fire alarm system or some type of suppression system. As of 1998, approximately half of the laboratories reported that they had a water sprinkler system. Other systems such as dry-chemical, clean

Specific Property Use	Fires	Fires Where Type of Material First Ignited Was		
		Gas	Liquid	Solid
Chemical or medical laboratory	91	3	24	3
General research laboratory	54	3	11	1
Physical materials testing laboratory	46	3	9	4
Electrical or electronics laboratory	23	1	1	0
Agricultural laboratory	14	0	4	0
Radioactive materials laboratory	5	1	1	1
Personnel or psychological laboratory	3	0	1	0
Unclassified laboratory	16	0	3	0
Unknown-type laboratory	41	1	5	1

Fig. 23–1.* Types of fires in laboratories** *(Reprinted with permission from the* **Fire Protection Handbook, 19th Edition, Copyright © 2003, National Fire Protection Association, Quincy, MA 02169)*

suppression and inerting agents, may be installed for the entire facility or for specific hazards.[1]

The causes of fires in laboratories vary (fig. 23–1). Also, some processes performed in laboratories may result in an explosion. For this reason, explosion protection could be installed in facilities. The intent of an explosion protection system is to prevent total destruction of a facility by venting the rapidly released gases to atmosphere. These vents must be large enough to handle the greatest anticipated overpressure in a facility. Just an overpressure of 0.5 to 3 psi could destroy a structure.[2]

DESCRIPTION OF PROCESS

Laboratory work could include just about anything and cannot be completely described in this chapter. There are thousands of processes that take place in a laboratory. Almost every industrial procedure or material started in a laboratory. Most people are familiar with some of the many laboratory tests carried out for medical purposes such as testing tissue for cancer cells or growing a culture to determine a type of infection. Some materials-testing laboratories subject the materials to extreme stresses to determine failure points. These pressures could be pneumatic or hydraulic. Of course, the temperature of tested materials is sometimes elevated. Chemicals could be used in laboratories to test materials or to make new compound. Materials could be irradiated to kill or alter the growth of organisms.

Radiation could also be used to heat materials or for uses such as X-rays to check for cracks or internal defects. In a quality control laboratory, materials may be intentionally burned, crushed, pierced, or overpressurized. There is no limit to the processes you may discover in a laboratory.

Given that the processes and materials are so varied in laboratories, it is imperative that thorough prefire incident planning be conducted. Laboratories may not be fully open on the intentions of the tests being conducted. They may be very protective of proprietary information and processes. Research and development laboratories cannot have their trade secrets revealed to other businesses. Fire protection and emergency management personnel must be sensitive to this need and ensure that they protect the companies' trade secrets. At the same time, it is vital to determine what hazards may be encountered by emergency responders. Hazards must be identified ahead of time to limit the risk to responders, and a vulnerability analysis must be conducted to determine who or what might be exposed.[3] Once the hazards (both materials and processes) are known, firefighters will be better prepared to respond to such incidents.

As mentioned earlier, laboratory testing is expensive, and manufacturers use this testing to refine processes, improve products, and optimize manufacturing procedures. Firefighters should realize that experiments may take months or years to complete. The inadvertent destruction or simple contamination of these experiments should be considered during investigative and emergency operations. Simply walking through a clean room laboratory in firefighter protective clothing could ruin years of work. Special safety clothing, such as an acid suit, may also be required in these areas.

Facilities with hazardous materials should also be placarded. One such system is the NFPA 704 symbol, which identifies health, flammability, reactivity, and special instructions.[4] Other common placards are for radiation, noise, explosives, and biological hazards. Placard symbols should be visible on typical response routes and on the internal doors and work spaces. Many other placards systems may be used, such as the Department of Transportation or locally adopted systems.

COMMON HAZARDS

As has already been mentioned, the potential hazards in laboratory facilities are wide and varied. There could be massive quantities of hazardous materials or substances that will kill you in small doses. Most commonly, there will be laboratory (small) quantities of hazardous materials to deal with. Some of these substances could kill you quickly, and others, relatively slowly. You will have to survey the facility, note the hazards, and do some research. Walking through the facility with one of the employees will help you to understand the processes and potential hazards. An excellent way to start is by reviewing the facility's material safety data sheets (MSDS). The facility may also use a system such as the Hazardous Materials Information System (HMIS). This is a computerized information management system designed to ensure the safe transportation of hazardous materials by air, highway, rail, or water.[5] Firefighters should reference this system during preincident response planning. Another type of plan that firefighters may already be aware of is the Hazardous Materials Management Plan (HMMP). Facilities required to develop these plans also must share them with the responsible fire department.

One of the common features that can potentially delay response forces are security systems. Security systems present the hazard of preventing firefighters from entering the affected areas or potentially isolating forces who gain entry. If security doors lock behind firefighters entering affected areas, they may not be able to be assisted by other responding forces. Communications may also be hindered by features such as lead lined walls, which were specifically designed to withstand energy waves, such as gamma radiation, and will also stop radio communications.

Flammable liquids and gases are common in laboratories. The quantities will vary depending on what the laboratory uses them for. There may be large cylinders or liquid storage tanks associated with these facilities. Inside of the facility and in the work spaces, there could be more quantities of these hazardous materials. The general rule for storage and use of hazardous materials in laboratories is that only the minimum amount necessary to perform the task should be out of the container or in the work area. Experimental work rarely requires large amounts of flammable or combustible liquids; however, it might require many different types of typically small amounts. Some types of hazardous materials that you may encounter include acids and bases; toxins; organic peroxides; pyrophorics, such as sodium, which is water reactive; or phosphorus, which must be kept wet; and oxidizers.

Radioactive materials and radiation-producing equipment could be used in laboratories. For highly radioactive material handling, hot cells and caves are used to handle materials remotely. The operators are shielded by walls and special glass that prevent passage of radiation.[6] Radiation-producing machines may also be encountered. These machines may have a small amount of radioactive material or a radiation-emitting device such as cold cathode discharge tubes emitting X-rays, and gamma-emitting substances such as colbalt-60, cesium-137 and technetium-99m are common. Gamma radiation has great penetrating power. Remember that even when materials are penetrated by gamma radiation, it does not make them radioactive.[7] Other types of radiation include intense light such as a laser, microwaves, or radio frequency. All types of radiation present their own specific hazards that must be planned for.

The three different types of radiation are alpha, beta, and gamma. Hazards from these types of radiation are covered at all levels of hazardous materials training. Time, distance, and shielding are the primary methods to protection personnel. Limit the time of potential exposure, maintain maximum distance from potential sources of radiation, and use shielding, such as concrete wall, when available.

Cryogenics, a liquefied gas stored at temperatures below –130°F, can be found in some laboratories. These extremely cold materials may be in either small or large quantities. Substances such as liquid nitrogen could be used to rapidly freeze a substance or material. Materials could also be tested at very high or low temperatures (fig. 23–2). Other materials may be encountered that are extremely hot, such as molten sulfur or aluminum. Placards should also identify these materials.

***Fig. 23–2.* Aircraft being tested in the climatic Laboratory at Eglin Air Force Base** *(Courtesy of the United States Air Force)*

Live animals can present a hazard to firefighters in testing laboratories. These animals may be used to test medicines or even cosmetics. They could be infected with diseases that could affect humans. Various animals are birds, monkeys, baboons, dogs, apes, mice, or just about any other species. Not only could they potentially harm you by giving you a disease, but they could also cause physical injury. These animals could bite, scratch, trample, or crush firefighters. Well-defined plans must be completed to handle animals in laboratories. In addition to preventing them from harming workers, firefighters, and the populace, there should be plans to evacuate the animals in case of a fire. This may not be possible in every situation, but it should be considered during the planning process.

Live organisms such as viruses and bacteria may be present. As mentioned, some laboratories will grow live organisms that could cause sickness in humans (fig. 23–3). The organisms could be in a culture dish, on human or animal tissue, or on a live laboratory animal. Sometimes these organisms are frozen for future study.

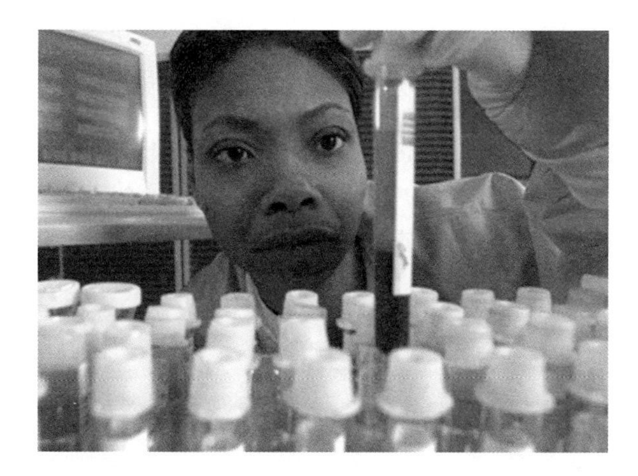

Fig. 23–3. **Medical laboratory work contains potential pathogens. (Courtesy of the United States Air Force)**

Wind tunnels can be found in some laboratories. These tunnels can be large or small. Wind speed can be low speed to supersonic. In any case, you would not want to enter one of these tunnels until the equipment had been turned off. There will probably be shutdown devices for emergency scenarios such as a fire or failure of the test subject. Ensure that lockout and tagout procedures are followed.

As you have read, the hazards in laboratories could be just about anything. All of the potential contents for nuclear, biological, and chemical exposures could be present. That being said, it may be anticipated that laboratory facilities could be the target for terrorist events. (Read more about terrorism in chapter 10.) The release of industrial radiological materials, chemicals, or biological organisms could create a complicated scenario for emergence responders to manage.

FIREFIGHTING TACTICS

The firefighting tactics associated with laboratories are similar to other industrial facilities. Hazards within the laboratory facilities must be identified during preincident response planning. In addition to the hazards, proper shutdown procedures must be known ahead of time. Special emergency shutdown procedures and systems may require multiple steps. Some laboratory hazards will be too dangerous for even the best equipped hazardous materials response teams to enter. These situations should be managed to limit the exposure to the community and to contain the incident in the affected facility. Personnel who could have been exposed to exotic materials will have to be controlled as well. Many laboratories have their own teams who respond for initial handling of spills. These teams will be specially trained and equipped to mitigate expected hazardous materials incidents in their facilities. Work with the laboratory hazardous materials response teams and learn how your team can either assist their team or be assisted by their team. Coordination with the plant safety personnel is critical during your preincident response planning. The importance of working with laboratory hazardous materials response teams cannot be stressed enough. Strict fire department hazardous materials response procedures could overly complicate an incident that could be quickly mitigated by a hazard specialist team from the facility or industry.

Collection points and a means of decontaminating firefighters and workers will have to be set up for appropriate materials. Firefighters entering the vicinity of radioactive materials must be equipped with radioactive dosimeters. Field monitoring should be conducted in all directions from the potential radiological release. Monitors and self-contained breathing apparatus must always used for potential radiological incidents.

Other types of detection instruments may also be needed. The use of flammable or toxic gas meters may also be required, depending on the types of hazardous materials in a facility. On-site detection systems may also be available. The laboratory may also have special detection equipment to handle exotic materials. A heat gun or thermal-imaging camera can help to determine which machinery surfaces are hot or cold.

Biological hazards or pathogen microorganisms may be present. As discussed, these hazards may be from live animals, human tissue, or cultures. Again, these hazards and proper personal protective equipment (PPE) should be identified ahead of time. Appropriate PPE could be NFPA 1999–type protection or other highly specialized gear. You must realize that the specialized PPE for a biohazard may not be appropriate or compatible with firefighting PPE. Where a contamination potential exists, disinfection procedures must be established to destroy recognized pathogenic microorganisms.[8] After exposure, or potential exposure to biological hazards, firefighters should follow their department's blood-borne pathogen procedures. Potentially exposed personnel may need to be quarantined as well. A physician should always be consulted to evaluate the potential that personnel have been exposed and to determine the appropriate course of action.

As with almost all other types of occupancies, the most effective means of fire control and extinguishment will be through a fixed or semifixed system. In many cases, the facility will be covered by an automatic sprinkler system. Specialized hazards or hazard areas may have their own fixed fire suppression system. Negative pressure ventilation systems may be installed and should continue to operate in the event of a fire and the actuation of the fire alarm or suppression system. Where hood and exhaust systems are installed, they should prevent the accumulation of flammable or toxic gases, and if a fire occurs, contain that fire to the immediate area.

CONCLUSION

Laboratory facilities present the full realm of hazards to firefighters. These facilities may be small, as part of another facility, or massive, taking up acres of land. Exotic materials and even live animals may be present. Exposure to these hazards could kill you quickly or over time. Your best bet is to understand the hazards by conducting facility preincident response planning. Take advantage of engineering features such as exhaust and fire suppression systems. Work closely with any specialized laboratory response plans already in place and build on their procedures. Laboratory facilities are generally safe and are staffed by highly trained workers but can present significant hazards because of their exotic operations.

NOTES

[1] National Fire Protection Association (NFPA). 2003. *Fire Protection Handbook,* 19th ed., vol. I. Quincy, MA: NFPA, pp. 13–262.

[2] Kennedy, P. M., and Kennedy, J. 1990. *Explosion Investigation and Analysis, Kennedy on Explosions.* Chicago, IL: Investigations Institute, p. 40.

[3] Noll, G. G., Hildebrand, M. S., and Yvorra, J. 2005. *Hazardous Materials: Managing the Incident,* 3rd ed. Chester, MD: Red Hat.

[4] National Fire Protection Association (NFPA). 2001. NFPA 704, *Standard System for the Identification of the Hazards of Materials for Emergency Response.* Quincy, MA: NFPA, p. 14.

[5] United States Department of Transportation, Office of Hazardous Materials Safety, 2006. *About HMIS.* Retrieved September 23, 2006, from http://hazmat.dot.gov/enforce/spills/abhmis.htm.

[6] NFPA. 2003. NFPA 801, *Standard for Fire Protection for Facilities Handling Radioactive Materials,* p. 21.

[7] United States Environmental Protection Agency (EPA). 2006. *Understanding Radiation, Gamma Rays.* Retrieved September 23, 2006, from http://www.epa.gov/radiation/understand/gamma.htm.

[8] NFPA. 2002. NFPA 471, *Recommended Practices for Responding to Hazardous Materials Incidents,* p. 27.

24

Marine Operations Including Shipyards

INTRODUCTION

Marine operations, while not specifically an industrial process, present unique challenges to the fire service (fig. 24–1). The unusual characteristics of a vessel's power plants, construction, and access to the vessel during a fire warrant this subject being covered in our text, just as aircraft facilities and aircraft themselves are covered in chapter 29. When responding to industrial facilities located on a waterway, it should be kept in mind that the response to the industrial facility may be the result of an incident involving a marine vessel that is at the facility, either loading or offloading product. In addition to vessel fires and their hazards, we will also examine shipyards because of their industrial and marine nature.

Fig. 24–1. **Marine fires present unique challenges to municipal and industrial firefighters** *(Courtesy of R. Callis)*

Fire departments throughout the United States that are located on one of the many coasts, bays, rivers, or tributaries may be called to marine-related incidents at industrial facilities. Not all of these departments have given thought to their possible response to a marine incident, though, or the varied problems that they may encounter on arrival. This chapter will highlight the various problems and unique characteristics of marine operations, including land-based support facilities. We will specifically look some of the vessel types that may be found at industrial facilities and offer specific and generalized tactics for control of fires on these vessels. Additional information and details can be found in publications such as *Marine Firefighting for Land-Based Firefighters,* published by Fire Protection Publications, and the National Fire Protection Association's NFPA 1405, *Guide for Land-Based Firefighters Who Respond to Marine Vessel Fires.*

TYPES OF VESSELS

The following vessels are those that may typically be found at industrial facilities because of the products carried and the various processes that may be using these products. In addition, vessels that may be present at terminal facilities are also included.

Roll-on/roll-off

Roll-on/roll-off (Ro-Ro) vessels are characterized by large doors in their ends or sides to allow vehicles to drive off or on while the ship is at berth (fig. 24–2). These vessels are similar to large parking garages and carry all types of vehicles, including tractor trailers, and may hold up to several thousand automobiles.[1] The Ro-Ro can have a very high distance from the waterline to the weather deck (*freeboard*), which can have negative effects for firefighting access to the ship. These vessels are found at port facilities used for loading and unloading products.

Fig. 24–2. **Roll-on/roll-off** *(Courtesy of R. Callis)*

Petroleum tankers

Petroleum tankers are designed to carry crude oil or finished petroleum products (fig. 24–3). Their sizes can range from 200 to 1,200 feet in length, with carrying capacities from 15,000 to 3,680,000 barrels (bbl).[2] It should be remembered that a barrel of oil is 42 gallons. Convert 3,680,000 bbls to gallons, and we can see the potential fuel load of over 154 million gallons. Petroleum tankers usually have various tanks on board to store their products. These tanks have systems on board to make the vapor space inert in the cargo tanks to minimize the chance of explosions. Pump rooms contain the pumps used to load and unload the specific product. In some instances, individual tanks may have pumps used for loading and unloading the product. Loading/unloading manifolds are located on the main deck for connections to shore side connections. There may be spillage at any point in the system, which can result in flammable vapors and an ensuing fire. These vessels may be present at fuel storage depots, refineries, or other industrial facilities that may use the products they carry.

Fig. 24–3. **Petroleum tanker** *(Courtesy of R. Callis)*

Chemical carriers

Chemical carriers are tankers that transport a variety of liquid products, including oils, solvents, sulfur, and other commodities (fig. 24–4). Many of these commodities are classified as hazardous materials or *dangerous goods*. The fire officer should ascertain what chemicals are on board the vessel by consulting with the ship's crew and the vessel's *Dangerous Cargo/Goods Manifest*. Fixed fire suppression systems may be present that are compatible with the products being carried, for example, fixed-foam fire monitors. These vessels may be located at industrial chemical plants that use, store, or refine the products.

Liquefied flammable gas carriers

These vessels transport *liquefied petroleum gas* (LPG) or *liquefied natural gas* (LNG) (fig. 24–5). These gases are liquefied to reduce their volume for transportation. The liquefaction takes place using a cryogenic process in which extremely low

Fig. 24–4. **Deck area of chemical carrier** *(Courtesy of R. Callis)*

temperatures (at least –130°F) convert the gas to a liquid. The temperatures vary depending on the materials being liquefied. The reduction factor that takes place is 600 for LNG and 270 for LPG. Large, spherical, insulated tanks are used to store the product on the ship; however, other configurations may be used where the tanks are not as readily noticeable, such as when spherical tanks are used. Vessels carrying these products are equipped with water spray fire systems in the deck area, which are used for exposure protection during a fire. The spray will assist with fire confinement and protect exposed metal surfaces from the effects of the radiant heat and/or direct flame impingement. Dry-chemical fire-extinguishing systems may also be present in some areas of the vessel. This type of vessel may be found at processing plants, storage facilities, or in the case of LNG, at liquefaction or regasification plants.[3]

Fig. 24–5. **LNG carrier** *(Courtesy of R. Callis)*

Bulk cargo vessels

These vessels are found at port facilities and can be either liquid bulk (see earlier, petroleum tankers and chemical carriers) or dry bulk carriers. Coal, wood chips, iron ore, grain, cement powder, salt, sugar, and sand are some of the products carried (fig. 24–6). These products are loaded into cargo holds directly and possess hazards unique to the products. For example, grain will posses the same hazards as storage silos, including engulfment, spontaneous combustion, and dust explosions (see chapter 15). Fires in these products may involve *burrowing*, or creating a deep-seated fire, and the only effective method of extinguishment may be the removal of the product to an area where it can be spread out for extinguishment. Remember to monitor and control dust conditions.

Fig. 24–6. **Bulk cargo vessel loading salt** *(Courtesy of Val Pamboukes)*

Break bulk cargo vessels

Break bulk cargo vessels are designed so that many different products can be carried in their holds (fig. 24–7). The products are usually carried on pallets or are separated by wood *dunnage* that is used to support the products carried. The materials transported are usually loaded and unloaded by cranes, either contained on the ship or on shore at the industrial or port facility.

Fig. 24–7. **Break bulk cargo vessel loading at terminal**

Container vessels

Container ships carry a variety of products in *intermodal containers* or tanks (fig. 24–8). Containers are usually steel-, aluminum-, or fiberglass-covered boxes with wooden floors. Little indication from the outside of the container will give clues to what is contained inside. Hazardous materials containers will usually be placarded. Container vessels use TEU (twenty-foot equivalent units) to measure their capacities.[4] Any combination of lengths may be used to determine a TEU, for example, one 40-ft container equals 2 TEU, while one 20-ft container equals 1 TEU. Tank containers are supported by steel framing that may be a box frame or beam frame. The majority of these containers are 20 ft long. The loading arrangement and location of containers on a ship may create problems for fire department personnel when responding to a fire in a container. Containers may have to be offloaded to ensure access to the container on fire. Offloading will take cooperation from the crew and shore-based facility personnel. Container cranes or unloading systems will have to be utilized.

Fig. 24–8. **Container ship** *(Courtesy of R. Callis)*

Barges

A barge is a smaller, flat-bottomed vessel used to transport products on smaller waterways. They are usually not self-propelled and are pushed or pulled by other vessels such as tug- or towboats (fig. 24–9). Barges may contain any product that can be carried on larger vessels such as containers, liquid products, or dry bulk materials. Many times, barges are found alongside industrial facilities on rivers either supplying fuel to the facility or loading/offloading product.

Fig. 24–9. **Tugboat and barge** *(Courtesy of R. Callis)*

Tugboats and towboats

Tugboats and towboats are small but very powerful vessels that are used to move another vessel, barge or groups of barges. They do this by either pushing or pulling. Both tugboats and towboats have low *freeboards*.

Tugboats are primarily used to assist larger vessels in docking and moving through confined waterways or narrow channels (fig. 24–10). They are also used to move barges. Many tugboats contain fire monitors that can assist with firefighting efforts on other vessels. Tugboats classified as firefighting vessels may have fire water flow capacities from 10,000 to 42,000 gpm (fig. 24–11).[5]

Fig. 24–10. **Tugboat** *(Courtesy of R. Callis)*

Fig. 24–11. **Firefighting tugboat** *(Courtesy of R. Callis)*

Towboats are used primarily to push a group of barges on rivers or intercoastal and inland waterways (fig. 24–12). If the engine room is flooded on a towboat or tugboat, it usually means the vessel will not remain afloat. If the vessel begins to take on seawater over the side because of the low freeboard, unless all watertight doors and hatches are secure, it will mean rapid *listing*, flooding, and sinking.

Fig. 24–12. **Towboat** *(Courtesy of R. Callis)*

VESSEL CONSTRUCTION

A vessel is constructed using a framing system that starts with the *keel*, which runs the length of a vessel's bottom. Structural frames are added that rise from the keel and are perpendicular to it. These structural frames are placed fairly close together and are further supported by longitudinal stiffeners that run the length of the vessel. The outer shell of the vessel is attached to this framing system and further gives the vessel strength. Depending on the type and size of the vessel, the framing and outer shell may be steel, wood, aluminum, or other composite material.

The framing system of large vessels includes numbering to provide a ready reference to locate spaces within the vessel and for determining direction once inside. Frame numbers generally begin at the *bow* and get larger toward the *stern* of the vessel. However, other numbering systems may be present, for example, stern to bow (fig. 24–13), middle toward both ends, or numbers preceded by a letter indicating that the vessel was lengthened by inserting a new section in the middle. The incident commander (IC) must be aware of the numbering system in use for the particular vessel.

Fig. 24–13. **Frame numbers (220 and 235) running from stern to bow**

The frame numbers are painted on the sides of these large vessels and can be used by ICs as reference numbers when reading the ship's *fire control plan* or general arrangement drawings. Firefighters inside of the vessel can find these frame numbers used in identification marking systems for passageways, cabins, fire stations, and the like, and can communicate their location to the command post for accountability or rescue purposes.

Ships are divided into vertical sections by bulkheads (walls). Bulkheads are usually attached to frames and may be numbered with the frame number or sequentially from the bow or the stern. These bulkheads also provide strength to the vessel and are used to separate spaces within the vessel. Bulkheads may carry different ratings and may provide protection against fire spread and flooding. Watertight bulkheads provide protection against flooding, while fire-rated bulkheads divide a vessel into vertical fire zones.

FIREFIGHTING CONCERNS

Shipboard fires and other emergencies, while not frequent occurrences, require knowledge of specialized firefighting as well as the hazard(s) involved. We have previously discussed some of the vessel types that municipal firefighters may encounter at industrial facilities, and in this section we will discuss the strategies and tactics involved. When fighting a shipboard fire, life safety is still our first priority. Within life safety, our firefighters' safety is our first priority. We should not commit firefighters unless they are fully aware of the hazards involved and have been specifically trained in marine firefighting and the conditions that they will encounter.

Hazards encountered

Some of the hazards that will be encountered during ship fires and are common to most if not all ships are as follows:

- Complicated interior arrangements
- Limited access
- Unfamiliar surroundings
- Flammable vapors
- Large fuel quantities
- Rapid heat development and high heat conditions
- Heavy smoke conditions
- Steam generation from fire department streams
- Locked access due to security provisions on board the vessel
- Narrow passageways
- Steep ladders and stairs (also called ladders) (fig. 24–14)
- Deck openings
- Confined spaces with oxygen-deficient atmospheres
- Limited radio communications
- High electrical currents
- Moving and rotating machinery and equipment
- Vessel movement (which can cause doors and hatches to close unexpectedly, causing injury or entrapment to firefighters)
- Large open areas

- Hazardous materials
- Compressed gases
- Oxygen-deficient spaces due to gaseous inerting

Fig. 24–14. **Steep ladders**

Unified command

As with all firefighting operations, an incident command system (ICS) must be implemented. However, with a shipboard fire, a unified command system is a must. There will be many players involved with the mitigation of a marine incident. These persons and organizations include (fig. 24–15) the following:

Coast Guard. The Coast Guard Captain of the Port has responsibility for port security and safety. The senior Coast Guard representative on the scene will act as the on-scene representative of the Captain of the Port. In this capacity, he or she will have overriding authority on all issues concerning port security and safety. He or she will also be able to assist with obtaining of resources necessary to effectively mitigate the incident as well as providing technical assistance.

Facility representative. The facility representative, for example, plant superintendent, or foreman, will have information on the status of systems within the facility that can be used for firefighting efforts such as fixed fire protection systems, water supplies, and semifixed systems. They will also have information on the process being undertaken at the time and whether the situation will affect plant operations to the extent they become an additional hazard. If the vessel was loading or unloading, the facility representative would have information on this aspect of operations. In addition to the facility representative, the facility fire chief should also be included.

Vessel's officers. Consultation with the vessel's officers will be necessary when making decisions regarding firefighting on board a ship. Just as has been discussed in other chapters concerning industrial facilities, certain information regarding the status of the vessel's systems will need to be taken into account by firefighting personnel. The familiarization of the ship's systems is another factor where the vessel's officers will be invaluable. They will be able to assist with the reading and interpretation of the ship's fire control plan. They will also assist with communications with personnel on board their vessels and provide additional information as necessary.

Vessel's owner representative. Many times, the captain of the vessel will be the initial representative of the vessel's owner. During the incident, the vessel's owner may send another representative, such as an agent or lawyer, to assist. These persons should be included in the decision-making process regarding the operations on board the ship.

Other parties. Other parties, such as the Environmental Protection Agency (EPA) representatives, salvage representatives, stability experts, and state and local emergency operations staff, may be part of the unified command system either as full partners or in advisory roles. Remember, at a marine incident, the fire chief cannot act alone. He or she will need assistance. Third-party marine incident specialists may be called in. These specialists are experts at extinguishing shipboard fires, stabilizing the vessel, and performing salvage if necessary.

Fig. 24–15. **Many participants will be involved at marine incidents, including the vessel's representative as well as the Coast Guard.** *(Courtesy of Val Pamboukes)*

Command post

The location of the command post should be in an area that is not on the involved ship. It should be large enough to accommodate the key players and have access to many forms of communication. The command post should have an area where the ship's and industrial facility's plans can be spread out and reviewed and where emergency response plans can be reviewed.

Strategic and tactical considerations

Fighting a marine fire at an industrial facility will involve the basic strategic plan used in other types of fires:

- Protection of life
- Prevention of extension and exposure protection
- Confining the fire
- Extinguishment

There are, however, items that need to be specifically addressed. These items are unique to this type of fire.

Fire extension. Due to the nature of the construction of most vessels, that is, steel construction, fire extension must be a strong consideration for fire commanders. Vessels will rapidly transmit heat via conduction, convection, and radiation on all six sides of the fire. The six sides include the four lateral sides, the area above, and area below. These areas must be identified and assessed, and fire boundaries established. When assessing these areas and establishing boundaries, the removal of exposed combustibles and the cooling of exposed metal should be accomplished. Ventilation should be secured, and electrical power to the affected spaces must also be disconnected. When cooling surfaces, use as little water as possible, and this water should be applied as quickly as possible. Only use as much water as is required to cool the surface and produce steam. Once steam is no longer being produced, cooling should be discontinued to prevent unnecessary water accumulation and the *free surface effect* that can affect stability.

Stability. Stability will have to be addressed throughout the incident. One of the first things the incident commander must do is to establish baseline stability readings. This can be done using *draft marks* on each side of the vessel (fig. 24–16), both *fore* and *aft* as well as readings from the inclinometer located on the bridge (fig. 24–17). Readings should then be taken at defined intervals, but at least every 30 minutes, and recorded in a log book. Stability specialists will need to be consulted to help determine these data. Dewatering operations will have to be considered at an early stage. Portable pumps will be needed in most dewatering operations when fire-water streams have been used. *Ballast* or cargo may have to be transferred to counteract list. This transfer will need to be performed by the ship's personnel using on-board pumping systems or other cargo-handling systems.

It needs to be remembered that every gallon of water added to a vessel by firefighting and/or cooling

Fig. 24–16. **Draft marks**

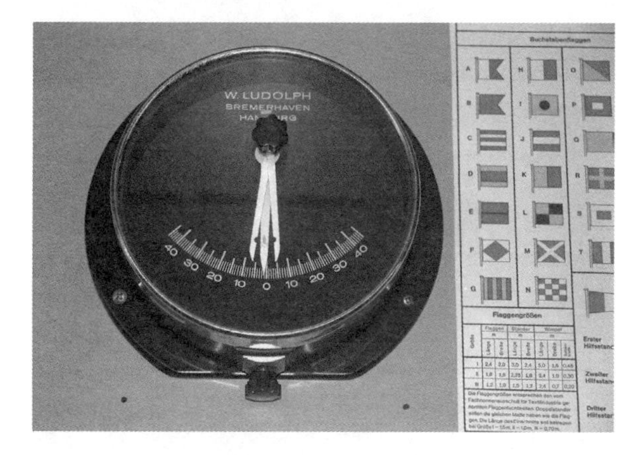

Fig. 24–17. **Inclinometer**

lines adds 8.3 pounds of weight. If this weight is not distributed evenly, it will cause the ship to list. This listing will place strain on mooring lines, cause doors and hatches on board the vessel to close or open, cause items that are not secured to move, and make the footing of firefighters hazardous. Eventually, it may cause the vessel to roll over.

Access. Access to the vessel will be limited during the best of conditions. There may only be one usable gangway available to firefighters. The gangway is narrow and may only be used by one or two firefighters at a time (fig. 24–18). Firefighters trying to board the vessel will be competing with those individuals trying to exit. For firefighter safety, a secondary means of egress must be established. In other instances, when the vessel is away from the pier, firefighters may have to board using the Jacob's ladder (fig. 24–19). Secondary means of access and exit can be accomplished using portable or aerial ladders. These devices must not be secured to the vessel. In case of vessel movement, damage will occur. It is best that ladders be monitored continuously. Aerial ladders and platforms that are touching the vessel can be severely damaged by the movement of the vessel. Once these ladders have been used, operators should move them a slight distance away from the railing but maintain a constant vigil of the area in case firefighters need to make a hasty retreat using the device. The gangway on the opposite side of the vessel that was used for primary access should be lowered also, and rescue boats should be available in case of emergency evacuation. Because of security concerns, the vessel's access to the inside will be secured either from the inside or outside. Security measures must be accounted for, and consultation with the ship's crew will be required to gain their assistance to open locks.

Fig. 24–18. **Gangway and Jacob's ladder** *(Courtesy of R. Callis)*

Fig. 24–19. **Firefighter using Jacob's ladder**

Logistics. Marine operations will involve comprehensive logistical planning. These events will require large numbers of self-contained breathing apparatus (SCBA) cylinders and possibly refill capabilities. Tools and equipment may have to be stockpiled on board the vessel so that firefighters who need the equipment can access it in a short amount of time. Specialized cutting torches for steel, thermal-imaging cameras, search ropes, additional radios, and additional hose and nozzles are just a few of the items that may have to be placed on the vessel for the use of firefighters. Foam supplies are another item that may need to be stockpiled on the ship during operations. High-expansion foam can be used in some instances involving cargo holds or engine rooms. Adequate supplies must be on hand prior to the start of foam operations. Remember, due to the large size of some vessels, delivery of any equipment requested will have a lag time because of the distances involved, even if the equipment is already located on shore at the incident. Incident commanders should anticipate the needs of the operation and have equipment standing by and available.

Water supply. Water supply may be a great concern to the firefighting forces at marine incidents. While vessels will have their own firefighting water supply, fire departments should use their own. This means that departments will use the industrial facility's hydrant system, the municipal hydrant system, or portable pumps or fire apparatus to draft from a water source where the incident takes place (fig. 24–20). Drafting sites will have to be carefully considered. Remember, the rise and fall of the *tide* may affect the drafting abilities of your pump or pumper. Portable pumps have been used in the past by placing them on tugboats or barges and supplying water to marine incidents, but this must be planned and practiced in advance (fig. 24–21). Fireboats are another source of water supply. Manifolds supplied by 3-inch hose and placed on the deck of the vessel have proven effective in the past to provide water to the firefighting forces on board. Five-inch supply has proven too heavy for this type of supply unless supported by the gangway. This latter method of supply will only interfere with access to the ship and may actually hinder operations.

Fig. 24–20. **Drafting using portable pump**

Fig. 24–21. **Portable pump placed on vessel to supply water for fire operations**

On vessels of 500 gross tons and greater there shall be at least one international shore connection to the fire main available to each side of the vessel in an accessible location. Suitable adapters must be available for furnishing the vessel's shore connections with couplings mating those on shore fire lines. Vessels greater than 500 gross tons must also be provided with at least one *international shore connection.*

These connections can be used if the local fire department also has one that matches their hose fittings (fig. 24–22). With the appropriate international shore connection, a municipal fire department would be able to connect to a ship's fire main to supply the water system. It is recommended that fire departments connect to the ship's fire-water system, just as they would connect to a standpipe system in a high-rise structure. This will support any firefighting operations that are using the ship's fire-water or water-based fixed system in case there is a loss of pumping ability by the ship's fire pumps.

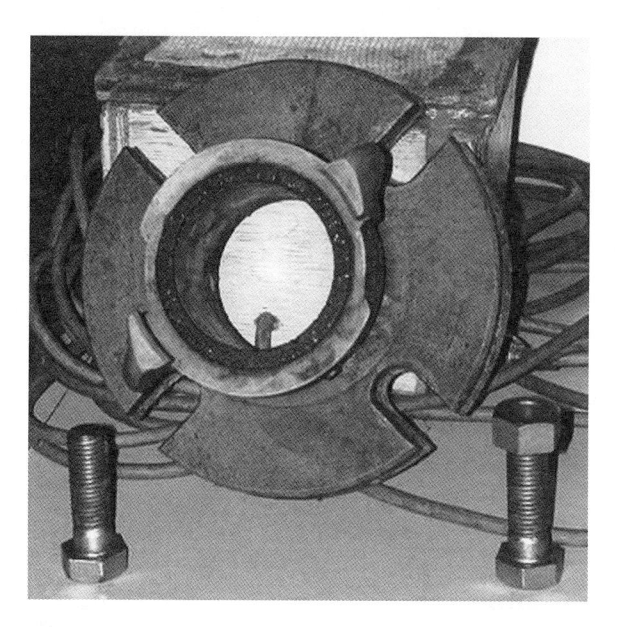

Fig. 24–22. **International shore connection (Courtesy of Val Pamboukes)**

Accountability. At all incidents accountability must be established for firefighters. At marine incidents, accountability will be a challenge due to the large scale of the operations. Accountability will initially be set up at the gangway to the vessel, where the majority of personnel board the ship. Where other access points are used, additional accountability officers will need to be stationed. Also, where personnel enter the ship from the deck, accountability will need to be established. Advise personnel that where they enter the vessel is where they should exit. If for some reason they exit at a different location, then they must report to the accountability officer at the original location that they have exited the ship.

Rapid intervention teams or crews (RIT/RIC). As with any incident, rapid intervention teams/crews must be established. Due to the size and complexity of these incidents, additional RIT/RICs must be established at any and all points of entrance to the vessel where personnel have entered. RIT/RIC equipment must be deployed with the team and the teams located on the vessel. Because of the additional danger of falling overboard, a water-side RIT/RIC should be established on a safety boat that patrols the outboard side of the vessel. This team may also be used to gather data on stability for the incident commander. This team should be equipped with the tools for water rescue, including throw ropes and life rings.

Fire control plans. In all ships, general arrangement plans will be available showing the fire stations; fire sections enclosed by fire divisions; and information on the fire detection, fire alarm system, sprinkler system, fire extinguishing appliances, means of access to different compartments, and the ventilation system including dampers and fan control positions. These plans shall be permanently exhibited for the guidance of the crew, but a duplicate set will be permanently stored outside of the *deckhouse* in a weather-tight container for the use of shore-side firefighting personnel (fig. 24–23).

Fig. 24–23. **Fire control plans in container on deckhouse bulkhead**

These plans should be used by fire commanders to assist with the strategic and tactical plans for mitigating the incident. The assistance of a person knowledgeable with the information contained on the plans should be sought by the incident commander.

Preincident response planning. It may be difficult to preplan every individual ship that you may come into contact with at an industrial facility, but departments can preplan for specific types of ships that may dock at industrial facilities. Departments can preplan for cargo type, ship construction, loading and offloading operations, and general marine firefighting. If plans can be obtained for every ship that is expected to call at an industrial facility, then the fire department is that much ahead of the learning curve because it will have detailed information for a particular ship. With today's security measures and operational security issues, this may not be supported by the management of vessels that arrive in your ports.

Specialized tactics. One of the specialized tactics that should be considered at marine fires is the use of foam. High-expansion foam generators for engine rooms or cargo spaces or using cellar nozzles supplied with aqueous film-forming foam (AFFF) solution for engine room spaces have been used in the past to successfully extinguish these difficult fires. A cellar nozzle lowered into an engine room can be a very effective tool in combating these hard-to-reach fires without committing firefighting personnel below decks. This technique of using cellar nozzles and AFFF has also been used to provide a foam blanket to large spills in pump rooms in tankers.

The use of thermal-imaging cameras during marine fire operations will be invaluable. The cameras can be used during interior operations in heavy smoke conditions, but can also be used on the exterior to locate hot spots that may pinpoint the fire's location on the interior. By comparing the hot spot's location with the general arrangement drawings of the ship, an idea of where the fire is located on the interior will be obtained. It can also pinpoint areas that will need boundary cooling.

The construction of a ship inherently creates ventilation difficulties because of many below-deck spaces, narrow passageways, and interior compartments with no direct access to the outside. Positive pressure ventilation (PPV), if used correctly, can help to ventilate the space as well as assisting the advancing firefighters as they move toward the seat of the fire. As with all PPV use, adequate and controlled openings must be provided opposite the fan; otherwise, the PPV can spread the fire or push fire to areas where firefighters may be operating, possibly with disastrous effects.

Use of on-board extinguishing systems must be considered. Systems such as steam flooding, CO_2 flooding, or inert gas systems may be used to successfully extinguish fires. Again, consultation and close cooperation with the vessel's crew is necessary. When using a system to exclude or deplete oxygen, ensure that the area to be flooded has been sealed. If the extinguishing agent is allowed to escape, then its effectiveness will be compromised.

Long-duration SCBA is a necessity. One-hour cylinders should be the minimum. Provisions for refilling on board the vessel should be considered.

Training

Fire departments that may respond to a marine incident should attend specialized training courses and seminars to be adequately prepared for such a response. Hose-handling techniques will be different, specialized tactics will be required, and command and control will be much different than at structural fires. The large scale of a marine event and the specialized knowledge required dictates that specialized training be conducted. The NFPA is currently developing the *Standard for Professional Qualifications for Marine Firefighting for Land-Based Firefighters* (NFPA 1005). Firefighters and officers who may be required to fight a marine fire should be trained and certified to this standard upon its approval and publication.

Training in marine firefighting is available at major fire training centers as well as through seminars and symposiums throughout the United States. Private nonprofit groups such as the Delaware River Marine Firefighting Task Force also offer courses in marine firefighting.

MARINE TRANSFER AREAS

Marine transfer areas are defined as the part of a waterfront facility between the vessel, or where the vessel moors, and the first shutoff valve on the pipeline immediately before the receiving tanks. These areas may have their own fire protection equipment such as fire monitors or fixed/semifixed foam systems, or under-pier sprinklers in case of spills and fires. The operation and reason for these systems should be known and understood by municipal departments responding to an incident involving marine transfer areas. In the case of a spill, the terminal operator may not be willing to activate such a system to control vapors. They should be encouraged to use foam to cover spills and suppress vapors. The municipal department may have to augment such systems with foam or water, depending on the system, weather conditions, or size of the spill. Weather, including winds and rain as well as sea conditions may either disperse or spread spills and affect foam application.

Fire protection and prevention issues within the marine transfer area include electrical power systems, transfer hoses and associated piping, lighting, communications, gas detection systems, and firefighting equipment/systems. These areas should be inspected and preplanned when performing preincident response planning for the facility. Because of their remote locations and waterfront setting, access for the fire department will be difficult. Narrow piers may be the only access to the area. Fireboats or portable pumps may be the closest water supply.

SHIPYARDS

Introduction

Our discussion of shipyards will include construction and repair facilities. By their nature, shipyards include many different industrial processes and structures. When a municipal fire department is called to respond to a shipyard, there is no guarantee as to what will be found on arrival. Every shipyard will present its own unique hazards and challenges to the firefighting force.

Fire hazards and concerns

The fire hazards associated with shipyards are numerous and can be challenging during fires. If we look at the work being performed, we will see that there may be many noncompatible processes taking place in close proximity to one another. There may be woodworking or fiberglassing taking place under the same roof as metalworking. Combustible dusts may have been created by work being performed, thus adding to the fire hazards present. These facilities usually have cutting and welding operations, with their inherent danger from open flames and sparks. High-voltage electrical systems will be present to run electric welders and large machinery. These facilities may have their own power generation stations (see chapter 30). In many cases, housekeeping is not the best, and the poor housekeeping will contribute to the fire's start as well as its spread.

Shipyards may use large quantities of paint and other finishing materials, which create a fire hazard. When using paints in large quantities, inevitably flammable solvents will also be present. These solvents can create flammable atmospheres if not used or stored properly.[6]

Shipyards may be located on a large area inland of a wide body of water or on a narrow plot of land along a limited width river or other tributary (fig. 24–24). These geographical locations will limit access by the local fire department (fig. 24–25). Locations within the facility may create water supply problems and necessitate long stretches of hose from remote locations or pumpers for drafting. Portable pumps, fireboats, or tugboats with pumping capabilities should also be considered as water sources.

Fig. 24–24. **View of shipyard from water side**

Fig. 24–25. **Locations of shipyards may limit access**

Since the original construction of a shipyard, the occupancy and use may have changed over the years due to economical reasons. Buildings may be subdivided and used for other types of occupancies. The clearances between buildings, combustibility of the structure, and proximity of structures housing incompatible processes are all factors that will have to be considered at shipyards. Many of the older shipyards have been subdivided and industrial parks created, changing the primary occupancy of the location. Only a small portion of the shipyard may actually be used for marine work. The buildings that were subdivided may contain occupancies for which the building was not designed, or whose original fire detection and extinguishing systems are not suitable for the new occupancy. This should be identified during inspection visits, and corrective actions taken by the fire department. All relevant information will need to be included in preincident response plans. The subdivision of the shipyard may create access problems as well as water supply problems. Hydrants that are part of the yard hydrant system, and even municipal water supplies, may be separated from the area by fencing or walls. During a fire, hydrants that should be available, based on the original design of the facility, may in fact no longer be available.

Fig. 24–26. **Vessel in dry dock**

Other problems that should be associated with shipyards are the state of the vessels in the yard being repaired as well as the access problems associated with these vessels. During repairs, a vessel's fire-extinguishing system, including the fire-water main and fire detection systems, may be out of service. Other systems such as the electrical or water systems may also be out of service. If the vessels are in a dry dock, then access will be very difficult (fig. 24–26). The only method to reach the vessel will be gangways from the dry dock to the vessel, which severely limits access for firefighting. Fuel or products may have been removed from on board the vessel. Some dry docks may be made of wood, adding to the exposure problem during a fire on board a vessel in the dry dock. In the case of tanks, they will have to be purged of vapors by either the shipyard personnel or the ship's personnel and made gasfree. Intrinsically safe equipment must be used and atmospheres monitored. Other fuel and products may be left on board the vessel, creating additional problems in case of fire. Tanks and/or cargo holds will be considered as exposures in case of fire. During repairs to vessels, incompatible trades operating in close proximity to one another may create a fire hazard, for example, open flame work being performed in the area where painting or the

Fig. 24–27. **Vessel in dry dock (Note the scaffolding.)**

Fig. 24–28. **Moving cranes may be present. Note the railway track for the crane and the position of the fire chief's vehicle.**

use of solvents is taking place. Also, the platforms of the scaffolding in the dry dock (fig. 24–27) are usually wood, and sparks from cutting operations have been known to set scaffolding on fire.

Typically, shipyards contain large machinery, including movable cranes and conveyors (fig. 24–28). Within structures, large forklifts fueled by flammable and combustible liquids or liquefied petroleum gas (LPG) should be anticipated. This machinery can move at any time due to operators not realizing firefighters are present, operator error, or equipment malfunction, causing a hazard to firefighters, or by moving and cutting hoselines. Machinery and equipment may also present rescue scenarios during a fire department response or be the cause of a fire themselves. Shipyards and marine repair also have many confined spaces and high-angle areas that will have to be confronted, either during firefighting or rescue.

Furthermore, shipyards will contain piers and bulkheads that are usually made of wood. Flammable floating debris can accumulate under or around these piers and bulkheads that can create a fire hazard. In case of fire, the areas under the pier or around the bulkheads may not be accessible to conventional firefighting streams. Streams operated from boats or from under-pier firefighting teams may have to be used. Cutting holes in the top of the pier, then inserting cellar nozzles is effective in fighting under-pier fires.

CONCLUSION

Shipyard and marine fires present not only the unique firefighting challenges that all industrial facilities present but also the unique challenges of vessel fires. As with any industrial fire, preincident response planning and drills are essential, as well as knowledge of the hazards present and specialized firefighting tactics, especially when a vessel is involved. These fires, whether they are at shipyards or on board ships, will usually present access and water supply problems for the fire department. It is through preincident response planning and training that any obstacles will be overcome so that a successful firefighting operation can be accomplished.

CASE STUDY

In July 2001 a fire occurred aboard a 950-ft long commercial container ship carrying military munitions at Sunny Point, North Carolina. The fire was located in the engine room of the ship and was initially fought by crews from a Department of Defense career fire department. The fire was located at a military installation, and the ship involved was chartered to the Military Sealift Command. Mutual aid was provided by surrounding volunteer departments throughout the 6 hours necessary to bring the fire under control and the many more hours necessary to finalize extinguishment and overhaul duties. Throughout the incident, 300 to 350 firefighters were engaged in the firefighting efforts, and one firefighter called the operations "organized chaos." Firefighters made repeated entries into the ship and required extensive rehabilitation after making an entry. Accountability of firefighting forces became difficult as fire departments from surrounding areas reported to the scene with different accountability systems. Eventually, one deceased victim was located in the engine room during search operations.[7]

Random thoughts

- Ship fires are extremely labor intensive. Early requests for additional personnel should be made early so that initial firefighting crews can be relieved.
- Large numbers of breathing air cylinders will be required because of the high heat and smoke conditions encountered at ship fires. Provisions for SCBA refill at the scene should be accounted for during the preincident planning activities.
- Long hose stretches will be necessary. Consider using 3-inch hose stretched to portable manifolds on deck to supply handlines.
- Thermal imaging cameras for locating the fire and for search activities is a must. At the incident described earlier, thermal cameras were used and proved extremely beneficial.
- Breathing air management techniques and accountability are required at ship fires. Because of the many departments that may respond, ensure that all personnel are familiar with the accountability systems in place in neighboring departments. It is preferred that all departments within a region use the same accountability system. If not, then during regional drills, the various accountability systems in use should be demonstrated so that departments become familiar with all the systems.
- When entering a ship, it is recommended to use search/guide ropes to assist with entry and exit from the ship.
- Units that are coming out of a ship after operations should brief the crews making entry. Fire location, conditions present, routes of entry, and the like, are all items that must be discussed. The exiting crews should draw a map on paper or on a bulkhead to visually demonstrate the fire area and its surroundings.
- Remember, a ship fire has six sides. All sides should be examined for extension. The steel construction of a ship contributes to a rapid fire spread.

NOTES

[1] HM Fire Service Inspectorate. 1999. *Fire Service Manual, Vol. 2: Operations, Marine Incidents.* London: The Stationary Office, p. 15.

[2] International Fire Service Training Association (IFSTA). 2001. *Marine Firefighting for Land-Based Firefighters.* Stillwater, OK: Fire Protection Publications, p. 60.

[3] IFSTA, *Marine Firefighting for Land-Based Firefighters*, p. 63.

[4] IFSTA, *Marine Firefighting for Land-Based Firefighters*, p. 67.

[5] American Bureau of Shipping (ABS). 1981. *Guide for Building and Classing Fire Fighting Vessels.* New York: ABS.

[6] National Fire Protection Association (NFPA). 1990. *Industrial Fire Hazards Handbook.* Quincy, MA: NFPA, pp. 507–508.

[7] The information that this case study is based on is from the article Disaster Averted at Sunny Point in *The State Port Pilot.* Retrieved July 20, 2001, from http://www.sppilot.com/front/shipfire.html.

25

Storage of Liquified Petroleum Gas

INTRODUCTION

The most common stored flammable gas is liquefied petroleum gas (LPG; see fig. 25–1). In developing fire protection methods and guidelines for liquefied flammable gases, the chief concern is a massive failure of a vessel containing a full inventory of LPG. The probability of this type of failure occurring can be mitigated, or at least controlled to a reasonable and tolerable extent, with appropriately designed and operated facilities, coupled with a local fire department/brigade response. Since most LPG fires originate as smaller fires that become increasingly more dangerous, this chapter will focus on fire protection methods and guidelines in relation to small leaks and fires in LPG spheres. Of greater importance to the firefighter is the more likely event of a leak from a pipe, valve, or other attached component leading to ignition, flash fire, pool fire, and eventually a pressure fire at the source.

Fig. 25–1. **LPG storage vessels**

Liquefied petroleum gas was first discovered in the 1900s. The applications and uses of LPG, which range from cooking and refrigeration to transportation, heating, and power generation, make it an all-purpose, portable, and efficient energy source. LPG consists of light hydrocarbons (propane, butane, propylene, or a mixture) with a vapor pressure of more than 40 psi at 100°F. At standard temperature and pressure, LPG is in a gaseous state. LPG is liquefied by moderate changes in pressure (i.e., in a process vessel) or a drop in temperature below its atmospheric boiling point. The unique properties of LPG allow for it to be stored or transported in a liquid form and used in a vapor form. LPG vapors are heavier than air and tend to collect on the ground and in low spots. After LPG is released, it readily mixes with air and could form a flammable mixture. As a release occurs, there will be an area closest to the release that is above the flammable range, an intermediate area that may be in the flammable range, and areas that will be below the flammable range. Mixing via natural currents and diffusion of LPG vapors affect the size and extent of these areas. If these processes continue, eventually the mixture is diluted to below the lower flammable limits (LFL).

Other characteristics of LPG include the following:

- LPG exerts a cooling effect as a result of vaporization due to releases at low pressure (called autorefrigeration).
- The density of LPG is almost half that of water; therefore water will settle to the bottom in LPG.

- Very small quantities of liquid will yield large quantities of vapor.
- When vaporized, LPG leaves no residue.
- When LPG evaporates, the autorefrigeration effect condenses the moisture from the surrounding air, causing ice to form. This is usually a good indication of a leak.
- LPG is odorless; therefore, agents such as ethyl mercaptan are added to commercial grades in most countries for better detection.

DESCRIPTION OF OCCUPANCY

LPG is derived from two main energy sources: natural gas processing and crude oil refining. When natural gas wells are drilled into the earth, the gas released is a mixture of several components. For example, a typical natural gas mixture may be (90%) methane or "natural gas," while the remaining percentage of components (10%) is a mixture of propane (5%) and other gases such as butane and ethane (5%). From there, the gas is shipped in tankers or via pipeline to secondary production facilities for further treatment and stabilization. From these facilities it is sent by bulk carrier or pipeline to various industrial plants and gas-filling facilities or used for power generation.

LPG is also collected in the crude oil drilling and refining process. LPG that is trapped inside crude oil is called associated gas. The associated gas is further divided at primary separation sites, gas oil separation plants (GOSPs) and or central processing facilities (CPFs). At these facilities, the produced fluids and gases from the wells are separated into individual streams based on their characteristics and properties and sent on for further treatment.

At refineries, LPG is collected in the first phase of refinement or crude distillation. The crude oil is then run through a distillation column, where a furnace heats it at high temperatures. During this process, vapors will rise to the top, and heavier crude oil components will fall to the bottom. As the vapors rise through the tower, cooling and liquefying occurs on *bubble trays,* aided by the introduction of naphtha. Naphtha is straight-run gasoline, and the heavier naphtha is generally unsuitable for blending with premium gasoline. Therefore, it is used as a feedstock in various refining processes such as in a reformer. These liberated gases are recovered to manufacture LPG.

In commercial applications, LPG is usually stored in large horizontal vessels called *bullets.* These bullets can range in volume size from 150 to 50,000 gallons (fig. 25–2). In industrial applications, LPG is typically stored in large vessels that are spherical or spheroid shaped. These are the large golf-ball shaped and oval vessels commonly seen at refineries and other similar occupancies. In this chapter, we will deal primarily with LPG spheres.

Fig. 25–2. **LPG bullets**

COMMON HAZARDS

To reduce the fire risk at LPG facilities, adherence to various design considerations and requirements such as layout, spacing, distance requirements for vessels, drainage, and containment control will help to limit the extent of fire damage. Additional considerations such as fireproofing, water draw systems, and relief systems are also important with respect to the integrity of the installation and the reduction of risk. These considerations address the various ways to prevent leaks or releases that may lead to a fire.

Equally important to the prevention of a leak or release is a properly designed, installed, and maintained fire protection system. These systems attempt to minimize or limit the fire damage once a fire occurs. In the event that a fire does occur, the levels of required fire protection are affected by several factors such as location and remoteness of the fire and the availability of water.

Pool fires are simply spill fires that have accumulated on the ground or solid surfaces to form pools of burning liquids. These can cause flame impingement (if ignition occurs) to the bottom or belly side of the tank. (See chapter 13 for details.)

When flammable gases leak from an LPG sphere and are ignited, the flames will form a jetting pattern. This condition is referred to as a *jet fire*. Jet fires are intense and often impinge on adjacent equipment and structures. The source of jet fire fuels should be isolated, and the jet fire itself should be allowed to burn out when possible. (See chapter 13 for details.)

A vapor cloud explosion (VCE) is one of the most dangerous hazards associated with these types of facilities. Normally, the response to this type of incident occurs after the fact and very rarely during or before. VCE is defined as an explosion occurring outdoors that produces damaging overpressures by the unplanned release of a large quantity of flammable vaporizing liquid or high-pressure gas from some form of containment. (See chapter 13 for details.)

The other dangerous hazard that exists at LPG storage facilities is the potential of a boiling liquid expanding vapor explosion (BLEVE). This sudden release of a large mass of superheated liquid from a pressurized vessel can create a devastating explosion. (see chapter 12 for details)

FIREFIGHTING TACTICS

Even with the proper installation of fixed fire protection systems, the importance of emergency response to LPG fires cannot be disregarded. LPG fires can escalate quickly, and a lack of manual activities by the fire department can lead to vessel failure. As part of this response, an up-to-date and complete preincident response plan is essential. The plan should include

- Facility name and location
- Map of facility
- Emergency phone numbers for key plant personnel
- Hydrant layouts and capacities
- Additional water supplies such as ponds and canals (Are they available in freezing weather conditions?)
- Hose lays and lengths required
- Multiple response approaches (wind dependant)
- Vessel inventories
- Fixed fire protection information
- Scenarios for both unignited and ignited leaks

Preincident response plans should identify the emergency response structure of the industrial plant as well as the incident/unified command structure that will be used. In some instances, a plant operations person may be acting as the incident commander (IC), with the municipal/town/volunteer department operating in a support role. Better to have this organized during the preincident response planning stage than on the fire ground during an actual incident.

Preincident plans must be exercised on a frequent basis and updated as necessary. Since recent events throughout the world have increased security in industrial plants, obtaining information for preincident response planning may be difficult. It is imperative that discussions with plant personnel remain open and take place frequently to maintain a team spirit and facilitate information sharing. Once information is obtained, it will be important for fire departments to maintain operational security for this information. Preincident response plans should be secured on the apparatus, and numbered copies tracked. These copies should be inventoried on a scheduled basis to maintain operational security.

When responding to an LPG facility, preincident response plans may indicate that responding personnel remain at a staging area, usually near the main gate or other location (depending on wind direction) until directions are received from plant personnel, especially in the case of a leak. The authors know of instances where vehicles were the source of ignition for a leak. In one case, although not an LPG incident, the fire department apparatus was the source of ignition for a natural gas explosion that destroyed a number of city neighborhood blocks. In another instance, a plant security vehicle investigating a reported leak was the source of ignition, causing a catastrophic loss to the facility.

On arrival, initial assessment of the situation is essential for the safety of personnel. During a fire, if there has been flame impingement on a vessel, especially on the vapor space with no water cooling of the area or fireproofing on the area of flame impingement, vessel failure could be imminent. If these conditions have been present for 10 minutes or more from the initial impingement (not the fire department arrival), then an immediate evacuation is recommended. It is important to remember that the initial time of the notification to the fire department may not be the time the incident started.

If it has been determined that a fire is safe to approach, an incident action plan (IAP) should be developed. This plan does not have to be initially written, but it should contain the following objectives:

- Cooling of exposed storage vessels
- Water application rates and water supplies available
- Shutting down fuel supply
- Monitoring of surrounding area using combustible gas indicators (CGIs)
- Evacuation of nonessential personnel
- Evacuation routes for responders and plant personnel in case of emergency

To determine if cooling water is required, the anticipated radiant heat flux from an adjacent tank, maximum tank shell temperatures if the vessel shell is not cooled, and other specific risk management guidelines must be analyzed (fig. 25–3). API 2510A, *Fire Protection Considerations for the Design and Operation of Liquefied Petroleum Gas (LPG) Storage Facilities* contains a procedure to identify the point at which cooling water should be applied based on the size of the pool fire and the distance between the vessel and the center of the fire. Additionally, an analysis of the relief valve parameters is necessary to maintain certain internal vessel pressures. Although computer models are available to more accurately anticipate the heat fluxes, this procedure helps to determine if a more detailed study is required. Basically, the procedure considers the radiant heat flux from a pool fire, assuming a 20-mile-per-hour wind and determines if the heat flux on the adjacent tank requires cooling.

Fig. 25–3. **Radiant heat on a pressurized storage vessel** *(Courtesy of Williams Fire and Hazard Control)*

In general, there are three primary methods that can be used to apply water for cooling or extinguishment to LPG vessels exposed to fire. The three methods are water deluge, water spray, and fixed monitors. Further, portable equipment such as ground and trailer-mounted monitors can be used, but should not be considered a primary means of water delivery. This is mainly due to the potentially extended setup times, logistics, and requirement of human intervention that is not necessarily reliable.

During the preincident response planning stages, methods of water applications for incidents should be evaluated. The first method involves the use of a water deluge system and some form of water distributor (fig. 25–4). This could include high-volume spray heads, perforated pipe, or a distribution weir. An underflow or overflow weir is a form of distribution weir that allows water to be evenly distributed over the surface area of a sphere by water flowing up the piping network, over the top of the sphere, and out of the weir. This type of water distributor is commonly used but is prone to corrosion from standing water and clogging; and requires increased preventive maintenance. Also, weirs may not be as effective on bullets and are often greatly affected by wind. The remaining components of this method are similar to other deluge installations. The typical deluge system contains a supply piping network, deluge valve and trim, and a branchline-piping network near the top of the sphere. Newer installations are usually activated automatically, whereas older installations are commonly activated manually. The decision as to which activation method to use requires evaluation of spacing, available protection, exposures, and other factors.

The principle behind the use of a deluge or weir system for LPG sphere protection is that the geometric shape of the sphere and gravity work together to an advantage. As water is applied to the top of the vessel, the shape of the sphere and the force of gravity facilitate the flow of the water as it covers the surface area of the vessel. This type of protection is very effective to facilitate an even distribution of water over the surface area.

Fig. 25–4. **Water spray on a sphere** *(Courtesy of Williams Fire and Hazard Control)*

Caution should be exercised, however, because paint, corrosion, dust, and other environmental influences can cause changes in the surface of the sphere, resulting in uneven water distribution. Further, settling and other conditions inside the weir can also cause uneven water flow over the sphere's surface.

The second method of application is the use of water spray systems. These systems comprise a piping network of spray nozzles that distribute water over the surface area of an LPG sphere. The spray nozzles are positioned to form a grid pattern that facilitates the complete coverage of the sphere's surface area. Larger orifices and piping should be considered to help reduce blockage due to scale and marine growth buildup. It is also important to properly size strainers to prevent blockage. Inspection of strainers should be part of the preventive maintenance program.

Vapor, heat, or flame detectors mounted in the vicinity of a vessel can complete automatic activation of these systems. Vapor detection provides early detection and warning, but activation of water application systems must be confirmed through flame detection. Flame detection provides quick activation, but use caution when positioning these detectors to prevent false activation from sunlight. Consideration need also be given to the installation of UV/IR combination detectors to reduce the false indication rate. These devices require testing and preventive maintenance programs. An evaluation of the facility is necessary to determine the correct type and location of devices. (see chapter 8 and 9 for details)

Fixed monitors, the third method of water application, permit the use of fixed hydrant-mounted monitors or individual monitors connected to the fire main to apply water to the fire area (fig. 25–5). In this case, water application is accomplished by operators manually opening valves to allow the flow of water to the LPG sphere. This procedure exposes operators to high heat fluxes and places them dangerously close to vessels under fire conditions. It is important to carefully study the plant and vessel layout if this method is elected. Proper placement, location, and quantity of fixed monitors must be reviewed and field-tested to ensure that proper application and even distribution of water to all parts of the vessel are accomplished. In some cases, remote activation and operation are suggested when proper spacing of monitors is not a possibility. Annual testing and preventive maintenance are necessary to ensure parameters have not changed and that coverage is still adequate.

Fig. 25–5. **Fixed monitors protecting spheres** *(Courtesy of Williams Fire and Hazard Control)*

The last method available is the use of portable monitors and hoses. This method uses hand-carried portable or trailer-mounted monitors deployed by the fire department. Although not one of the three primary methods of water application, preparations and planning for this type of application should not be forgotten. Quantity of monitors, monitor flow calculations, and predetermined hose lays should be reviewed prior to an incident to ensure that adequate capabilities are available. This method is considerably more dangerous than previously mentioned methods due to exposing personnel to the hazards and risks associated with LPG firefighting.

When utilizing the four water application techniques discussed previously, a combination of techniques provides ample fire protection, such as the use of a deluge or water spray system and portable monitors. A combination of a water deluge/distributor with a fixed water spray system with portable monitor backup from the fire department provides excellent coverage.

A water application rate for these fixed fire protection systems depends on the type of fire situation. When a vessel is exposed to only radiant heat without direct flame contact, a density of 0.1 gpm per square foot of vessel surface area is the minimum. If direct flame contact, or impingement, occurs, a density larger than 0.1 gpm, up to 0.25 gpm, per square foot of vessel surface area is the minimum.

When dealing with fixed or portable monitors, 250 to 500 gpm is the minimum flow that should be initially considered. However, field verification and flow testing is necessary to ensure that adequate and proper coverage is provided. Placement of monitors must also be field-verified against approved plans to ensure acceptable spacing and access.

In the event of a liquid pool that has not ignited, water spray to control/dilute the vapors is warranted. Water should not contact the spilled material where possible; this would increase the vaporization. Use combustible gas indicators to determine the extent of the vapors. Always remember to position fire apparatus upwind and uphill.

It should be noted that the control of the incident cannot be accomplished by the fire department alone. Expert advice, cooperation, and remedial actions by plant personnel will be required. Preincident planning and familiarization visits are the times to begin to foster this cooperation and identification of plant personnel who will be needed to assist the fire department.

CONCLUSION

Since most LPG fires originate as smaller fires that become increasingly more dangerous, three primary methods to apply water in a quick manner can help reduce the risk of LPG vessel failure. The deployment of portable monitors and hoses, although not one of the three primary methods of water application, is an important backup to the primary methods. LPG fires can escalate quickly, and a lack of manual suppression activities by the fire department can lead to vessel failure. It is necessary, however, to take control of the fuel source before attempting to suppress the fire. In any case, a preincident response plan, along with proper training and drills, is important to reduce the risk of injuries and promote a quicker and safer response.

26

Steel and Aluminum Manufacturing

INTRODUCTION

Like other large, complex occupancies, such as pulp and paper (chapter 19) and automotive (chapter 34) manufacturing facilities, steel and aluminum manufacturing facilities contain large, wide-open structures with multiple hazards (fig. 26–1). Steel and aluminum facilities, much like pulp and paper plants, are some of the oldest manufacturing facilities in existence. Iron, the raw material for steel, has been used for centuries. This chapter will discuss a few of the popular methods of reducing iron for steel manufacturing and the general process of aluminum smelting. The chapter will also address some of the finishing processes that are common to the industry. The intent is to provide the reader with some basic knowledge of the process and process terminology to aid in both emergency response planning and firefighting tactics.

Fig. 26–1. **Overview of steel plant**

DESCRIPTION OF PROCESS

The process of steel manufacturing can be broken down to two basic processes: *basic oxygen steelmaking* (BOS) and *electric arc furnace* (EAF). After the raw or crude steel is manufactured, it is then cast, rolled, and/or treated to meet customer demands. Aluminum smelting, similar to steel manufacturing, uses an *electrolytic* process to smelt the raw materials to produce aluminum.

Steel manufacturing—Basic oxygen steelmaking (BOS)

Basic oxygen steelmaking (BOS) accounts for 60% of the world's total output of steel.[1] The BOS process is the dominant steelmaking technology. There are multiple manufacturing variations using the BOS process, such as top blowing, bottom blowing, and a combination of the two. This chapter will focus on the more popular top-blowing variation.

Raw materials

With either the BOS or EAF process, the basic raw material for steel manufacturing is iron ore and scrap metal (up to 30%) from junkyards (fig. 26–2). Iron ore is delivered to the steel plants

Fig. 26–2. **Scrapyard at steel plant**

Fig. 26–3. **Yard storage of taconite**

Fig. 26–4. **Unloading of taconite**

via rail, truck, or barge from iron mines in various parts of the world. The mining of low-grade iron ore, typically referred to as *taconite* (fig. 26–3), goes through a long process, even before arriving at the steel-manufacturing facility, commonly referred to as a steel mill. In general, taconite is mined, crushed, separated, mixed, concentrated, and palletized for shipment to the steel mill. Iron is then used in a blast furnace to produce so-called hot metal and is one of the key ingredients in the manufacture of steel. Another major component is scrap metal from various suppliers, which is sent to the steel mill for recycling (see chapter 18). The scrap metal can be anything from cars, barrels, cylinders, rebar, or dismantled process vessels, tanks, or pipes. Some scrap, referred to as home scrap, is generated on-site, including off-specification cast products such as flat steel, rebar, and billets. The storage of taconite and scrap are typically outdoors, in large storage yards. The storage yards can cover tens to hundreds of acres. Movement of these raw materials is usually via large front-end loaders and conveyors belts. Other components of steel include the following:[2]

- *Fluxes.* Fluxes serve two important purposes; they combine with SiO_2 (silicon dioxide) from the hot metal to form a slag. Then, slag absorbs and retains sulfur and phosphorus from the hot metal made in the blast furnace. *Lime* and *dolomite* are the most common fluxes.
- *Coolants.* Limestone and scrap are added to the furnace in the event that overheating is problem.
- *Alloys.* Bulk alloys, in the form of ingots or blocks, are charged from overhead bins into the ladle. Common alloys are *ferromanganese* and *silicomanganese.*

The differences in the BOS and EAF processes begin at this stage. They are mainly related to how the taconite is formed to make either hot metal for the BOS process or direct reduced iron (DRI) for the EAF process. The BOS uses a blast furnace to produce hot metal, in contrast to most EAF processes, which typically use an *iron reduction process* to produce the DRI.

Iron ore processing—Blast furnace

The BOS process, using a blast furnace to produce the hot metal, chemically reduces and physically converts iron oxides into liquid iron called hot metal, which is about 93% pure iron. The blast furnace (fig. 26–5) is a huge, steel stack lined with refractory brick, where taconite pellets (iron ore Fe_2O_3), limestone, and other components are dumped into the top via conveyors, as preheated air is blown into the bottom. The raw materials require 6 to 8 hours to "cook" and descend to the bottom of the furnace via gravity. The bottom of the blast furnace is where the raw materials become the final product of liquid slag and liquid iron. These liquid products are drained from the furnace at regular intervals. Once a blast furnace is started, it will continuously run for 4 to 10 years with only short stops to perform planned maintenance.

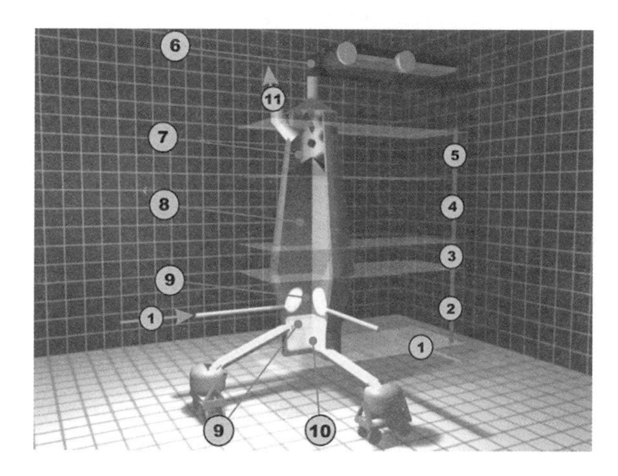

Fig. 26–5. **Blast furnace** *(Courtesy of Robert Blazek)*

1. Hot blast from Cowper stoves
2. Melting zone
3. Reduction zone of ferrous oxide
4. Reduction zone of ferric oxide
5. Preheating zone
6. Feed of ore, limestone, and coke
7. Exhaust gases
8. Column of ore, coke, and limestone
9. Removal of slag
10. Tapping of molten pig iron
11. Collection of waste gases

Steel production—Basic oxygen furnace

The product of the BOS process is molten steel with a specified chemical analysis, depending on customer requirement and intended uses. The steel usually has a temperature of 2900°F–3000°F during the final stage of the process. From here, it may undergo further refining in a secondary refining process or be sent directly to the continuous caster where it is solidified into semi finished shapes: billets or slabs. The place where the BOS process takes place is commonly referred to as the melt shop. The BOS process is similar to baking a cake or cookies, in that it is a batch process, and all the raw materials have to be at the right proportion and temperature. Once the temperature and chemical analysis of the hot metal from the blast furnace are known, a computer in the central control room will determine the optimum proportions of scrap and hot metal, flux additions, lance height and oxygen blowing time. This is referred to as charging the furnace. When everything is ready, the beginning of the BOS process, called the *heat,* starts. The basic oxygen furnace (BOF), now charged, is tilted about 45°, and scrap metal charge is dumped into the top BOF. The hot metal is immediately poured directly onto the scrap from an overhead crane. Fumes, fire, smoke, and sparks are emitted from the BOF's mouth and collected by the pollution control system. This is a very violent reaction and extremely noisy. Then the vessel is tilted back to the vertical position, and various fluxes are dropped into the BOF from overhead bins or manually added via cranes, while the lance is lowered to a few feet above the bottom of the vessel. The oxygen lance is water-cooled with a multihole copper tip. Through this lance, oxygen of greater than 99.5% purity is blown into the mix. This process is also very noisy, and caution should be used while in the area during stirring times. Once the heat is completed and the BOF is ready for tapping, the preheated ladle is positioned in the ladle car under the furnace. The BOF is slowly tilted, and molten steel emerges from a port near the top of the BOF. This is called the taphole. The taphole is generally plugged with material that prevents slag entering the ladle as the vessel turns down. As the BOF is tilted, molten steel burns through the plug. Slag is a by-product of the steel-manufacturing process and comes from remaining iron oxides, impure scrap components, and desulfurization. Operators use caution to ensure that slag does not mix with the molten steel. After tapping steel into the ladle, the ladle car and molten steel are now ready to head to the caster for transforming into billets or slabs. To remove additional excess slag, the BOF is turned upside down, and the remaining slag flows into the *slag pot*. The slag is then carried away as a waste by-product and sold for use in concrete or other manufacturing processes and uses.[3]

Steel manufacturing—Electric arc furnace (EAF)

The raw materials are the same in both processes, as well as the yard storage of taconite and scrap metal. The difference with the EAF, which accounts for approximately 35% of the manufacture of steel, is that it mainly uses a mixture of scrap metal and DRI as the furnace charge. DRI is a virgin iron source that is relatively uniform in composition and virtually free from *tramp elements*. In basic terms, the oxygen has been removed, and a very nearly elemental form of iron remains. DRI is *pyrophoric* because iron does not like to be in this form, and therefore it tries to reoxidize. This will be further discussed in the common hazards section of this chapter.

Fig. 26–6. **Reformer (side view)**

Fig. 26–7. **Reformer (top view)**

Iron ore processing—Electric arc furnace

The process of converting taconite to DRI can be accomplished by several different methods. The two most common of methods are the Midrex® and HYL® processes.

The Midrex technology of converting taconite to DRI uses a reformer (fig. 26–6) to produce reducing gas, which consists of about 95% combined hydrogen plus carbon monoxide, is heated to about 1500°F–1700°F (fig. 26–7). The reducing gas is fed through the bottom end of a furnace or reactor, while taconite is loaded or charged at the top of the reactor by conveyor belt. The taconite, via gravity, is fed continuously from the top of the reactor, passing uniformly through the preheat, reduction, and cooling zones of the reactor (fig. 26–8). The gas flows upward in the reactor against the descending taconite. This is where the oxygen is removed from the taconite, and it is then transformed into a form of element

Fig. 26–8. **Midrex reactor column**

iron (97%). At the top of the reactor, the partially spent reducing gas (approximately 70% hydrogen plus carbon monoxide) exits the reactor and is recompressed, enriched with natural gas, and transported to the gas reformer. The reformer changes the mixture back to 95% hydrogen plus carbon monoxide, and it is then ready for reuse by the direct reduction process. The cooled DRI (roughly 97% pure iron) is discharged through the bottom of the reactor onto a conveyor belt, after which it goes through a series of product screens for removal of dust, fines, or other fugitive elements, and is treated with a cementitious coating to help minimize the danger of spontaneous ignition during storage. The DRI is now stored in large silos under a nitrogen blanket to further prevent spontaneous ignition.

The HYL process, commonly known to be a safer technology than the Midrex, follows the same basic process. Like Midrex, the HYL process uses reformed natural gas from a gas reformer (fig. 26–9). Prior to the gas reformer, natural gas is mixed with excess steam (above *stoichiometric limits*) and is then passed over nickel-based catalysts. After the reformer, the water vapor (steam) in the reformed gas is removed to produce a hydrogen-rich reducing gas. The HYL process uses four reactors in the reducing section (fig. 26–10). The process is the same as the Midrex; taconite is moved via conveyor to the top of the reactor and gravity-fed into the reactor, while the reformer gas flows counter to the taconite flow. The reduction process of the taconite, transforming it to DRI, occurs in the first and second reduction stages. The third stage is used for cooling the

Fig. 26–9. **HYL Reactor with reformer**

Fig. 26–10. **HYL Reactor with reformer (close-up)**

DRI from temperatures above 1800°F prior to discharging from the bottom of the reactor. The final stage permits finishing touches on the DRI to further produce a more stable product with a reduced pyrophoric risk. At the top of the reactor, the partially spent reducing gas is returned for reformulation and use again in the next reactor. The DRI is now sent through product screens and stored in silos under a nitrogen blanket.

Steel production—Electric arc furnace

The EAF process basically begins in the same manner as the BOF, in the melt shop. The EAF is charged with scrap metal, DRI, fluxes, and other components, depending on the desired chemistry and metallurgy (fig. 26–11). The EAF, however, is a closed batch process rather than an open batch process of the BOF. The EAF has a roof and three electrodes, which are direct three-phase AC movable graphite electrodes on a hydraulic-operated arm that penetrate through the roof. After the EAF is charged, the roof swings over the EAF and is locked in place. The electrodes are slowly lowered just above the scrap, and an arc is struck that provides the heat for melting the scrap. Melting is accomplished by supplying energy to the furnace interior. This energy can be electrical or chemical. Radiation is the primary heat transfer method, but the current resistance through the metal also causes the charged materials to heat and melt, which is known as electrical energy. Chemical energy is supplied via several sources including oxy-fuel burners and oxygen lances. Much like the BOF, oxygen lancers are used to stir the melt and purify the mix. This is an extremely loud and violent reaction, and caution should be used when in the melt shop furnace floor during this time. It is common to have flames, smoke, and other products of combustion emanate from the EAF. The arcing process is very profound, and some people may experience physical distress. Dust and fugitive emissions are collected and sent to baghouses for environmental control. The steel usually has a temperature of 2900°F–3000°F during the final stage of the process. When the temperature, metallurgy, and chemistry are within the desired parameters, the operator in the control room begins the tapping process. Unlike the tapping process in the BOF, some EAFs are bottom tapped. When the furnace is ready to tap, a plug is removed or often referred to as the slide gate, and the molten metal flows into the ladle car waiting underneath. Other EAFs are tapped in a similar nature as the BOF. The slag that is produced is during this process is removed and sent for environmental recycling (fig. 26–12).

Fig. 26–11. **EAF being charged from overhead crane**

Fig. 26–12. **Slag pot**

Casting of molten steel

From the melt shop, raw or crude steel is produced. This is not the final step in the process but an intermediate step. Whether the molten steel is produced using the BOS or EAF process, we must now cast the molten steel into a form that is acceptable for rolling and/or some form of heat treatment. This is completed using a caster (fig. 26–13). A caster is a water-cooled piece of equipment that uses a mold and a series of rollers to form the steel. Most casting is done in the continuous mode. Casters can produce either *billets* or *slabs*. Billets are used for long products such as rebar or as stock for producing nails, bolts, and other types of fittings. Slabs are used for producing steel sheets, car parts, household goods, and more. The only basic difference in a continuous caster producing billets or slabs is the end product. A caster works on this basic principle: Molten steel from the melt shop is transferred to the caster via a ladle car. The ladle car then is place over a device called a *tundish car*. This is basically a reservoir over the mold and the top part of the caster. It is lined with refractory brick to protect it from the extreme heat of the molten steel. This permits the precise control of the flow of the molten steel.

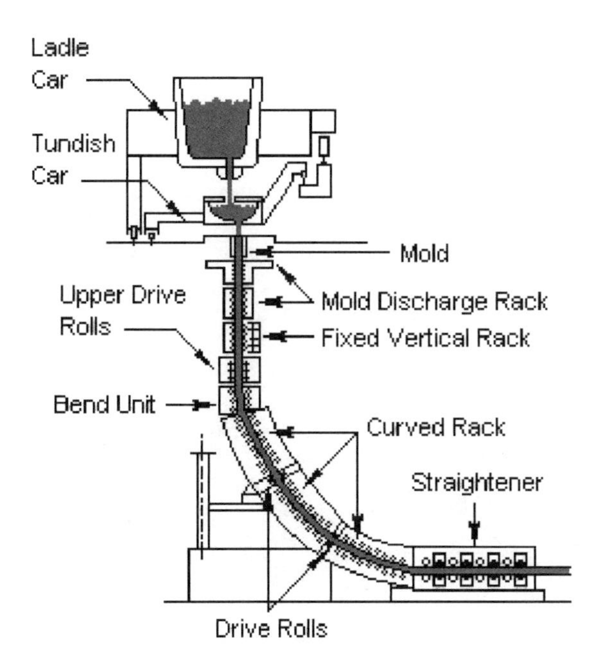

Fig. 26–13. **Continuous caster diagram**

Once the flow of molten steel enters the tundish car, it then flows into the mold. The mold is where the final product is determined, either a slab (thick or thin) or billet. From the mold and throughout the remaining casting process, water is used to cool the molten steel. The steel now exits the mold in a controlled flow via the mold discharge rack and enters the fixed vertical rack. The steel, now cooling, enter a series of rollers, and the shaping process continues. Up until now, the process has been vertical and gravity fed. At this point in the casting process, the steel enters the bending roller unit and the curve rack. This bends the now cooler steel and changes the direction from vertical to horizontal. Since gravity is no longer the driving force, drive rollers roll the steel. The steel, still red hot, but in a more solid form, enters the last stage of the cast, the straightener, and finally exits the caster. As the billet or slab exits the caster, it is cut into predetermined lengths via a gas cutter. This gas cutter is a robotic unit with several gas-cutting torches. The robotic gas cutter clamps onto the billet or slab and follows the steel down the process, cutting the steel to lengths. At the end of the process, a large crane picks up the billet or slab and moves it to a temporary storage area for final cooling and future use in the rolling mills. Some steel mills directly sell the billets or slabs for third-party rolling and/or finishing, while other steel mills have on-site rolling and finishing mills.

Rolling mills

Some of the common forms of rolling mills include the bar (long products), hot strip, and cold strip (both are flat products) rolling mills. These different types of mills produce rebar, wire rod, plates, and coiled steel.

A bar-rolling mill consists of a reheat furnace (fig. 26–14) to reheat the steel billets, which allows them to be rolled. The basic process begins with a crane that hoists steel billets onto a conveyor, where an operator will load a preheat furnace with the billets. When the billets are heated to red hot (about 1000°F), the billet is then sent down a series of rollers and roughing stands at high speed. A roughing stand (fig. 26–15) is a piece or series of equipment that roll the steel, similar to using a rolling pin to roll out dough to make cookies. The size of each subsequent roughing stand decreases in preset increments. The pins or rollers oftentimes are in the shape of disc with a concave surface. One is on top, and the other on the bottom. As the billets reach the first roughing stand, the billet begins to elongate. The process continues at high speed and passes through a series of sequentially decreasing diameter roughing mills toward finishing mills. A finishing stand or mill is the same as the roughing equipment, but smaller tolerances between stages are permitted. This is where the bar is finished to within millimeters of the desired size. The billet further elongates and begins to form bars, rebar, wire rod, or seamless tube. At the end of the bar-rolling mill, a series of cooling racks is used to air-cool the bar prior to transporting it to a staging area for either warehousing or sales.

Fig. 26–14. **Reheat furnace**

Fig. 26–15. **Roughing mill**

A hot strip mill has the same beginning process as the bar-rolling mill, with the exception that a hot strip mill uses slabs of steel instead of billets. The slabs are reheated in a reheat furnace and sent down a set of rollers to roughing stand and finishing mills. The roughing and finishing stands are basically the same, but use larger rollers than the bar mill. The slab is then transformed into a flat piece of steel. After several passes through the roughing and finishing mills, the flat steel is then sent to the end of the hot strip mill, where it is wound on a downcoiler to form a steel coil (fig. 26–16). The hot striped coil of steel, also referred to as hot rolled coils, is now sent to a staging area for warehousing, sales, or for further processing in a cold strip mill.

Fig. 26–16. **Downcoiler**

The cold strip mill is one of the final stops for the coiled steel roll. In a cold strip mill, there may be a series of smaller finishing mills called skin-pass, tube, or color-coating mills (fig. 26–17). These are all used for small-scale production and are based on specific customer orders. Basically, a cold strip mill uses the hot rolled coils as feed stock. The hot rolled coils enter the cold mill at the pickling line (fig. 26–18). A pickling line, whether a push-pull or continuous, uses a pickling acid, typically sulfuric acid, to clean the surface of the steel from oils, rust, and other surface imperfections and prepares the surface for further finishing and processing (figs. 26–19 and 26–20). Many times, the pickling line involves using an acid regeneration boiler, similar to the *black liquor recovery boiler* (BLRB) (mentioned in chapter 19, on pulp and paper manufacturing). This acid regeneration boiler uses a chemical recovery process, commonly found in various other industries. In principle, the spent sulfuric acid is returned to the boiler where the acid impurities (rust or other deposits) are consumed and removed, while the acid is regenerated in the boiler, and new sulfuric acid is mixed with the regenerated acid to be sent back to the pickling line. This method reduces the amount of new acid that would be needed to operate this process. From the pickling line, the coils can be sent to the finishing or skin-pass mill for final reshaping to meet very specific customer demands or for further processing. A finishing or skin-pass mill is a large machine that uses a batch process (one coil at a time) for refining the shape (often called the coffin) and the thickness of the steel coil (fig. 26–21). These types of mills check the *camber* of the steel. Camber is the deviation of a side edge from a straight edge. Measurement is taken by placing a straight edge on the concave side of a sheet and measuring the distance between the sheet edge and the straight edge in the center of the arc. Camber is caused by one side being elongated more than the other. The thickness of the steel coil must also fall within very tight tolerances; otherwise, they will not properly fit on the customer's machine. After this process, the steel coils are either sent to the warehouse for storage or sales. Some coils continue down the process for continued finishing.

Fig. 26–17. **Finishing mill**

Fig. 26–18. **Pickling line overview**

Fig. 26–19. **Pickling line entrance**

Fig. 26–20. **Pickling line**

Fig. 26–21. **Finishing mill 2**

The next step in the process is annealing (chapter 28). *Annealing* is a heat treatment process that is used to heat metal close to its melting point and cool it over a long period of time. The result of this causes billions of tiny atoms to align themselves in such a way that a tough and hardened structure is produced in some types of metals. In this process, large annealing ovens can process three coils at a time (fig. 26–22). Some facilities might have 20 or more of these annealing furnaces.

Fig. 26–22. **Batch annealing ovens**

After the annealing process, the final step involves the galvanizing line. The galvanizing line consists of a large machine that unrolls, dips, and rerolls the coils of steel (fig. 26–23). This process is typically a continuous process and uses a molten zinc bath to dip the coils. The process starts by placing a roll of coiled steel on the feeder reel. The steel is then fed through a straightener, much like a roll of tape in a tape dispenser. As the coil unrolls, another coil is added to the machine at the end of the first coil, and both are joined together to form a large length of steel (fig. 26–24). The steel coil now travels through a series of rollers under tension. This machine can be very high, in excess of 50 feet or so (fig. 26–25). As the steel travels through the open-air machine, on the downward drop, the steel is immersed in a large, ground level, molten pit of zinc (fig. 26–26). The zinc is supplied in bricks, or ingots, that are about the size of a shoe box. The steel enters the molten zinc bath, then exits on a continuous set of rollers to be dried. At the end of the machine is a recoiler, which rolls the steel coil back in to the typical coil shape (fig. 26–27). The coil is now ready for warehousing or immediate sale (fig. 26–28).

Fig. 26–23. Galvanizing line overview

Fig. 26–24. Coils loading in the galvanizing line

Fig. 26–25. Unloading, continuous coils in the galvanizing line

Fig. 26–26. Molten pit of zinc

Fig. 26–27. Recoiling finished galvanized coils

Fig. 26–28. Warehouse of finished coils

Similar to automotive (chapter 34) and pulp and paper manufacturing (chapter 19), there are other parts of the plant that are common as well. There are offices, laboratories (chapter 23), power substations, mechanical workshops, warehouses (chapter 17), hazardous chemicals (chapter 11), and other nonprocess related plants areas.

Aluminum manufacturing

Similar to the manufacture of steel, but far simpler and more pure, the aluminum process begins the same way—at the mines. However, instead of iron ore or taconite, the aluminum process needs *bauxite* ore. Bauxite ore typically is mined on the surface in the same manner as taconite. Bauxite is mined; then the ore is crushed, washed, and screened for size classification, after which, the bauxite is separated to remove clay and silica. The raw bauxite is then ready for delivery to the aluminum plant.

Raw materials

The initial step in the basic aluminum manufacturing process to the mix and crush the bauxite in a solution of hot caustic soda in equipment called digesters. This process allows the alumina hydrate (the raw material for aluminum) to dissolve from the bauxite ore (fig. 26–29). After the mud (red in color) and remaining clay residue are removed by *decantation* and filtration, the hot caustic soda solution is piped to large tanks, where the alumina hydrate begins to crystallize. This alumina hydrate is then sent to a series of filters and dryers to dry under very high temperatures. During this final filtering and drying process, the alumina hydrate is transformed into a very fine powder referred to as simply alumina. The alumina is then transferred to the aluminum reduction plant, where electrolytic process will be used to produce aluminum.

Fig. 26–29. **Bauxite ore** *(Source: U.S. Geological Survey)*

Aluminum electrolytic process—Aluminum smelting

The aluminum electrolytic process separates the oxygen from the alumina to produce aluminum. To obtain the metal form of aluminum, the elements are separated via the electricity in the smelting process. This takes place using a similar method as the melt shop in a steel mill. The electrolytic process or aluminum smelting reaction happens in a large, carbon-lined steel cell, commonly referred to as the pot. The bottom of the pot acts as a *cathode* (positive electrode). Carbon blocks are suspended in the pot to serve as *anodes* (negative electrodes). In same basic principle as the EAF uses, direct electrical current is passed through the pot. Alumina is added to the pot, yielding a mixture called *molten electrolyte*. This mainly comprises the mineral *cryolite*. As the electrical current passes from the anode to the cathode, a reaction takes place. The passing current causes the oxygen in the alumina mixture to react with the carbon anode to produce carbon dioxide. As this reaction transpires, the alumina settles to the bottom of the port to be captured (in a similar manner to tapping the BOF or EAF), then to be sent for casting and fabricating.[4]

Casting and fabricating

Molten aluminum, a so-called cleaner metal than steel, is cast and fabricated in a similar method to casting and rolling in a steel mill. Prior to fabricating with the rolling mill aspect of aluminum production, the molten aluminum is treated to ensure and verify its cleanliness and purity levels. At this time, alloying process may take place, depending on customer demands. Alloying ingredients are added to increase the overall strength and additional special properties and qualities. The finished aluminum metal is then cast into ingots and shipped to customers. These ingots can range in size, shape, and alloy content depending on customer needs. Ingots, sold to other companies, are used to manufacture sheets, plates, foil, and soft drink cans, as well as other applications such as engineering equipment, aircraft, and other similar uses.

Some facilities will also have a fabrication plant as well as a casting plant. This alternative technique to casting of ingots is similar to the rolling mill operations of the steel mill. In the aluminum-fabricating facility, molten aluminum metal is cast directly into a semifinished form. In this case, the molten aluminum is used to directly form sheets, foil, or rod products for future sale to customers. In this process, the aluminum is sent either through a series of rollers, like a roughing and finishing mill to form the sheets or rods, or to an extrusion machine that forms bars of aluminum.

COMMON HAZARDS

As with most large and complex industrial facilities, there are many potential hazards. Steel and aluminum manufacturing facilities are no exception. Some of the more commonly found hazards include molten metal spills, pyrophoric iron dust, electrical cable trays, conveyor belts, hazardous chemicals, flammable and combustible gases and liquids, and scrapyard fires.

Molten metal spills

Molten metal spills create a dangerous situation. These situations arise from failures of furnace taps or slide gates on ladles in the melt shop operations of a steel mill or in aluminum-smelting operations. Molten metal, relatively safe in the furnace or pot, can cause multiple fires if containment is lost. If a molten metal breakout occurs, it will behave similarly to honey or syrup, that is, flowing toward the direction of least resistance. This could be toward small offices or pallets of materials, or into the cable trenches that are common in the area. No matter where it flows, or whether direct or indirect contact, almost anything in its path will spontaneously ignite. It is common during molten metal breakouts to have several small fires. Particularly hazardous is the involvement of the cable trays. If molten metal reaches the cable trays and they get ignited, this could not only shut down operations in the plant but also spread to other, noninvolved areas. Housekeeping and *continuity of combustibles* is key to preventing fire spread.

Pyrophoric metal dust

During reduction in the Midrex or HYL processes, dust is one of the major by-products. As mentioned earlier, DRI is produced via oxygen reduction in a reactor. It comes in the form of pellets and often lumps. DRI, if exposed to air, will begin to slowly reoxidize. This is the reason why it is stored in silos under a nitrogen blanket. The reoxidation reaction is *exothermic,* meaning heat relieving or heat generating. The rate at which this accelerates is related to the total surface area. DRI dust may create an explosion hazard as well. When DRI comes in contact with water, it may undergo a further chemical reaction, emitting hydrogen gas. This process is a surface process and does not take place within the DRI pellet or lump. The quality of the DRI is a contributing factor in this reoxidation and hydrogen gas generation reaction; therefore the higher the quality of DRI (closer to 100%), the more rapidly this reaction will take place. The basic chemistry relates to the need for the DRI, near elemental iron, to reform back into the iron oxide.

Aluminum dust also creates a dust hazard. In general, aluminum has a high ignition temperature and therefore is not normally considered a fire hazard. Dry aluminum powders are capable of exploding but only under very certain conditions. Aluminum processing and finishing produce dusts that are capable of sustaining combustion. These fires can also originate in the dust collection systems, screens, and fabricating operations.

Electrical cable trays

Due to the large amount of equipment, steel and aluminum mills have multiple process and control rooms, as well as motor control centers (MCCs) and process logic control centers (PLCs). Typically, these rooms control a particular section of the process such as the melt shop, caster, reheat furnace, casting, rolling, and fabrication. Because of the presence of all these control rooms, MCCs, and PLCs, electrical cables are found in large concentrations. The cables are usually grouped into large cable trays and located in raised floors, vertical along walls, overhead at the ceiling, or in horizontal underground tunnels (cable tunnels). These cable trays are exposed to the accumulation of dust and combustibles, adding to the fire hazard. Because of the nature of the cable tunnels, which may pass through other building(s) or process compartments, many times they contribute to the spread of fires. These are especially susceptible to molten metal breakouts.

Conveyor belts

As mentioned in chapter 15, conveyor belts also pose a fire hazard at these locations. These large conveyor belt systems can range in length up to 500 feet or more and operate in a series of several sections, either vertical or horizontal (fig. 26–30). In the event of a fire, the belt can break, and because of the delay in shutting down the conveyor motors, the belt may form a pile. If ignition occurs from overheating or friction from the motors and rollers, a large ensuing fire that is fed from a large pile of burning rubber conveyor belt can occur. Further, fires can occur from accumulated DRI dust and pellets while being transferred to the product screens, storage silos, and melt shop floor. It should also be noted that these conveyors could be 100 feet or more in the air, creating challenging attack obstacles.

Fig. 26–30. **Conveyor belt**

Hazardous chemicals

Hazardous materials and chemicals (chapter 11) are also present in most of the manufacturing process. These are especially concentrated in the melt shop, aluminum smelting, mechanical workshops, sulfuric acid regeneration plant, wastewater treatment plant, and laboratories. These chemicals include dolomite, lime, sulfuric acid, and others. A detailed survey should be included in your department's preincident response planning.

Flammable and combustible gases and liquids

Flammable gases are present in the front end of the iron reduction, steel, and aluminum productions. Natural gas is used in the reformers of the DRI plants and blast furnaces, and as robotic cutting torches in the casting operations. Annealing and other heat treatment process hazards also involve the presence of flammable or special atmospheric gases (see chapter 28 for more details).

With the presence of large hydraulically operated equipment comes a large amount of combustible oil in the holding tanks. This is common throughout all processes. In some cases, these large storage tanks are under the related process equipment with limited access via a tunnel. If these holding tanks catch fire or are exposed to a fire, this could create a very hazardous situation. Because there is also a presence of oil pumps, the likelihood of a pump seal or gasket failure can lead to a quenching oil spray fire. These are particularly dangerous if the built-in safety controls fail to operate.

Scrapyard fires

Scrapyard fires are one of the most common types of fires at steel plants. These fires can occur in the storage area of the incoming scrap or in areas where scrap shredder waste is stored. Although not particularly dangerous with respect to a large loss of life or property, fires in these areas are difficult because of the limited water supplies and the potential of unstable loads. These fires may also involve the shredder unit, which grinds whole cars and separates the components into metal, plastic, foam, and unknowns. These shredder equipment fires (see chapter 18) are more a nuisance than anything else. Limited water supply is expected.

FIREFIGHTING TACTICS

As previously discussed, a well-planned and tested emergency response plan is needed when responding to these large industrial facilities. Preincident response planning should be carried out in these heavy industrial facilities to identify the hazards, which have the potential to develop into a major incident. In the event of a fire at this type of facility, it is advantageous to have information readily available that relates to the common hazards, material safety data sheets (MSDS), process flow diagrams, fire safety provisions, access roads, staging areas, and location of water sources. Good loss prevention starts with the fire-safe design of the facility such as the installation of fixed suppression and detection systems and construction features.

Arriving at the scene of a fire at a large steel- or aluminum-manufacturing facility can be very confusing, considering the multiple gates, process buildings, and magnitude and size of the plant (fig. 26–31). Proper preincident response planning and several site visits will help eliminate some of this confusion. Depending on the type of incident, most often, establishing a reliable water supply and supporting the fixed systems (if available) is a priority.

Fig. 26–31. **Large open spaces and high ceilings**

After arriving at the scene, establish a command post, if one has not already been set up. Try to first determine the process area that is involved in the incident. By doing this, you can begin to list the possible hazards common to that part of the plant.

Before firefighters enter these process areas, they should be wearing full personal protective equipment (PPE) and self-contained breathing apparatus (SCBA). Additionally, they must have communication equipment or the ability to communicate to the command post, hand lights and forcible entry/exit equipment. A local escort, perhaps a member of the local fire brigade, should also accompany the initial attack crew. Complex interiors created by equipment, multiple floors, and dead ends should be reviewed on the preincident response plan before entry. When firefighters enter large facilities with high ceilings such as these, even a serious fire may not appear dangerous. Products of combustion, such as smoke and carbon monoxide, can quickly build up and incapacitate unprepared firefighters.

Because of the large areas covered by these plants, locations of standpipes and hose cabinets are important. Similar to warehouse (chapter 17) and automotive (chapter 34) facilities, the use of high-rise or hotel packs will be beneficial to use. Connecting into the standpipe (preferably a Class I) will help reduce, although not necessarily eliminate, extremely long hose lays and expedite manual firefighting efforts. Because these buildings can be rather large and long (several hundred to several thousand feet), it may be possible to drive inside and through an uninvolved section of the plant and set up operations on the unexposed side of a fire wall. It then may become necessary to set up attack lines from the truck and reduce to smaller handlines to extend the effective range and attack the fire from there. Confinement of the fire to one portion of the structure or piece of equipment may be possible. In many cases, these facilities are built with fire barriers and fire walls. The fire may burn in a section of the facility, but automatic and manual efforts to confine the fire to the immediate area of origin may be the best tactic.

Overhead cranes, holding in process or raw materials, can become weak in a fire and result in a release of the load. During events where overhead cranes are present, firefighters should exercise caution to ensure that the crane and the crane's load are secured in a safe position. Avoid operating under a crane with a load.

Fires involving the cable trays, whether overhead or in tunnels, can spread to other areas of the facility. If a fire occurs in the cable trays, fires can spread to the control room, causing the loss of plant operations and control. If this happens, a very dangerous situation can arise. Therefore, all efforts should be made to protect, isolate, and contain cable tray fires to prevent involving control rooms motor control centers in the fire. Also, cable tunnel fires are extremely dangerous. Similar in nature to basement fires, these cable tunnels can be a few feet underground, collect heat and other products of combustion, and cause a loss of communications because of the thick concrete walls. This is especially true if your department uses a newer digital radio system. Different forms of communications should be considered prior to fighting a fire in a cable tunnel. Cable tunnel fires also create a challenge with respect to supplying water to handlines. These cable tunnels can be several hundred feet long, and it may become necessary to set up attack lines from the truck and gated down to smaller handlines to extend the effective range. If Class I or III standpipes are provided, using high-rise or hotel packs will eliminate the need for several hundred feet of fire hose and expedite the attack.

Molten metal spills arise from premature failures of furnace taps or slide gates on ladles in the melt shop operations of the steel mill or aluminum-smelting operations. If responding to this type of incident, it is very important that you do not plunge a straight hose stream into the released molten steel. This will create a large high-pressure gas expansion explosion, commonly referred to as a steam explosion. In brief, this is a rapid expansion of water to steam in the confinement of the molten steel. If this occurs, molten steel will be sprayed everywhere, on people and combustibles alike. In this case, a fine mist from the hoseline on the front-running edge of the spill, as well as the top of the molten steel, will cause an increase in the cooling rate, forcing the molten steel to form a crust layer and stop the fluid flow. Continue this operation every few minutes until the steel cools and hardens, at which time it can be removed by a chisel or front-end loader.

DRI, if exposed to air or water, it will begin to slowly reoxidize (exothermic) and release hydrogen gas. In this case, it is important to remove the smoldering DRI dust fines or pellets to an outside environment. This can be accomplished using hand tools or small front-end loaders. Once in the open air (or if already outside), use copious amounts of water to cool the materials. Using foam has not been found to be better than water. Other successful agents include Met-L-X, a dry powder specifically designed for Class D metals. If you cannot remove the DRI fines or pellets from the area, the use of SCBA is required due to the production of hydrogen gas and the fact that the reoxidation process may reduce the percentage of oxygen in the atmosphere, depending on the size of the incident. Fires involving aluminum dust or fires should also be handled carefully. These fires are also best handled using sand, dirt, or a Class D extinguishing agent. Caution should always be exercised when determining the use of water for aluminum fires. Water is an excellent cooling agent because of its ability to absorb heat. Water has the highest heat-absorbing ability of most common substances. Under proper conditions and application, water has been used successfully to lower the temperature of burning metals to below their ignition point. The authors have had great success in using water for the extinguishment of many, many DRI fires involving dust, fines, and pellets.

Large conveyor belt systems can range in length up to 500 feet or more and operate in a series of several sections, either vertical or horizontal. A conveyor belt fire can generate large amounts of smoke and heat, similar in nature to a tire fire. PPE, SCBA, and a good water supply are the best weapons in extinguishing these fires. Further, electrical and mechanical isolation of the conveyor is important prior to entering in and around the rollers.

Flammable gases are present in the front end of the iron reduction, steel, and aluminum productions. Natural gas is used in the reformers of the DRI plants and blast furnaces, and as robotic cutting torches in the casting operations. Annealing and other heat treatment process hazards also involve the presence of flammable or special atmospheric gases (see chapter 28 for more details). The firefighting tactics are similar for fires involving flammable and combustible gases and liquids. Remember that we do not want to extinguish a gas fire before we are capable of isolating the supply. When fighting fires involving flammable and combustible liquids, the use of quality Class B foam is important. (See chapters 8, 31, and 33.)

CONCLUSION

Preincident response planning should be carried out in steel and aluminium facilities to help identify some of the common hazards and different firefighting strategies and tactics needed to combat such fires. The sheer size and magnitude, combined with the complex operations of these facilities, necessitate the need for frequent site visits. Many of these larger facilities have an on-site fire brigade, which may include a full-time department. On-site training with these local plant brigade members will aid in bridging the gap between the municipal and industrial fire responders and is highly encouraged.

NOTES

[1] Stubbles, J. 2000. *The Basic Oxygen Steelmaking (BOS) Process.* Mason, Ohio: Author, p. 1.

[2] Stubbles, *The Basic Oxygen Steelmaking (BOS) Process,* p. 2.

[3] Stubbles, *The Basic Oxygen Steelmaking (BOS) Process,* p. 4.

[4] Habashi, F. 1993. *A Textbook of Hydrometallurgy.* Quebec, Canada: Metallurgie Extractive, p. 194.

27

Printing Establishments

INTRODUCTION

If you look around the room where you are sitting, you will probably see many items that have been through a printing press. When we think of the printing process, we normally think of newspapers, books, and magazines, but much more has gone through the process. Our clothes may have the design or color applied using the printing process. The packaging of everyday products will usually have been through a printing process. Even the labels on goods that we buy for consumption have been through the printing process.

The printing process is actually composed of many different processes, each with its own unique problems and hazards. In the end, though, we have characters or designs on paper, cloth, metal, synthetic, or other surfaces. The processes used in printing include lithography, gravure, flexography, letterpress, screen printing, and other plateless-printing processes. Plateless printing is becoming a major force in the industry, with the continuing application of computer-controlled printing processes.

DESCRIPTION OF PROCESS

Printing processes may extend over a large area of a plant and include a variety of printing methods, storage of raw materials, finishing areas, and finished product storage. Printing equipment can range from the size of a large copier to machines that are 8 feet tall and 100 feet long. The basic printing process transfers an image from an original form to various other materials.

The following are some of the printing methods used in the printing process:

Lithography (offset)

Lithography uses a printing plate on which the image areas are neither raised nor indented in relation to the nonimage areas. The image and nonimage areas are on the same plane of the plate are defined by differing physiochemical properties. In short, the ink sits flat on the surface of the paper. Lithography printing is based on the fact that water and oil do not mix. The print-receptive areas are chemically treated to make them oleophilic (oil loving), and the nonprinted areas are chemically treated to make them hydrophilic (water loving). Ink is then applied to the oleophilic areas during printing. Water is applied to the hydrophilic area of the printing plate and prevents ink from migrating into the nonimage area.

Other types of plates are known as bimetal plates, where metals are used because of their affinity or nonaffinity to water or oils.[1] Bimetal plates are composed of two metal layers. One has an affinity for ink, which will be the printed area, and the other has an affinity for water. Paper or other raw material is passed through a series of plates and rollers to transfer the images.

Rotogravure

Rotogravure printing uses a sunken or depressed surface for image transfer. This printing looks almost the same as offset printing, but with a much higher quality. Unlike other types of printing where the inked surface is raised, gravure has the image areas depressed. The image areas are etched into copper cylinders or wraparound plates by either a chemical or an electromechanical process.[2] The chemical process is rare and has been largely replaced by the electromechanical process. Following etching, the cylinder is completed by the application of an electroplate of chromium, which improves the durability of the plate. The electrolyte used in chrome plating consists of chromic acid, sulfuric acid, deionized water, and small amounts of organic additives. The cylinder rotates in an ink bath with the excess ink wiped from the surface or raised areas. As the raw materials pass through a plate cylinder and an impression cylinder, the image is transferred on to the raw material.

Letterpress

In letterpress printing, the paper or material being printed comes in direct contact with the type.[3] A small press job may be set from individually cast and hand-set type in a form for the particular job being processed. The metal type can be remelted to form new type as needed. There are three types of letterpresses in operation today: the platen, flat-bed, and rotary presses. The platen press consists of two flat surfaces called the bed and the platen. The type or image plate is locked into the bed. The platen provides the smooth backing for the paper or other material that is to be printed. The plate is inked, and the platen presses the paper or other material against the plate, producing the impression.[4]

Flat bed cylinder presses use vertical or horizontal beds. The plate is locked into a bed, and the bed is passed over an inking roller, then against the material to be printed. An impression cylinder forces the material against the plate, completing the image transfer. This type of printing is relatively slow, so it is being replaced with other, faster processes.

The most popular press used in letterpress printing is the rotary press. This requires curved, image-carrying plates. In some instances, rotary presses will use heat-set inks and are equipped with dryers, usually the high-velocity hot-air type.

Flexography

Flexography combines the features of letterpress and rotogravure printing. The process uses relief plates as in letterpress, and low-viscosity fast-drying inks, as in *rotogravure*. The inks may be either fast-drying solvent or water-based. The plates are usually made of low-cost rubber or photopolymer. Flexography has been gradually becoming a major factor in the printing industry because it is a relatively simple operation, and it is easily adapted to the use of water-based inks.

Flexographic presses that are used for publications will generally incorporate infrared dryers to ensure drying of the ink. Publication flexography will be used during the production of newspapers, comics, newspaper inserts, and catalogues. Flexography is also used in packaging printing such as folding cartons, labels, and other forms of packaging materials. Flexographic printing can recycle most of the spent inks and wash water, depending on the printing equipment used.

Engraving

Engraving is the type of printing used to print currency. It is primarily used where very fine detail is necessary. The image is raised above the surface of the paper and yields the sharpest image of all traditional printing methods.[5] A plate is engraved with the image needed. Plates are made of steel when very long runs are necessary, and made of copper when short printing runs are needed. Ink is applied to the plate and fills the cavities. The plate is then wiped clean, leaving the image area, that is, recessed areas, filled with ink. Intense pressure is used to transfer the image to the material.

Thermography

Thermography is used to create special effects in printing stationary, invitations, greeting cards, and the like. It creates a raised printed surface resembling genuine engraving but without expensive engraved plates. Specialized ink is applied to the surface using conventional printing techniques such as letterpress or offset, and the wet inks are dusted with a special powder. The powder on the noninked surface is removed using suction, and the surface is passed under a heater, which then fuses the ink and powder. The printing will then swell, causing a raised printing surface.

Screen printing

Screen printing employs a porous screen of nylon, silk, Dacron™, or stainless steel that has been mounted on a frame. An image is produced on the frame by applying a stencil, which leaves the image area porous and the nonimage areas impervious to ink. The screen is placed over the material to be printed, and the ink is applied to the screen. A squeegee or blade is drawn across the screen to force the ink through the porous areas and onto the surface to be printed.

Using screen printing allows a greater thickness of ink to be applied to the material than other printing techniques. Years ago, this process was manually performed, but today automated presses—flat-bed and rotary—are used. The screen-printing process can be used on almost any material.

Plateless printing

Plateless printing differs from other methods of printing in the sense that it does not use printing plates or any other form of physical image carrier. Plateless printing relies on sophisticated computer software and hardware that controls the printing elements. Plateless-printing processes include electronic printing, ink jet printing, magnetography and thermal printing.

Electronic printing. Electronic printing includes both xerographic and laser printing. In both of these processes an image is recorded on a drum in the form of an electrostatic charge. This charge is transferred to the material to be printed, usually paper. The toner, which consists of a conductive fine powder, is then spread on the paper. The toner is attracted to the areas of the paper that have been electrostatically charged, and the image remains. The paper is then heat treated to affix the toner to the paper.

The difference between laser and xerographic printing is the method in which the image is formed on the drum. For xerographic printing, light reflected from the image to be printed is projected on to the drum through a camera lens. Laser printing has the image inputted in digital form from a computer. A laser then projects the image onto the drum. Laser printing is used for short-run, in-plant, and quick printing applications. It can also be used for proof copies of printed material, which will eventually have final printing performed using one of the other conventional forms of printing. It may also be used to produce camera-ready documents that will be printed using one of the traditional forms of printing.

Ink jet printing. Ink jet printing operates by spraying a pattern of individual ink droplets onto the material to be printed. Computers control the input of the dot matrix image. Ink jet printing has the advantage of speed but loses some resolution in the process.

Magnetography. Electronic printing and magnetography are similar in process except that magnetography uses a magnetic instead of an electrostatic photoconductor. The cost of the toner, which is relatively high, and the slow speed of the process are also negative aspects of this type of printing.

Thermal printing. During the process of thermal printing, a chemical reaction occurs when portions of a thermal-coated paper are subjected to heat. This process is employed in facsimile machines and other office machine applications. Newer facsimile machines use plain paper instead of the thermal-coated paper.

Postpress operations

In addition to the direct printing operations, there are additional processes involved known as postpress operations. These operations include the following major activities: cutting, folding, assembling, and binding. The major chemicals that are used are generally adhesives that are used in binding and other assembly operations. During postpress operations, it may be that not all of the following operations are performed. This is dependent on the requirements for the finished product.

Cutting. Machines that cut the printed product vary in size and capacity. These machines are known as paper cutters and typically operate on the guillotine principle. Electric or hydraulic-powered guillotines are in use as well as hand-operated versions in some facilities. When moving the stacks of paper, which can be very heavy, air may be applied to the underside of the stack to reduce the friction between the table and the paper. Some printing presses have integrated cutting equipment built into the press equipment. The scraps resulting from the cutting operation will be collected and then baled for recycling. There may be power-operated machines that collect the scraps. Housekeeping will be necessary to prevent fires in these areas.

Folding. Folding can be used for simple print jobs such as pamphlets or to fold pages of a paper, book, or magazine into groups (signatures) that will be used to form the finished product. As in cutting, certain printing processes may also have integrated folding included.

Assembly. Prior to binding, all the printed elements of a final product will have to be brought together to form the finished product. The first step in the assembly process is known as *collating.* Collating is the gathering of individual sheets of paper instead of the groups (signatures). A term known as *gathering* is used when the signatures are brought together during assembly. When signatures are combined, they are inserted within each other to form the final book. Assembly may be manual, semiautomatic, or fully automatic. In manual operations, workers handle the products and assemble by hand. Some machinery may be used to move the products. In semiautomatic processes, the products are hand fed into machinery, which then performs the assembly. In fully automatic processes, all of the work performed has been automated. In all systems, machinery such as conveyor belts may be involved to move products from one area to another.

Binding. The final printed product is held together with binding. In adhesive binding, the paper is clamped together in a press, then a liquid glue is applied to the edge using a brush. One type of glue that is used is usually a water-soluble latex. This is used when making pads, so that the papers can be easily removed from the pad. When binding paperback books, stronger glue is necessary. A hot-melt glue is used, and a piece of gauzelike material is inserted into the glue for added strength.

The side-binding technique uses a fastening device that is passed through the pile of paper. Loose-leaf binding as well as stapling are examples. Another form of side binding is the spiral notebook type. Metal or plastic spiral bindings may be used. Side-sewing binding involves drilling the binding edge with holes, clamping, and then passing a needle and thread through the holes to complete the process. Semiautomatic and automatic machines may be used for this process.

Saddle binding is used when assembling one or more signatures along their folded edge(s). This type of binding is used in magazines where wire stitches are placed in the fold of the magazine. This type of stitching can be performed automatically in line during the postpress operations.

Finished material storage. The finished products are usually stored on skids or pallets and covered with plastic. Industrial lift trucks as well as overhead traveling cranes are used to move the products while on the skids or pallets.

COMMON HAZARDS

The hazards associated with the printing process are varied. During the letterpress process, especially in older machines, an ink mist may be emitted that can adhere to ceilings and walls. If it is allowed to accumulate, it may become a fire hazard because of the oils contained in the inks. Letterpress and other types of printing machines may also leak oil and grease, especially older models that have not been well maintained. This oil and grease may be absorbed into the floor, which will allow a fire to spread more rapidly if one does start. Concrete or other such noncombustible flooring materials will not allow the oil or grease to be absorbed but can contribute to the fire hazard by allowing oils and grease to accumulate. It may also cause a slipping hazard to responding firefighters.

The use of solvents for cleaning the plates and machinery parts also creates a fire hazard. Rags used for cleaning must be stored properly. Improper storage of oily rags has been the cause of fires in the past, not only at printing facilities but also at other industrial facilities.

Flammable ink may be used in printing, and this ink may be transported through piping to the printing presses. A leak in the piping may create a spray that will be easily ignited. The source of supply must be shut down and the surrounding areas cooled. Bulk storage of flammable ink creates hazards, as indicated in other chapters on bulk storage of flammable liquids.

Acids such as nitric acid, ferric acid, or sulfuric acid are used in some of the printing processes to etch the printing plates. The acid is usually reclaimed, but during an emergency, it may come into contact with responders. The use of the acid in the process, as well as its storage location, must be familiar to emergency responders.

The storage of raw stock and finished products can be a hazard to firefighters if ignited. The material may absorb water during firefighting operations and increase the *live load* of the building. Incident commanders (ICs) must take this into account during firefighting operations. Storage of raw materials may impede access to areas of the printing facility, also. Aisle widths may be reduced depending on the amount and configuration of the storage. Rolled stock may *exfoliate* during fire conditions and contribute to fire spread. Expansion of paper materials may actually push a wall out, causing a localized or complete collapse of a wall or building.

Another hazard that may be encountered by firefighters is the use of high-voltage machinery. Care must be exercised during operations to ensure that responders do not come into contact with the high-voltage electricity that may be on the premises.

The mechanical hazards of the machinery must not be overlooked. Loose clothing may be drawn into rotating machinery . Other machinery such as presses or cutting edges may cause amputation if inadvertently activated during firefighting activities. Ensure that all power to machinery is off during fire operations. Use lockout/tagout procedures.

The storage of forklifts or other vehicles in printing establishments must be acknowledged. Liquified petroleum gas (LPG)-, diesel-, or gasoline-fueled equipment may be present in storage areas as well as loading areas.

Finished product storage may contribute to the fire load and live load of the building if it has the ability to absorb water during fire operations. These finished products may be plastic, paper, cloth, or just about any other material to which the printing process could be applied. Plastics and foams that are used to wrap the finished products or themselves are finished products may contribute to the toxic smoke conditions present during a fire.

FIREFIGHTING TACTICS

The firefighting tactics employed at these occupancies should be based on standard firefighting procedures that are in use in your department. As with all industrial firefighting, preincident response planning is a necessity. Knowing the processes involved, the machinery present, and its specific hazards will assist in successfully extinguishing fires. Normal departmental tactics may be adjusted based on the hazards that will be encountered. Many departments have the first units to investigate the incident bring a pressurized water extinguisher with them to the fire location. It may be wise to bring a multipurpose dry-chemical extinguisher instead, especially where high voltages or oil-fed fires may be encountered. The use of fire alarm enunciator panels and reports from operators will help in locating the fire.

Fixed systems may be required because of building and fire codes based on occupancy classifications, and their presence should be identified during preincident response planning. As always, fire department support of fixed fire suppression systems should be a priority. These systems may be water sprinklers or other systems such as manual carbon dioxide (fig. 27–1). If they have been activated, allow them to perform their intended function. When safe to do so, shut down the systems and perform final extinguishment, overhaul, and salvage. Support of these systems should be one of your first priorities in a fire situation, after life hazard.

Fig. 27–1. **Carbon dioxide extinguishing system discharge** *(Courtesy of Ansul)*

During serious fires, the IC must be aware of the loads present on the structure. The added weight of machinery on the fire floor or floors above may necessitate the early withdrawal of firefighting personnel. The weight of the water being applied to storage areas must also be taken into account. Cellulous materials, for example, can absorb great quantities of water, and as mentioned, could potentially cause a structure to collapse. The condition of the building structure should be noted as well. Older buildings that have housed printing operations may have a weakened structure due to the vibration of the machinery as well as the saturation of oils and inks (fig. 27–2). Accumulations of dust may also be present in some areas, causing a potential dust explosion under the right conditions (see chapter 15).

Fig. 27–2. **Older buildings that house printing operations may have been weakened over the years due to vibrations as well as saturation of oils and inks.**

The storage of hazardous materials must be considered and noted during preincident response planning. Material safety data sheets (MSDS) should be included in your department's preincident response plans and also obtained from plant operators when arriving at the scene.

CONCLUSION

As with any industrial response, printing establishments must be treated using sound firefighting tactics, standard operating procedures, and preincident response planning. Familiarization visits, as well as fire prevention inspections, can facilitate the preincident response planning aspects of an emergency response and assist the IC and other fire department personnel when responding incidents at these occupancies. The printing process may appear mundane, but these fires present significant safety hazards to firefighters.

NOTES

[1] National Fire Protection Association (NFPA). 1990. *Industrial Fire Hazards Handbook*, 3rd ed. Quincy, MA: NFPA, p. 384.

[2] NFPA, *Industrial Fire Hazards Handbook*, p. 385.

[3] NFPA, *Industrial Fire Hazards Handbook*, pp. 382–383.

[4] Printers' National Environmental Assistance Center. *Print Process Descriptions: Printing Industry Overview.* Retrieved April 6, 2006, from www.pneac.org/printprocess.

[5] Invitesite.com. *Descriptions of Printing Processes.* Retrieved April 6, 2006, from www.invitesite.com/white_pages/printingprocesses.html.

28

Quenching and Annealing

INTRODUCTION

Oil-quenching and annealing operations are heat treatment processes that both harden and temper metals. They use a variety of media, ranging from gases to liquids, and a multitude of methods to obtain the desired results. Some of these operations are small in nature, while other, larger commercial operations can be quite big. Some of these types of facilities exist as stand-alone processes, where the metals are manufactured someplace else and are shipped to a metal-processing facility for further processing before a finished product is produced. However, other locations not only manufacture the metals but also further treat them before sending to a finishing plant to become aircraft parts, eating utensils, or building components.

DESCRIPTION OF PROCESS

Oil-quenching processes, which involve the use of oil, and annealing processes, which involve the use of a special gas, are two of the most common methods of tempering or hardening metals. Both contain different types of process equipment and techniques. We will first discuss oil quenching, followed by annealing.

Oil quenching is a process that uses oil in large tanks to heat-treat metal (fig. 28–1). It is essentially a controlled cooling or quenching of heated metals by immersion in a liquid-quenching medium. This process hardens and tempers metal by imparting metallurgical changes to its surface to create a more hardened material that is not possible without this treatment. Due to the combustible nature

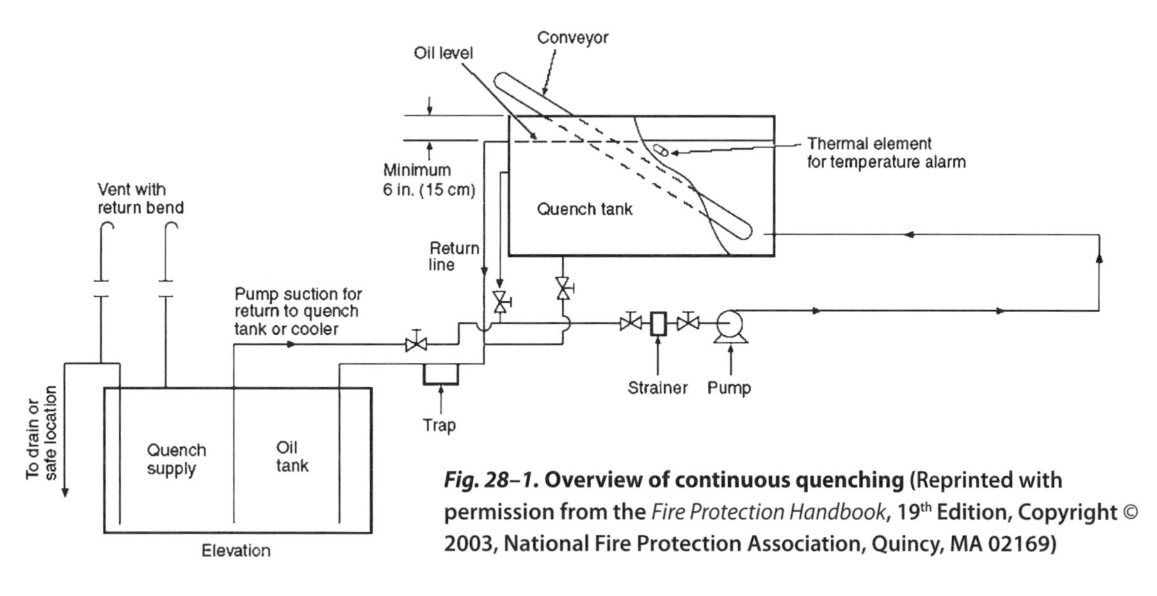

Fig. 28–1. **Overview of continuous quenching (Reprinted with permission from the *Fire Protection Handbook*, 19th Edition, Copyright © 2003, National Fire Protection Association, Quincy, MA 02169)**

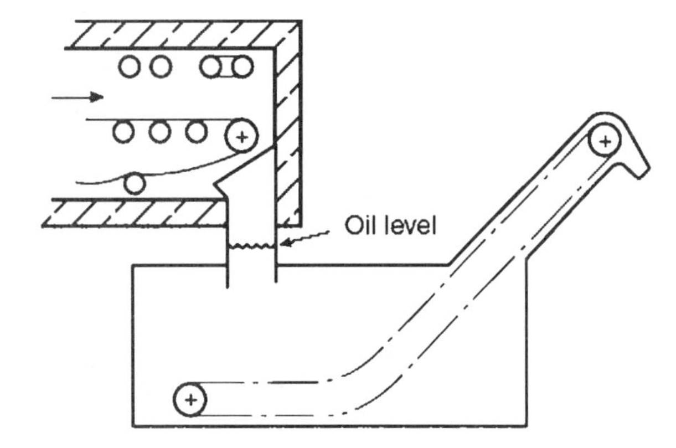

***Fig. 28–2.* Overview of chute-type quenching**
(Reprinted with permission from the **Fire Protection Handbook**, *19th Edition, Copyright © 2003, National Fire Protection Association, Quincy, MA 02169)*

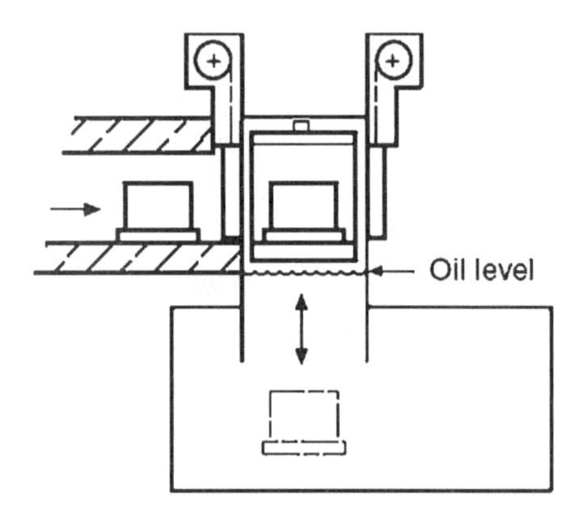

***Fig. 28–3.* Overview of dunk-type elevator quenching**
(Reprinted with permission from the **Fire Protection Handbook**, *19th Edition, Copyright © 2003, National Fire Protection Association, Quincy, MA 02169)*

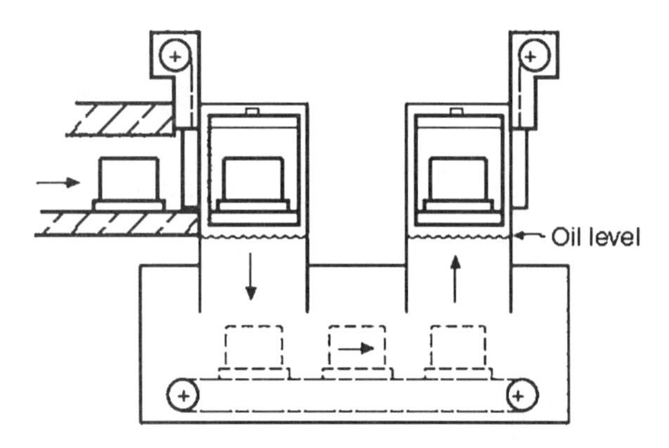

***Fig. 28–4.* Overview of transfer-type elevator quenching** *(Reprinted with permission from the* **Fire Protection Handbook**, *19th Edition, Copyright © 2003, National Fire Protection Association, Quincy, MA 02169)*

of quench oils, the process presents serious fire hazard. There are two basic types of oil-quenching processes, batch and continuous (automatic or semiautomatic) quenching.[1]

In a batch oil-quenching process, a large tank with quenching oil is heated to a predefined temperature. In many cases, the quenching oil is some type of mineral oil or a mixture of mineral oil and animal or vegetable oils. The operating temperature ranges from 100°F to 400°F, with an average flash point associated with these types of oils in the range of 500°F. The quenching tank should be provided with overflow drains, liquid level control, and enough freeboard design to prevent splashing. *Freeboard* is defined as the distance from the quenching oil surface to the top of the tank. Above the quenching tank, you will usually find a crane or some sort of lifting device that is used to pick up the metal material and maneuver to the tank. The crane size varies with the specific operation and can range from a few tons to tens of tons.[2]

In the basic operation, a supply of metal stock, which can be in many forms, such as rolled or already cut to a specific design, is staged in an area where the crane or lifting device has access to it (fig. 28–2). The crane lifts the device and quenches the material at a specified rate into the tank. Sometimes, because of the pretreatment of the material, flashing can occur at or near the oil surface. This is typical of many operations. Figures 28–3 and 28–4 illustrate other types of quenching operations. It is important to note the level of housekeeping in the area to prevent the buildup of combustibles as well as the continuity of combustibles to prevent fire spread. After the quenching operations take place, it is important to allow the material to cool to preserve the proper *crystalline* structure on the surface. Cooling is usually done by water immersion or sprays. The rate of cooling is closely monitored to achieve the desired outcome.

Continuous operation, whether automatic or semiautomatic, involves more than just a crane and a quenching tank. Continuous operation involves elevators, conveyors, hoists, and cranes, or some combination of these to dip the material in the tank, move it through the quenching process, and remove it from the oil tank. This process is more complicated than a batch operation and involves larger tanks, thus increasing the amount of oil present and increasing the fire hazard. In a continuous process application, oil is supplied via a quench supply tank, where oil is pumped to a tank and recirculated back to the quenching oil supply tank. A conveyor or similar means is used to dip the material into the tank and return it to the surface. From there, the conveyor takes the material to an area to begin the cooling process, similar to the batch application. As with the batching operation, flash fires can occur at or near the surface during entry of the material.[3]

In both cases, the temperature of the oil is an important factor as it relates to the quenching process. Control of the oil temperature is designed within specified limits relating to a particular heat-treating process. This is based on the desired outcome, final metal chemistry, and the intended use of the final product. If the temperature controls fail, overheating of the oil will increase the hazard. It is important that the cooling system used to maintain proper temperatures does not enter the oil-quenching tank or any other water (fire sprinkler or hose) as well. If an excessive amount of water is introduced into the tank, a boilover can occur if material in process is immersed in the oil tank.

Safety controls are installed on various parts of the system to ensure that the operation shuts down during unsafe or abnormal conditions. These safety controls include emergency drains, oil temperature controls, tank level indicators, loss of pump pressure, and other manual overrides to prevent partial immersion conditions. Partial immersion conditions are discussed in the next section.

Annealing is another heat treatment process that is used to heat metal close to its melting point and cool it over a long period of time. The result of this causes billions of tiny atoms to align themselves in such a way that it produces tough and hardened structures in some types of metals. In this hardening process (using controlled atmospheres), steel is heated to the critical temperature to make the steel more suitable for coating, rolling, or hardening. There are different types of annealing such as stress relieving, full anneal, and *spheroidization.* In these cases, annealing or tempering is used to soften and make the steel more ductile, as well as reducing any internal stresses. The processes can be a batch process or continuous process, similar to the oil quenching.

In a batch annealing process, the metal material can either be in raw form, such as a roll, or an already cut and designed piece, such as an aircraft part or hand tool like a screwdriver. A batch annealing furnace can be small and simple, similar to a large baking oven, or very large bell-shaped equipment that can house up to three rolls of coiled steel on side (fig. 28–4). Either process involves some sort of natural gas to the equipment as a fuel for the heat source.

Similar to the quenching process, a crane or hoist is used to load the material into the batch annealing furnace (fig. 28–5). Once enclosed in the

Fig. 28–5. **Loading the batch annealing furnace**

furnace, gas is supplied, and the heating process begins (fig. 28–6). This is a lengthy process and could take up to 6 hours for the cycle of heating and cooling (fig. 28–7).

Fig. 28–6. **Crane lowering furnace housing**

Fig. 28–7. **Batch annealing furnaces in operation**

In a continuous annealing process, a conveyor moves the desired material to the annealing furnace, where automatic doors open, and the material is permitted to enter. Some smaller annealing furnaces, especially in automatic continuous operations that use controlled atmospheres, use a flame curtain surrounding the entrance and exit of the annealing furnace to help reduce the amount of oxygen rushing in and also help maintain the internal temperature as the doors open and close. They also burn off any combustible gases that may escape during the time the doors are open. In some facilities, there could be dozens of these furnaces operating at the same time.

Controlled atmospheres, another form of annealing, use special generators to create various gases for hardening the metals. The generators create various types of atmospheres inside the furnaces to achieve a desired effect. Some of these atmospheres include

- Inert: Use of nitrogen, argon, and helium.
- Endothermic gas: Generated by *cracking* methane gas into 40% nitrogen, 40% hydrogen, and 20% carbon monoxide.
- Hydrogen: Either commercially purchased or generated.
- Exothermic gas: Methane or propane burned in a highly controlled ratio produces this gas, which can have 1% to 21% carbon monoxide.

COMMON HAZARDS

The most common hazard associated with the quenching operations relates to the large amount of flammable liquids present. However, with the annealing processes, the most common hazard originates with some sort of flammable or toxic gas, whether it is simple methane or one of the specialized atmospheric gases.

In quenching facilities, one of the more common hazards is a partial immersion. During the quenching process, if the crane or hoist fails, and the material is not fully immersed, a flash fire can erupt. If housekeeping is not adequate, this flash fire can spread to other areas via spilled oil. The large amount of combustible oil in the quenching tank or the holding tank also presents a hazard, that is, if this oil catches fire or is exposed to a fire. Oil pumps are also present, and a pump seal or gasket failure can lead to a quenching oil spray fire. These fires are particularly dangerous if the built-in safety controls fail to operate. Because of the nature of these types of facilities, a fine mist of oil is sometimes present over most of the area, including areas that are remote to the actual quenching operations. Good housekeeping must be maintained to reduce this risk. Many of these units have a local carbon dioxide system available in the event of a fire (fig. 28–8). These systems are usually manual, and the nozzles are located at the surface. Operators should be trained in and practice using these systems.

Fig. 28–8. **Carbon dioxide discharge on dip tank**
(Courtesy of Ansul)

Storage of the oil can also be a hazard, specifically, the method and amount of storage. If possible, storage of the oil should be outside in an unattached structure or clearly segregated from the actual process.

Annealing process hazards involve the presence of flammable gases. As mentioned earlier, even in annealing, where metals are softened, the furnaces are supplied with some sort of flammable gas. This gas is in itself a hazard, but now it is brought into the building and used as a heat source. Therefore, all safety precautions normally followed for dealing with flammable gases should be followed. Some of the more common dangers include equipment explosion due to the presence of an explosive atmosphere inside the furnace, crane malfunction, and operator error.

In the event of failure of the safety control, the right mixture of air and flammable gas could cause an explosion during the ignition phase of the process. This usually involves a single furnace with collateral damage to the surrounding areas. Crane malfunctions have been known to happen. Large overhead cranes moving around the area while large bell-shaped batch annealing furnaces are operating can spell disaster. If a crane, while carrying a coil of steel, hits an annealing furnace while in operation, it can lead to an explosion or fire. Also, if a crane or hoist were to drop a metal coil or part onto the fuel control systems or piping, this could also lead to the same consequences. Operator error, as in most industrial facilities, also tends to be a problem. Lack of process understanding, complacency, or errors in judgment can lead to a devastating effect.

FIREFIGHTING TACTICS

As with all firefighting tactics and discussed many times in this book, a well-planned and tested emergency response plan is needed. Preincident response planning must be carried out in cases where a building is identified as having the potential to develop into a major incident. Quenching or annealing facilities are no exception. In the event of a fire at this type of facility, it is advantageous to have information readily available that relates to the structure, fire safety provisions, location of water sources, presence of large amounts of oil storage, gas supply safety shutoffs, and any other information that may be relevant. Good loss prevention starts with the firesafe design of the plant, and firefighters should become familiar with the design features. The use of sprinklers, noncombustible construction, and proper housekeeping are a preferred choices for these types of facilities.

In the event of a fire, location of an adequate water supply is important. Supplying the sprinkler system is always important. Begin to locate power and gas cutoffs if applicable.

Before entering a quenching facility, try to determine the following:

- What type of fire (spray or tank)
- Type of oil involved
- Safety shutoff locations
- Is there product in the tank? Partial Immersion?

Remember that a response to a quenching facility means the presence of both oil, sometimes in large volumes, and oil mist in the area could lead to more than rapid fire spread. Ensure that adequate water and perhaps foam are available. If a tank fire has occurred, remember your tank fire tactics from chapter 31. Do not plunge your fire stream into the oil tank, but rather "bounce" the foam/water solution off the sides of the freeboard to cover the surface. If possible, have an operator remove the product from the tank using the protection of a foam blanket. Don't forget that the presence of a manual carbon dioxide system can aid in the extinguishment of the fire. In the event of a spray fire, simply cool the area where the leak is coming from, and get plant personnel to isolate the power to the pump. Continue to cool the area to prevent further rupture of the pipe or flange.

If responding to an annealing facility, the initial response issues remain the same; preplan, water supply, and utility shutoffs. Before entering an annealing facility, try to determine the following:

- Process description: batch or continuous?
- Type of furnaces: controlled or special atmosphere?
- Location of utilities: gas and electric shut-offs?
- Nature of event: fire or explosion?

Response to an annealing facility is different from a quenching facility. As mentioned earlier, a quenching facility mainly deals with oil and not flammable gases. Responding to an annealing facility with the presence of a flammable gas is similar to responding to a pipeline or chemical process fire; flammable gas may be escaping via a failure or damage to a piece of equipment or pipe. With these types of emergencies, foam will not help. Because extinguishing a gas fire is not desirable, your basic priority is to cool the involved and surrounding area until the gas supply can be safely shut off. Do not try to extinguish a gas-fueled fire, even at an annealing facility, until you are certain that the gas supply can be isolated.

Responses involving partially submerged products should immediately be removed from the tank. Leaks from tanks should be isolated as quickly as possible. In the event that a piece of equipment is damaged, and a gas leaks occurs without ignition, remember the tactics from chapters 12 and 13. When vapor clouds are present, special precautions must be taken to protect personnel from a potential explosion. Strategic decisions regarding vapor clouds are complicated, and it is important to suppress the vapors and control ignition sources. All of this information should be identified in your department's planning efforts.

CONCLUSION

Fires and other incidents involving quenching and annealing operations can be very challenging. This is mainly attributed to the presence of oil in large amounts or flammable gas inside closed environments. Knowing the type of process, associated process hazards, and the nature of the event will aid in the determination of firefighting tactics during your response.

NOTES

[1] National Fire Protection Association (NFPA). 2003. *Fire Protection Handbook,* 19th ed., vol. I. Quincy, MA: NFPA, p. 6–175.

[2] NFPA, *Fire Protection Handbook,* p. 6–177.

[3] National Fire Protection Association (NFPA). 1990. *Industrial Fire Hazards Handbook,* 3rd ed. Quincy, MA: NFPA, p. 723.

29

Aircraft and Aircraft Facilities

INTRODUCTION

Aircraft rescue and firefighting is in many ways similar to structural firefighting, but there are some significant differences. Typically, in a structure, there will be several paths of egress for the occupants that were designed to remain open and be used at any time during an emergency. Passenger aircraft, however, have a large number of occupants in a relatively confined space. Paths of egress can only be used while the aircraft is on the ground and has stopped moving. Highly trained flight crews facilitate the egress of passengers through strong commands and hands-on direction. Passengers are channeled to the nearest exit and directed to leave the plane through gestures and verbal commands.

Aircraft are compact carriers of personnel, fuel, and hazardous cargo, and they are constructed of exotic composite materials. There are numerous other types of aircraft that are described in more detail later in this chapter. Special planning and training are required for all firefighters to manage these incidents effectively. One place to start your planning is with the aircraft facilities.

Fig. 29–1.* An aircraft being assembled in a large hangar** *(Reprinted with permission from the* **Fire Protection Handbook, 19th Edition, Copyright © 2003, National Fire Protection Association, Quincy, MA 02169)*

AIRCRAFT HANGARS

Aircraft hangars are designed to shelter one or more aircraft. Because of the high value of aircraft, most hangars are protected by fire suppression systems. Hangars are designed to National Fire Protection Association (NFPA) Code 409, *Standard on Aircraft Hangars,* and other building codes such as the International Building Code (IBC; fig. 29–1). Suppression systems are designed to control fuel and support equipment fires that threaten aircraft, adjacent aircraft and the hangar structure. Many of these structures will have high bays, large doors, and expansive floor areas. Typical construction will be steel vertical and horizontal structural members.

The safest and most expedient entry into aircraft hangars is through an open main hangar bay door. If the main doors are closed on arrival, the incident commander (IC) may order the doors opened. The code or standard requires that the power remain on for the doors even after the activation of the suppression system. Firefighters may have to enter the hangar through personnel doors to operate the electrical door controls for the bay doors. Many hangars are also equipped with exterior hangar door control switches. Extreme care must be taken when operating hangar doors since personnel and equipment could be crushed by the rolling doors. Firefighters may be injured or hoselines cut if the doors are moved without coordination with all personnel on the scene.

Hangars may also be equipped with smoke management systems. There may be fans that operate with the suppression system or panels that melt when exposed to heat. Actuation of the exhaust fans may be manual. If an uncontrolled fire burns inside of a hangar, the roofing material may become involved. This would indicate intense heat inside of the hangar and a strong possibility of structural damage. The possibility of roof collapse must be considered before committing personnel.

Another type of aircraft hangar may be a tension membrane structure. These structures are commonly known as RUBB buildings after a major building manufacturer. Modern membrane structures are surprisingly resistant to the elements and even fire. The fabric will burn away but not support combustion. There are several documented cases where a severe fire occurred inside of the structure and burned away the membrane exposed to fire, but the metal frame structure survived intact.

Engine-testing facilities

Engine-testing facilities are designed to operationally test aircraft engines. Some engine-test facilities are designed for the entire aircraft to be inside and others for the engine to be placed in a test stand without the aircraft. These facilities are used to troubleshoot engines and simulate thrust conditions as if the aircraft were flying. At some facilities, engines can be tested at full throttle and even after burner for sustained periods.

Engine-test facilities are typically constructed of concrete. These facilities may also be referred to as a *hush houses,* because they have a sound-dampening exhaust chamber to channel the engine exhaust and greatly reduce the noise produced. A control room is provided for the engine mechanic to visually monitor and control the engine during testing.

Automatic and manually operated fire suppression systems are typically required to rapidly extinguish a fire should it occur. Common fires are from unburned fuel in the engine exhaust, which could be quickly extinguished by aircraft maintenance personnel by increasing the engine thrust and blowing the fire out. An exhaust fire may also occur as the engine is being shut down from unburned fuel. External fuel and hydraulic fires could also occur under the engine nacelle. The *nacelle* is the external engine covering. Less frequently, an engine could catastrophically fail during testing. As indicated, these structures are built of concrete and designed to withstand high heat. As with other types of aircraft fires, the main risk of fire in the engine test facility is from the associated fuel and hydraulic systems.

Extreme care should be taken if entering the engine test portion of the facility because of the potential for extreme heat. If an aircraft engine is relatively intact, consideration should be given to using a clean agent such as carbon dioxide or Halotron™. If dry-chemical or foam is used in an aircraft engine, the engine will have to be completely refurbished. That being said, foam and dry-chemical should be used, however, if the fire is extensive and could jeopardize the aircraft or test structure.

Aircraft engine maintenance and storage facilities may also be encountered. These facilities are less hazardous and probably protected by suppression systems. The most likely type of suppression system is a closed head sprinkler system. Only small quantities of aircraft fuel and oils will be present. General structural firefighting methods will be effective in these facilities.

Fuel cell maintenance and aircraft paint facilities

Aircraft fuel cell maintenance and aircraft painting facilities are similar in design. Fuel cell maintenance and aircraft painting is performed in facilities that look like fairly typical aircraft facilities.

The difference in these facilities and regular aircraft facilities will be the air exhaust systems and electrical classification of the building. Most aircraft hangars have electrical systems that are classified as explosion proof to some degree, but fuel cell maintenance and painting facilities are designed to a much higher standard. Explosive vapors may be present at any level in these facilities. Great care has been taken in the design of these facilities to eliminate potential sources of ignition. High volumes of air move through the facility by intake and exhaust fans to keep vapors below the flammable limits. Gas detectors are also used to constantly monitor the atmosphere in the facilities. Entry to these facilities is strictly controlled to ensure that personnel do not carry sources of ignition into the potentially explosive areas. Stray electrostatic charges from maintenance personnel have ignited vapors in these facilities.

In the aircraft fuel cell maintenance facilities, maintenance technicians open the aircraft internal fuel cells. Even though the aircraft is defueled, there is still standing fuel in the cell. Maintenance personnel use absorbent materials and drying methods to remove as much of the hazard as possible. Some fuel cells are large enough for a person to enter and require entry to perform the maintenance action. Many of these fuel cells are classified as permit-required confined spaces. Efforts must be made to check these spaces to ensure that maintenance personnel are not inside. Remember that any equipment that you take into these facilities must be intrinsically safe.

Fuel cell maintenance and painting facilities will most probably be protected by a fixed fire suppression foam system. Firefighters may encounter quantities of paint and solvents in the aircraft-painting facilities, as well as small quantities of fuel in the fuel cell maintenance facilities.

COMMON HAZARDS

Aircraft pose many hazards to firefighters. The various types and classifications of the different aircraft pose unique concerns. Consider the following types of aircraft that could be involved in an incident in your jurisdiction. Some types include helicopter (cargo and personnel), light aircraft, passenger aircraft, cargo aircraft, military aircraft, and variations of all the above. Hundreds of aircraft crash every year around the country. It is only a matter of time before an incident occurs in your response area.

Some other common hazards associated with aircraft are aircraft engine exhausts and intakes. Generally, firefighters should not approach within 25 feet from the engine intake and 150 feet to the engine exhaust. Personnel may be sucked into aircraft engines or severely burned by aircraft exhaust. Wind velocity from the exhaust may be strong enough to turn over vehicles.

Wheel and brake fires are common. Wheel and brake assemblies become hot from friction caused by the aircraft brakes. Heat transfers from the brake assembly into the wheel and tire. The wheel assemblies contain fusible plugs designed to blow out at high temperatures. Sometimes, however, heat builds up so rapidly that the sidewall of the tires fails with explosive and deadly force. Firefighters can be seriously injured or killed from exploding wheel assemblies. Water, foam, carbon dioxide, and other cooling agents should not be applied to wheel assemblies if the tires are still inflated. The agent of choice is dry-chemical because it does not rapidly cool the brake assemblies. Dry-chemical should be repeatedly applied to burning wheel assemblies until the fire is extinguished and the assemblies are below the ignition temperatures of the hydraulic fluid and rubber tires. Firefighters must avoid the sides of the wheels and approach from the front or the rear. Wheel assemblies are also made of combustible metals such as magnesium. These fires may be extinguished using a Class D dry-powder fire extinguisher.

Large quantities of fuel and cargo are carried on large-frame aircraft. Tens of thousands of gallons of fuel could be carried on these aircraft. Firefighting vehicles must not approach from downhill unless absolutely necessary. Firefighting vehicles have caused ignition of spilled fuels and have been destroyed by fuel fires flowing from burning aircraft. Fuel spills must be secured with foam, and the foam blanket must be reapplied as necessary. Wind, heat, and rain will have a negative effect on the foam blanket. In addition to the fuel, almost any type cargo is carried on aircraft. Everything from personal belongings to explosives may be encountered. If available, the airbill should be referenced. The airbill is usually found in the cockpit, but also may be attached to the outside of the packages.[1] The pilot is the responsible person for the airbill. The control tower may have additional information from the aircrew concerning the number of personnel on board, as well as fuel, weapons, and cargo.

Military aircraft pose additional and different dangers. Military aircraft come in various configurations. Some are very similar to commercial aircraft and pose no greater risk. However, it is prudent to consider all military aircraft as carrying munitions or hazardous cargo. Further, the military has been developing and deploying unmanned aerial vehicles (UAVs) such a the Predator in recent times. These UAVs may also carry live munitions.

The military uses the US Department of Transportation (DOT) and United Nations (UN) system to categorize munitions. Hazards from military munitions range from small arms ammunition (Class 1, Division 4) to bombs and missiles (Class 1, Division 1). Initial withdrawal distance should be determined by utilizing the DOT Emergency Response Guide (ERG). Further specific guidance should be obtained from military personnel once the aircraft cargo or munitions load is known.

Aircraft munitions are unlikely to detonate when involved in a crash. This being noted, firefighters cannot rule out a detonation either from the impact of the crash or possible resulting fire. Weapons can be classed as hard bombs, such as 500-lb guided and nonguided bombs, and thin skinned, such as missiles and rockets (figs. 29–2 to 29–5). Other subcategories such as bomblets are dispersed from a canister type of weapon. Often encountered at the scene of crashed military aircraft is machine gun ammunition such as 20- and 30-mm shells. Even these relatively small weapons can kill firefighters. Special weapons such as chemical munitions are not part of the active weapons inventory.

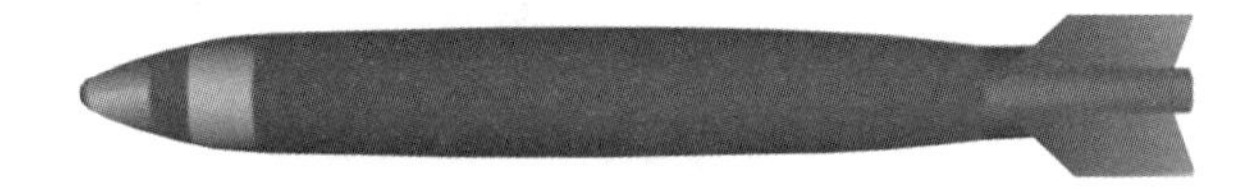

Fig. 29–2. **An example of a hard-cased bomb. Note the yellow band on the nose indicates it is a live weapon.** *(Courtesy of the United States Air Force)*

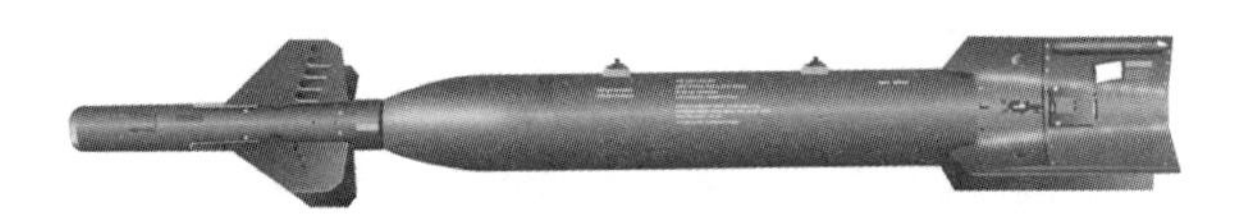

Fig. 29–3. **An example of an inert guided weapon (note the blue band). A live weapon of this type** *would contain high explosives but no rocket motor. (Courtesy of the United States Air Force)*

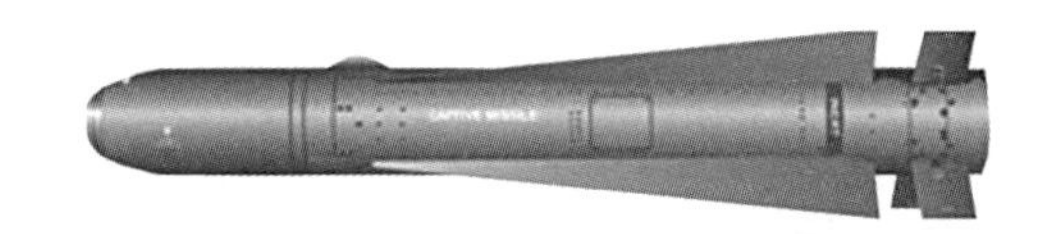

Fig. 29–4. **An example of an inert missile. A live weapon would contain high explosives and a rocket motor.** *(Courtesy of the United States Air Force)*

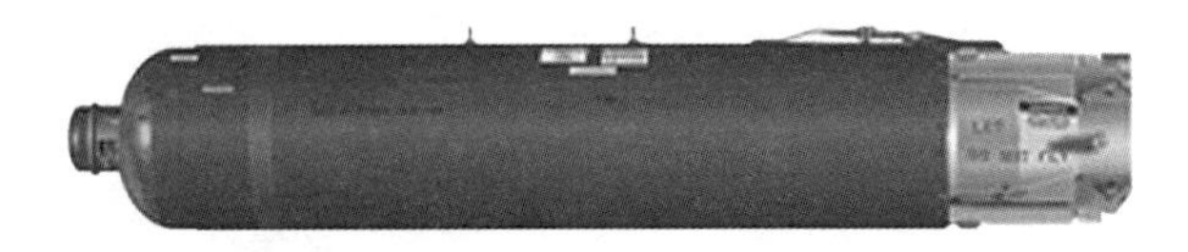

Fig. 29–5. **An example of a canister type weapon designed to dispense multiple smaller munitions.** *(Courtesy of the United States Air Force)*

Conventional military weapons contain high explosives. The munitions could detonate as a result of an aircraft crash, but it is unlikely. An explosion is more likely when weapons are involved in fire. Hard-case weapons are more resilient to external heat than thin-skinned weapons such as missiles. Firefighters should not approach military aircraft crash sites until it is confirmed that there are no weapons on board the aircraft.

Conventional munitions explosives material may become liquid when exposed to heat and flow out of the weapon's case. Care must be exercised by responding firefighters to ensure they do not drive over or step on the resolidified explosive materials. These resolidified explosives may be pressure sensitive and detonate when compressed by a tire or boot.

In addition to the detonation hazard presented by military munitions, there is also a projection hazard presented by missiles and rockets. When missiles or rockets are involved in fire, they can actually launch. Extreme care must be taken to ensure that emergency vehicles and personnel remain clear of the front and rear of the aircraft. Firefighters must remain clear of the rear of the aircraft because of the hazards presented by the hot exhaust from the missile engines. Color coding is used on munitions to help identify active or training munitions. For example, blue (either for the entire weapon or a band) is used to identify inert munitions. Yellow and black banding is used to mark munitions with live warheads or rocket motors. Extreme caution must be exercised even with blue practice munitions since they may contain spotting charges or even live rocket motors. The local military response agency can assist by providing orientation of specific weapons and their color coding.

The low likelihood of the detonation of military armament is illustrated by the June 16, 2005, crash of a military attack aircraft in Yuma, Arizona. An AV8 Harrier Jump Jet crashed into a residential neighborhood. Some of the smaller machine gun rounds did explode, but the four 500-lb bombs did not. Emergency officials prudently evacuated a 1-mile area around the crash. If weapons are encountered they, should not be moved. Explosive Ordinance Disposal (EOD) should be called to locate and disarm any munitions encountered. Typically, an EOD unit may be notified through a local military base. The senior fire official should use extreme caution and strongly consider evacuation of the immediate area around any munitions. Fighter and attack aircraft carry a wide variety of weapons (fig. 29–6).

Fig. 29–6. **An A-10 attack aircraft firing its 30-mm Gatling gun. It can carry almost any type of weapon.** *(Courtesy of the United States Air Force)*

Aircraft incidents involving nuclear weapons are possible but highly unlikely. Furthermore, the possibility of a nuclear weapon detonating in a crash is almost impossible. If a nuclear weapon is involved in a crash, there is a chance of radioactive material being released from a burning weapon. Radiation in the form of alpha particles could contaminate firefighters and equipment. Specific hot, warm, and cold zones must be established. Vehicles or personnel who have worked in the hot zone should not leave the warm zone without being decontaminated. Everything inside the warm and hot zones must be considered contaminated until proven otherwise. There is also a quantity of radioactive material associated with nuclear weapons and conventional explosives material. Time, distance, and shielding must be considered when dealing with nuclear materials. The closest federal government agency (e.g., United States Coast Guard, Federal Bureau of Investigation, military unit, etc.) personnel will assume control of the nuclear weapons accident once they arrive.

As with all firefighting operations, firefighter safety must not be jeopardized when there is nothing to gain, or if there is a possibility of an explosion. The incident command system (ICS) must be implemented on all fires. The area must be sectored and tactical objectives established.

Other hazards associated with military aircraft include aircraft ejection systems on fighter and some bomber aircraft. These ejection systems are designed to jettison the canopies and propel the pilot and seat several hundred feet into the air in less than a second. The pilot must initiate the ejection sequence by pulling an ejection handle. In some cases, the aircrew members may fail to eject and will crash with the aircraft. Extreme caution must be exercised when approaching a military aircraft. The rocket motor for the canopy ejection system may present an extreme hazard to firefighters, because a disoriented pilot could initiate the ejection sequence as firefighters are working to remove him or her from the aircraft.

All fire departments should have an up-to-date electronic or hard copy of United States Air Force Technical Order 00–105E-9. This technical order (T.O.) contains information on all United States military, NASA, governmental, and NATO aircraft, as well as many civil aviation aircraft. The T.O. covers hazards associated with the aircraft, aircraft entry, aircraft shutdown, and aircrew and passenger removal. This T.O. can be downloaded from www.afcesa.af.mil/cex/cexf/cex_firemgmt.asp, or you can write or call the following:

HQ AFCESA/CEXF
139 Barnes Drive, Suite 1
Tyndall AFB, FL 32409
1–888-AFCESA-1

Fire departments located close to military installations should get hands-on training on aircraft operating in their jurisdiction. If your department has a mutual aid agreement with a military or other government organization, it should provide training for your personnel on its aircraft and possibly live training fires (fig. 29–7). Joint exercises should also be conducted regularly to make sure that all participants understand their roles. Military and other governmental craft may present significant explosive and other types of hazardous materials hazards. In addition to the mass detonation and explosion hazards, there may be radiation, composite fibers, and exotic materials such a hydrazine, which is used as a fuel for auxiliary power units (APUs) and emergency power units (EPUs). If firefighters are assigned to perform aircraft rescue and firefighting duties, they are required to be trained for these duties according to NFPA 1500, *Standard on Fire Department Occupational Safety and Health Program.*

Fig. 29–7. **Foam used for live training fire** *(Courtesy of the United States Air Force)*

Unfortunately, we have seen catastrophic incidents involving NASA craft. The crash of the space shuttles Challenger and Columbia spread hazardous materials over large areas in several states. T.O. 00–105E-9 is an excellent source of emergency information on NASA craft. The shuttle contains numerous hazardous materials such as hydrazine, liquid hydrogen, liquid oxygen, and solid rocket fuel.

As a result of the attacks of September 11, 2001, more military fighter aircraft are flying with live munitions to counter a potential hijacking or hostile aircraft. Aircraft may also be equipped with dummy munitions for pilot training. Even though they are dummy or training munitions, they may be equipped with spotting charges or live rocket motors. They should always be handled by trained munitions experts.

***Fig. 29–8.* High-expansion foam discharging in a large aircraft hangar** *(Courtesy of the United States Air Force)*

FIREFIGHTING TACTICS

Hangars and maintenance/testing facilities

The general rules of supporting fire suppression systems as one of the first priorities applies to hangars as well as aircraft maintenance and testing facilities. Effective operation of the suppression system is the most effective method of extinguishing these fires. On large structures such as aircraft assembly and component storage facilities and large aircraft hangars, there may be no fire department connections. Many of the larger structures will be supported by dedicated water and foam systems that require water quantities beyond the capabilities of most fire vehicles. Aircraft hangars that have aircraft not considered purged of fuel (even defueled aircraft may contain hundreds of gallons of fuel) should be protected by some type of foam suppression system. Even though these systems may not be equipped with fire department connections, the systems may require manual actuation of the fire-water pumps or foam system. Various types of foam suppression systems include open head deluge, high-expansion foam, and closed head sprinkler systems. Actuation of these systems could be through heat detectors, smoke detectors, or flame detectors (fig. 29–8). There are many types of foam, including high-expansion, aqueous film-forming foam (AFFF), fluoroprotein (FP) and film-forming fluoroprotein (FFFP). AFFF and high-expansion foam systems are predominant.

Purged aircraft (such as aircraft that return for major upgrades) and new aircraft (never fueled) may be protected by water sprinkler systems. Closed head water sprinkler systems are more likely to be equipped with fire department connections for support of the suppression systems.

Firefighting in these large structures can be complicated by their size and complexity. Aircraft maintainers are trained to remove aircraft from burning structures; however, aircraft under maintenance may not be able to be removed even if time permits. These aircraft may be partially disassembled or on jack stands, so removal may not be possible.

NFPA 409 requires that power remain on aircraft doors after the suppression system actuates. If there is no power for the doors, extensive effort may be required to open these doors, sometimes including aircraft tugs and rigging. To gain access to aircraft, these doors will have to be opened. Initial entry to aircraft hangars and large facilities may have to be through personnel doors in order to access the main hangar door controls. Crews entering to actuate fire suppression systems or to open the doors should be protected by handlines. Furthermore, entry crews must wear self-contained breathing apparatus (SCBA). Because of high ceilings and large volume, the atmosphere inside of these structures may at first appear safe even in the presence of a large fire. Fire may rapidly spread because of fuel release, oxygen equipment involvement, structural collapse, or explosions.

Uncontrolled fires in these structures could result in structural collapse without warning.

Explosions may occur for many reasons, most notably from oxygen cylinders, fuel tanks, and even associated welding cylinders. Any closed vessels or piping could BLEVE (boiling liquid expanding vapor explosion). Aircraft involved in a fire could release highly toxic fumes and carbon fiber materials. These carbon fiber materials will disable electronic equipment such as radios and are hazardous to firefighters through inhalation. Again, all firefighters involved in aircraft incidents must wear SCBA.

One type of facility that firefighters may face is an engine-test cell. Firefighters should never enter the testing portion of these facilities until the engines are shut down. Extreme danger is presented from both the engine intake and exhaust. Again, specialized suppression systems may be present and represent the best method for extinguishing these fires. These systems may be either foam or gaseous suppression systems.

Preincident response planning and training should take place on these specialized aircraft facilities. As mentioned, there are many hazards and complexities associated with these facilities—high-volume water systems, dedicated fire pumps and foam systems, and aircraft hangar doors, to name few. Since many of these facilities are associated with active airports, just accessing the facilities may take special identification and access badges. Since 9/11, airfield security measures have been further tightened. Security personnel who would have once simply waved emergency vehicles through checkpoints have been warned that terrorists may use these vehicles in suicide bombings. Airports are required to develop disaster plans that cover security procedures for personnel to enter the airport area. Your department should know what these procedures are and practice them during the annual exercises. Special entry and exit gates, as well as staging areas, are part of these plans. Unannounced emergency vehicles will be suspect and could result in the use of deadly force and the injury to security and fire protection personnel.

Fighting aircraft fires

Aircraft fires present many challenges to the structural firefighter. As mentioned earlier, there is likely to be a large number of people involved. If at all possible, passengers will try to escape on their own power. Rapid extrication of several hundred casualties would be extremely labor- and time-intensive. Furthermore, the survival of passengers in a burning aircraft rapidly diminishes after 3 minutes. Structural firefighting crews would arrive most likely after the 3-minute threshold. Imagine the complexity of rescuing 300 injured victims from an aircraft disaster.

For example, in a best-case scenario, 50 firefighters are on the scene of a major disaster requiring the rescue of 300 victims. Many of the 50 firefighters will have duties such as driving the response vehicles, supporting the firefighting operations, commanding the scene, acting as safety officers, and being part of rapid intervention teams, and maybe even triage and medical treatment. This leaves very few firefighters to actually fight the fire and extricate the victims. The most critical action a fire protection team can take at the scene of an aircraft incident involving a large number of casualties is to protect the paths of egress so that passengers can escape on their own power. More passengers will be saved using their own means of escape than could ever be saved by firefighting and rescue forces. Protecting the paths of egress and opening hatches and doors are the most likely actions firefighters can take to save the largest number of victims in a mass casualty situation (fig. 29–9).

***Fig. 29–9.* Aircraft sometimes crash relatively intact.**
(Courtesy of the United States Air Force)

The type of fire involving an aircraft will most likely depend on the cause. In other words, if the aircraft impacts the earth, there is a high likelihood that there will be a large initial fire—but not always (fig. 29–9). If the fire is a result of a maintenance accident, the fire will likely start small but could grow rapidly. Large fuel fires are best fought with mass applications of foam by aircraft rescue and firefighting (ARFF) apparatus. Dry-chemical fire extinguishers and vehicle-mounted systems should be used to enhance the effectiveness of AFFF for pressurized fires and for three-dimensional fires (fig. 29–10). Remember that AFFF alone will not extinguish a three-dimensional fire, and dry-chemical will not seal the surface of spilled fuel. Dry-chemical and AFFF should be used together.

Fig. 29–10. **Dry-chemical agent is critical on three-dimensional fires.** *(Courtesy of the United States Air Force)*

ARFF firefighters are required to wear proximity (aluminized) fire protective clothing that reflects most of radiant heat, compliant with NFPA 1976, *Standard on Protective Ensemble for Proximity Fire Fighting* (fig. 29–11). Structural fire PPE meeting NFPA 1971, *Standard on Protective Ensemble for Structural Fire Fighting*, requirements will normally provide adequate protection but to a lesser degree when combating large fuel fires. NFPA 1500, *Standard on Fire Department Occupational Safety and Health Program*, requires firefighters with a primarily ARFF mission to be equipped with proximity PPE. Structural firefighting PPE can be used by municipal firefighters to combat aircraft fires. The main disadvantage when using structural PPE is the limited radiant heat protection when combating the large fuel fires in the initial stages. The proximity clothing is far superior in reflecting radiated heat. When combating large fuel fires, firefighters wearing structural PPE should be aware that their PPE could become very hot. A good initial indication will be the heat sensation felt through the SCBA face piece.

Fig. 29–11. **Firefighters in proximity fire protective clothing** *(Courtesy of the United States Air Force)*

With familiarization and training, firefighters should be able to open commercial aircraft doors externally (fig. 29–12). What happens when a municipal firefighter uses a ladder to open the door? If the ladder is not positioned correctly, the aircraft's chute will push the ladder away from the door as the chute expands. If the aircrew is able, they will open the doors from the cabin and deploy the evacuation slides.

Fig. 29–12. **Nighttime aircraft fire where firefighters must enter the aircraft using ladders** *(Courtesy of NFPA)*

Consideration must be given to the hazards mentioned with military aircraft. Because of the hazards presented from the forward-firing munitions and the hot exhausts formed by the missiles, personnel should only consider approaching at a 45° angle from the front or the rear. The sides of brake and wheel assemblies must also be avoided.

Helicopter fires can be fought in a similar manner as small aircraft fires. Approach to a helicopter must be in the sight of the pilot if the aircraft is still intact. Main rotors and tail rotors must be avoided. Cabin and cargo fires are possible, and aircraft fuel fires should be expected after a crash. Doors are normally able to be opened from the outside, and windows easily removed.

After the exterior fire is under control, interior aircraft fires should be treated in most cases like structural fires. Firefighters must not waste their limited resources applying foam or water to the shell of the aircraft if the fire is inside. Firefighters must make entry or use suppression methods from piercing nozzles such as a Snozzle® (fig. 29–13). Ventilation should also be initiated to improve the atmosphere for any trapped passengers. High-volume smoke fans will rapidly clear smoke out of the passenger compartment.

Fig. 29–13. **Commuter jet crash at Teterboro Airport New Jersey and the use of the Snozzle piercing nozzle** *(Courtesy of Crash Rescue Equipment Service Inc.)*

Other considerations for aircraft fires include oxygenated and cargo fires. The passenger and aircraft oxygen system must be shut off. Firefighters will not be able to extinguish oxygenated fires. Remember that aircraft can be used to carry any type of cargo. Hazardous materials could include many items such as chemicals, explosives, etiologic hazards, and compressed gases. The airbill should be checked as soon as possible. Personnel who have been in the vicinity of an aircraft accident or fire should be decontaminated. Jet fuels, carbon fiber, hydraulic fluid, and biohazards from human remains must be considered. In addition, if personnel have been exposed to severely injured or deceased victims, they should take part in a critical incident stress debriefing program.

CONCLUSION

Aircraft facilities and aircraft present many challenges to firefighters. The technical facilities, large floor space, and potential for large fuel loads must be considered even in the hangar. The exotic construction materials and fuels, as well as military weapons and ejection systems, can be deadly. Municipal firefighters must properly prepare to handle these emergencies both in the absence of and alongside airport firefighters. Communications must be established with state and federal agencies. Mutual aid training and drills involving mass casualty scenarios should be practiced on a regular basis. Now is the time to develop the capabilities so that municipal firefighters can competently deal with aircraft and aircraft facility emergencies.

NOTES

[1] Noll, G. G., Hildebrand, M. S., and Yvorra, J. 2005. *Hazardous Materials: Managing the Incident,* 3rd ed. Chester, MD: Red Hat, p. 254.

30

Electric Power Generation

INTRODUCTION

If we look at our lifestyles, we will notice that almost everything we use or see in our daily activities involves electrical power. Electricity powers our coffee pots, electric shavers, televisions, computers, and the advertising signs we see in our towns and cities. Most everything in our lives, including our safety and security, depends on the use of electric power. We have become totally dependant on electricity.

This electricity is produced in power generation plants, and several energy sources can be used for producing the electricity (fig. 30–1). These sources include fuels such as coal, oil, nuclear energy, natural gas, wind, and water.

Although there are similarities in how the electricity is produced, the fuel used for this production has its own inherent fire risks and challenges for firefighters. This chapter looks at the processes used in the production of electricity, the hazards associated with these processes, and the various fuels used in this production. The production of electricity using nuclear power was covered in chapter 14.

Fig. 30–1.* Electrical generating plant** *(Reprinted with permission from the* **Fire Protection Handbook, 19th Edition, Copyright © 2003, National Fire Protection Association, Quincy, MA 02169)*

DESCRIPTION OF PROCESS

In the simplest terminology, the electrical generation process uses a turbine to rotate a shaft connected to a generator that produces electricity. Turbines may be driven by power sources such as steam, wind, solar, water, or other fluids. Most commonly, steam is used to operate turbines in fossil fuel power plants or nuclear power plants (chapter 14); and water is used in hydroelectric power plants.

Where steam is used to turn the turbines, large boilers are required to produce the steam. The fuel used in the boilers at a particular facility is usually stored on-site. It is conveyed from the storage area where it is prepared for induction into the boiler. Coal is pulverized for induction; and fuel oil may be heated if necessary, as in the case of numbers 4, 5, or 6 oil, then pressurized. The fuel is burned in the boiler to create steam. Pulverized coal is blown into the furnace from fuel nozzles at the four corners, and it rapidly combusts, forming a large fireball at the center. This heats the water that circulates through the boiler tubes.

Steam created by the boiler can be at extremely high pressures, up to 5,000 psi, and extremely high temperatures, (1,000°F). This steam is piped to the turbine, where it is directed against blades inside the turbine, causing them to spin. The turbine actually consists of a series of steam turbines interconnected to each other and a generator on a common shaft. There is a high-pressure turbine at one end, followed by an intermediate-pressure turbine, two low-pressure turbines, and the generator. As steam moves through the system and drops in pressure, it expands in volume, requiring larger diameter and longer blades in each succeeding turbine to extract the remaining energy. It is so heavy that it must be kept turning slowly even when shut down (3 rpm), so that the shaft will not sag even slightly and become unbalanced.

The generator produces the electricity by a process known as *electromagnetic induction.* This process involves moving an electric conductor through a magnetic field, which causes an electric current to be induced into the conductor. From the generator, the electricity is raised to transmission voltage by the use of a step-up transformer.

According to the National Fire Protection Association, more than 50% of the electricity generated uses coal as fuel, with the remaining percentage spread among nuclear, natural gas, water, and other fuels including petroleum, *geothermal,* wood, waste, wind, and solar.[1] The obvious hazards involved in electricity production are the fuels used and the process used to create the heat required for steam generation. These hazards, as well as others, will be discussed in detail later in this chapter.

Large quantities of water are needed for the cooling process in the production of steam as well as for cooling large heat loads created by the generators. Because of the large quantities of water required, many power generation plants are located near a body of water, which is then used for cooling. Some generators are cooled using hydrogen. While there are similarities involved with the process of power generation, individual generation plants differ based on local climate, power generation demands, size of the units, and so forth.

Modern turbines incorporate a condenser, which returns the steam to a liquid form after it passes through the turbine. Cooling towers are used in conjunction with the condenser (fig. 30–2). Cooling towers circulate cool water around the piping containing the steam, causing the steam to condense to liquid form.

***Fig. 30–2.* Electric power generation plant. Structures with steam emissions are the cooling towers.**

In hydroelectric generating plants, water from a reservoir flows through gates and then through the turbines spinning the shafts connected to the generators. The energy extracted from water to spin the turbines depends not only on the volume but also on the difference in height between the source and the water's outflow. This height difference is called the *head.* The water leaving the turbines is returned to the river downstream of the generation plant. In some cases, this water is returned to a reservoir where it is pumped back to the original reservoir for reuse in the power distribution process. This type of hydroelectric plant is known as a pumped-storage plant.

As we are well aware, environmental considerations and the nimby (not in my backyard) syndrome have placed many industrial facilities in remote areas. This is no different when initially locating electric power generation plants. Many of these facilities are in remote areas, and this may affect emergency response and firefighting operations if required.

The facility itself consists of a power block that houses the steam generator, turbine generator, and auxiliary equipment.[2] The buildings are usually large, open, and undivided structures consisting of a steel skeleton frame and metal panel walls. Other areas of a power-generating station consist of fuel storage and transformer areas (fig. 30–3). There may also be offices, labs, or other support buildings located on the property. Hydroelectric power plants may be located several stories underground and constructed of reinforced concrete.

***Fig. 30–3.* Electric power generation plant. Note the large fuel storage tanks.** *(Courtesy of Kevin J. Kolb, 2003)*

Control rooms are separate from the boilers, turbines, and generators, and they should have a means of egress separate from these other plant areas. In the event of a fire or explosion, operators may be trapped in control rooms and must be accounted for by the first arriving emergency responders.

To reduce emissions from electrical generation plants, particularly coal-fired plants, the flue gas system contains equipment to clean the exhausts. This helps to control the release of sulfur dioxide, nitrogen oxide, and particulate matter.[3] The equipment that is used to reduce emissions consists of flue gas desulfurization systems (also known as *scrubbers*), electrostatic precipitators, fabric filters (baghouses), and chimneys to release the off-gases. *Electrostatic precipitators* remove fibers, dust, and ash from the air.

COMMON HAZARDS

One of the first hazards associated with electric power generation is fuel storage and use. The fuels used to generate electricity and their associated hazards are described in the following sections.

Coal

Coal consists of carbon with varying amounts of hydrogen, oxygen, nitrogen, sulfur, and mineral ash. Various forms of coal include anthracite, which is a very hard and clean coal, to lignite, a very soft, brownish coal. In the past, bituminous coal was primarily used for coal-fired boilers. This is a soft coal that yields a high heat value. But, bituminous coal contains sulfur, which when burned emits sulfur dioxide (SO_2) emissions and eventually results in acid rain. To reduce sulfur emissions, many facilities that use coal-fired boilers have begun using subbituminous coal. These coals are mined in the western region of the United States, namely, Wyoming and Montana. A common source of this coal is known as the Powder River Basin (PRB) deposits.

A major hazard when storing and using coal is spontaneous heating of the coal. Spontaneous heating occurs from a slow oxidation reaction with the coal pile that causes heat to build up to the point of igniting the coal.[4] The softer the coal or lower the grade, the more likely that the coal will heat spontaneously. Freshly mined coal is also more susceptible to spontaneous ignition than coal that has been out of the mines for longer periods. High moisture content, small particle size, high sulfur content, and the extent that air is trapped in piles all contribute to the potential for spontaneous ignition. Coal storage piles should be located away from important structures at power generation facilities.

When stored for long periods, coal should be compacted. Compacted coal contains less air than uncompacted coal, which means it is less likely to have appreciable air movement through the pile, and thus a low chance of oxidation on the surface (which would produce the heat that leads to self-ignition). If the coal is not to be compacted, it should be kept moving to the boiler to reduce the storage time. This will help to reduce the spontaneous heating tendencies of coal.

Coal dust is also susceptible to flash fires and dust explosions. Facilities for handling coal include receiving areas, crushers, and conveyors (fig. 30–4). Dust may be produced at these areas and distributed as combustible dust-air mixtures. The mechanisms and power sources that operate in these areas may be the source of ignition for this combustible mixture if not properly designed, installed, and maintained. Friction between the conveyor belt and roller, belt slippage, or static electricity can be a source of ignition when dealing with conveyor belts. Automatic sprinklers may be present at areas above and below the conveyor belt (fig. 30–5).

***Fig. 30–4.* Coal conveyor** *(Reprinted with permission from the* **Fire Protection Handbook**, *19th Edition, Copyright © 2003, National Fire Protection Association, Quincy, MA 02169)*

***Fig. 30–5.* Coal conveyor. Note the automatic sprinkler piping.**

Normally, a dust explosion will not be isolated. A second, more violent explosion usually occurs immediately after the first because of additional dust being loosened from the initial explosion. This is called a *sympathetic* explosion.

Methane and other gases can be given off from coal storage when it is freshly mined or crushed. This gas can accumulate over or at the higher levels of storage silos, and it may ignite explosively. During spontaneous heating carbon monoxide gas will be created and should be anticipated by firefighting personnel. Remember that carbon monoxide is a toxic and combustible gas.

If ignition of coal does occur, a hazard exists because of the conveyor systems used to transport the coal. In some instances, these conveyors may spread the burning coal throughout the plant. Again, firefighting forces should consider this during their size-up and incident action planning.

Natural gas

Much of the electrical production today is fueled by natural gas. The normal hazards associated with natural gas should be considered, and the systems should be designed, installed, and maintained in accordance with recognized standards. The greatest hazards for firefighters will be the potential for leaks from the natural gas piping system within the facility.

Oil

In today's electric plant, very little of the energy production uses oil as a fuel source. Economics of scale limit its use in the large quantities. The major hazards involved would be within the boiler itself or the preheating of the fuel oil where it is needed. Large boiler fires similar to those in a residence building may be encountered, but on a much larger scale. These hazards are covered later in this chapter.

The storage of the oil will be in large tanks usually located aboveground. Chapter 31 addresses the hazards of aboveground storage tanks. Heavier grade fuels such as, number 6 fuel oil, require heating to keep it fluid enough, and so problems may arise with this type of heating system.

Nuclear

Nuclear fuels are covered in chapter 14. Note that nuclear energy is considered one of the safest means of power generation.

Alternative fuels

Alternative fuels used in power generation may include refuse-derived fuels (RDF), municipal solid waste (MSW), and biomass. Biomass fuels include those that are composed of forest and agricultural by-products, including wood chips, wood waste, sugar cane stalks, and the like. The hazards associated with these alternative fuels include the problems associated with bulk storage of combustible materials, dust explosions where applicable, process equipment fires, and fuel accumulations in areas not designed for safe storage.

MSW consists of normal waste associated with residential or light commercial occupancies. MSW can be burned without any specific processing as a fuel, or it can be processed into RDF. In this case, MSW is stored and shredded on-site, then conveyed to a further storage area before being used as a fuel. During the processing of MSW, flammable or explosive materials can be ignited. Shredding machinery should be located within a separate enclosure. Further hazards can result from the processing machinery itself because of the electric power, hydraulic lines, moving parts, and high pressures required.

Rubber tires may also be used as a fuel in the primary or secondary boilers. The process of conversion to a fuel includes storage and shredding on-site, further storage of the processed fuel, and conveyance of the fuel to the boiler. If the tires ignite, the fire will be very difficult to extinguish (see chapter 18). Hazards associated with this type of fuel are similar to those mentioned for MSW.

OTHER HAZARDS

Other hazards that are associated with electric power generation are covered in the following sections.

Hydrogen gas

Hydrogen gas is used as a cooling medium for many generators. Hydrogen has an explosive range of 4% to 75% volume of gas in air, and for this reason, it is very dangerous. Within the generator itself, hydrogen explosions are considered unlikely because the hydrogen is typically 100% pure and considered too rich to support combustion.[5] The hazards of this gas are the greatest during the off-loading and storage of the product at the facility. Hydrogen is received in bulk and may be found on-site in tube trailers or high-pressure cylinders. Leak detection equipment should be installed where required by applicable codes.

Rotating machinery lubrication

Rotating machinery present, including the steam turbines, requires large quantities of lubricating oil. This oil may be under pressures of 150 to 250 psi and may be combustible. When released under pressure when a piping is breached, the oil will be released with high pressure, and contact with a hot surface may cause ignition. This will result in a spray fire, three-dimensional fire, pool fire, or combination thereof. These fires will produce large quantities of smoke and heat. The fuel may continue to flow, feeding the fire until the machinery is shut down and completely stopped. Turbine generators may take from 20 to 40 minutes to coast down.[6] Fires in turbine generators may also affect structural components of the building they are housed in during fire conditions. Automatic sprinklers should be present at bearings, instrumentation and piping for turbine generators.

In addition to a breached pipe, other causes of fires may include excessive vibration, human error, internal machine failures, overspeed, or instrumentation or control failure. All of these may result in pool fires or spray fires. Foam application and isolating the fuel supply at the source are methods of effectively dealing with these type fires.

Boilers

The major hazard with boilers is a firebox explosion. An uneven or low-fuel flow will allow a portion of the boiler to be without flame. The fuel continues to be introduced into the firebox, vaporizing without ignition. It forms an explosive cloud within the firebox or at times outside the firebox. This condition is sometimes known as the "white ghost." When this explosive cloud reaches an ignition source, an explosion occurs. The ignition source can be either an open flame or a hot boiler surface.

Cooling towers

The condensers are located within a cooling tower. The hazard associated with a cooling tower consists of the combustible materials used in its construction. Although many cooling towers are constructed using noncombustible materials, there are those that are constructed using combustible materials for economic reasons.

Transformers

After generation, electricity must be converted into a suitable voltage for transmission. Large power transformers are used for this purpose, and they may be located adjacent to the turbine building. Transformers are cooled with mineral oil in quantities that range from 20,000 to 25,000 gallons. Internal faults can rupture transformer casings, which will release oil and most likely result in ignition of the oil. Fire walls, transformer separation, and automatic water spray systems are used in the fire protection of these transformers (fig. 30–6).

***Fig. 30–6.* Transformers protected by water spray system** *(Reprinted with permission from the* **Fire Protection Handbook**, *19th Edition, Copyright © 2003, National Fire Protection Association, Quincy, MA 02169)*

Large undivided areas

The large size of the equipment required for electric power generation and the requirements for maintenance and operational access necessitate large areas that are usually undivided. These areas will allow heat and smoke to travel unimpeded in most cases, hindering firefighting operations. Open grating catwalks are present in many facilities, and these catwalks may provide the only access to upper areas of the power plant. The use of these catwalks may be blocked by fire or thermal currents.

Location in rural areas

As previously mentioned, environmental or social issues may have situated the electric generating plant in a remote or rural area. These areas may be protected by smaller fire departments with limited staffing, resources, and water supplies. Situations such as this must be considered when developing preincident response plans and firefighting tactics.

FIREFIGHTING TACTICS

As with any firefighting operations, preincident response planning is the key to a successful operation. Familiarization visits begin the preincident response planning process. Power generation facilities must have emergency response plans. These plans should be consulted during an incident, and their information used in developing a fire department's preincident response plan. Other items of interest are the fuel type, fuel storage location, process operations, hazardous materials, water supply, fire suppression systems present, any hazardous materials present, and access concerns. Some power generation plants will have their own fire brigades, which should be in compliance with NFPA 600, *Standard on Industrial Fire Brigades* (see chapter 2 regarding industrial qualifications). One of the first priorities of the responding fire personnel is to confirm that all plant personnel have been accounted for. There may be personnel trapped in control rooms or other areas.

Hydrogen

Fires involving hydrogen gas should be allowed to burn until the fuel source can be isolated. There may be times when the fire may have to be put out in order to reach the shutoff valve, but this tactic should be a last resort. While the fire burns, exposures that are affected must be cooled. Care must be taken to avoid getting water on areas containing live electricity. Consultation with plant personnel is a must before any water is used in electrical power generation facilities. Hydrogen burns with a nonluminous flame, which may be invisible during daylight. A technique that has been used when approaching hydrogen fires is to hold a broom out in front while progressing slowly, and at the same time throw dirt ahead. The combustibles in the dirt will incandesce and thus locate the edge of the fire.[7] Hydrogen vapors, if not on fire, may be dispersed using fog patterns on nozzles. However, water should not be allowed to contact pooled hydrogen because it will increase the rate of vaporization.

Natural gas

Fires in natural gas facilities should be treated as standard policies and procedures dictate, usually by letting the product burn until the fuel source can be isolated. There may be times where the fire may have to be put out to reach the shutoff valve, but as with hydrogen, this tactic should be used only as a last resort. While the fire burns, the cooling of affected exposures will be required. Care must be taken to avoid getting water on areas containing live electricity. Consultation with plant personnel is a must before any water is used in electrical power generation facilities. In the event of a gas leak with no fire, vapor disbursement is warranted and all ignition sources should be shut down. If this cannot be accomplished immediately, then evacuation will be necessary.

Coal

When a fire occurs in a coal pile, the coal will usually have to be removed from the pile and spread out on the ground so that the burning coal can be extinguished using water (fig. 30–7). If water is used on the intact pile to attempt extinguishment, the water can accelerate the oxidation of the coal and exacerbate the situation. Further, if water is applied to a coal pile, silo, or bunker by driving

***Fig. 30–7.* Fire department concluding a drill with local electric plant operations personnel. Note coal spread out for ease of extinguishment.** *(Courtesy of Dagsboro Volunteer Fire Department)*

Fig. 30–8. **Piercing nozzles similar to this one can be used in the extinguishment of coal fires. Longer lengths will be required.**

the stream through the pile, the water may reach a hot spot, creating a steam explosion. This initial explosion may loosen coal deposits and create a dust explosion. Burning coal in a pile can create instability in the pile, and the pile may shift or collapse inward. Therefore, personnel should not operate on top of coal piles unless their weight is supported and distributed by using planks or other methods. Personnel should be connected to safety lines in this case. It is best, however, that this placement of firefighting personnel not be used.

Class A foams and *wetting agents* have been used in some cases as a *surfactant.* When used in this manner, Class A foams should be proportioned at 0.1% so that excessive bubbles will not be created during the application.[8] During the application, thermal cameras or infrared detectors can be used to determine hot spots and direct the foam to the seat of the fire. Piercing nozzles can also be used penetrate the coal pile and allow the extinguishing medium to reach the seat (fig. 30–8).

Micelle-encapsulating agents have been used to extinguish coal fires and are the extinguishing media of choice for PRB coal fires. They may also be used on Class A and B fires. The characteristics of these micelle-encapsulating agents are as follows:

- On Class B fires, the agents encapsulate both the liquid and vapor phase molecules of the fuel and immediately render them nonflammable.
- The agents reduce the surface tension of water.
- The agents interrupt the free radical chain reaction of the fire tetrahedron. They can be used on coal fires in concentrations of 0.5% to 1.0%.[9]

Inerting gas has been used to extinguish coal fires, but large volumes of the gas must be used, and the gas must not be permitted to escape. Carbon dioxide (CO_2) has been used for this purpose. CO_2 has a vapor density of 1.5 times that of air, so it must be injected high into the pile or silo. It will inert the air above the pile as well as drift down through the pile. At the same time, gas is injected into the lower part of the silo. The CO_2 will begin to inert the voids of the coal. Usually, CO_2 systems are used in combination with a detection system so that early detection and extinguishment of fires can take place. Even with the application of CO_2, manual extinguishment methods will still need to be applied, namely, coal removal and water extinguishment. Nitrogen has also been used to inert coal silos. Because of its density (similar to air), nitrogen needs to be applied to multiple points simultaneously and in great quantities.

A coal silo involved in a fire will have to be emptied so that extinguishment can be accomplished. Plant personnel should be consulted so that they may assist with the coal removal. During removal processes, high-expansion foam can be used to reduce dust formulation and the possibility of dust explosions. High-expansion foam can be applied to the top of the pile or upper portion of coal in the silo. When firefighters are working around coal fires, there will be high concentrations of carbon monoxide gas being liberated. Air monitoring must be performed and self contained breathing apparatus (SCBA) used. Rotation of firefighting teams, rehabilitation of firefighters, and SCBA refill and logistics must be considered by the incident commander (IC).

Conveyor belt fires can be extinguished using manual fire suppression techniques such as hose streams. The possibility that the fire may have been spread by the moving conveyor belt will require the entire run of the conveyor to be examined. Because of this possibility, many conveyor systems have automatic detection systems installed that will stop the conveyor and activate a water-based extinguishing system. When the detection system does not provide for the system to be shut down, then a deluge system will be installed along the length of the conveyor, with all deluge heads operating. If the conveyor is shut down automatically, then a localized closed head sprinkler system may be deployed. Fire departments will have to ensure the status of the extinguishment system as part of the initial size-up and evaluate the effectiveness of the system prior to its being shut down.

Turbine lubricating oil

The extinguishment of fires involving turbine lubricating oil needs to be coordinated closely with plant personnel. The minimization of the resultant oil spill is a primary concern. The shutting down of the system could cause more damage to the equipment, so the plant personnel will have to institute emergency shutdown procedures and give guidance to the municipal fire department when they respond. Fire departments can cool surrounding structures, again under the guidance of plant personnel, because of possible electrical hazards, and wait until the machinery is stopped. Foam may be required to extinguish pool fires and cover the pooled oil.

Electrostatic precipitators

If a fire occurs in one of these units, the unit should automatically begin emergency shutdown. When operating, the unit contains an oxygen-deficient atmosphere, so opening the unit to an air source may allow for a *backdraft* condition. When the entire unit has been shut down (fuel completely shut off and the air supply shut down), the unit may be opened and water applied for extinguishment.

Scrubbers

Fires in scrubbers have occurred when the lining is made of a combustible material. These fires are difficult to fight because of their inaccessibility and because of the smoke and heat produced. There will usually be an automatic sprinkler system installed in the scrubber area. Consultation with plant personnel is required during one of these fires.

Cable trays

Control rooms and the associated plants have a large number of wires and cables needed for control and monitoring of the various systems and processes. The cables are usually grouped into large cable trays and located in raised floors, vertically along walls, overhead at the ceiling, or in horizontal underground tunnels (cable tunnels). These cable trays may be exposed to the accumulation of dust and combustibles, adding to the fire hazard. Because the cable trays and tunnels may pass through other building(s) or support areas, these may at times contribute to the spread of fires. Firefighters should examine the run of these trays and tunnels when there is a possibility that fires have entered these areas.

CONCLUSION

Electric power generation contains more than a simple process. The support processes can be just as hazardous as the electric generation itself. Fires can be challenging, and successful extinguishment will only take place when there is full cooperation among plant operations personnel, plant fire brigades, and municipal fire departments. Joint training must be conducted, and preincident response planning is crucial. An educated fire department is a more efficient and safe department. Visit these critical power generation facilities and learn how the electricity we take for granted is produced, and how, in the event of an emergency, we can be better prepared to respond.

CASE STUDY

A large oil refinery may produce much of its own electrical power using the equipment described in this chapter. One such refinery was operating a steam turbine generator to produce electricity to run a number of its refining units. The turbine generator was out of service for refurbishment of internal parts that had been damaged. On completion of the repairs, the turbine was started, but it kept shutting down because minor adjustments had to be made to previously installed parts. The turbine was restarted and appeared to be running normally. It continued to operate normally for 3 days with no problems. The turbine was checked at 0400 hours by plant operators, and conditions appeared to be normal. At 0517 hours a fire occurred at the nondrive end of the turbine in the vicinity of the lube oil filters. This fire was the result of a release of lubrication oil that subsequently contacted hot steam pipes, providing the heat for ignition. The ensuing fire was large and ultimately consumed the surrounding area and two levels of the structure in the immediate area of the turbine.

Steam was being directed to the turbine and presented noise and scalding hazards for responding firefighters, and the turbine was still turning, feeding oil to the fire. The plant operations personnel immediately notified the fire department and started shutdown of the process. Remember, the shutdown of electricity to the refinery units will create other problems within the refinery, so multiple tasks had to be completed by the plant operations personnel: shutdown of steam, isolation of electricity, and routing of electricity from other generators to the refinery units.

The response of the fire department was on the order of 7 minutes from receipt of alarm to arrival. On arrival, plant personnel directed the fire department to begin extinguishment of the structure fire and cooling of exposed areas. The noise from the steam was deafening to all personnel on the scene, making communications nearly impossible. Firefighters were unable to enter the structure until the plant was shut down. This took nearly 20 minutes, during which time the fire department continued operating fixed and portable master streams on the fire and surrounding areas. The emergency shutdown of the steam was activated by plant personnel, but failed to operate because the cables were in the vicinity of the fire and were affected early by the fire, rendering the system inoperative. Steam to the unit was manually shut down. Once the steam was shut down and the power isolated, handlines were advanced into the structure where it was found that only small fires remained in pools of oil around the turbine, wiring within the turbine and generator, and cable trays supplying the area. These were extinguished using handlines and portable dry-chemical extinguishers. The total time for extinguishment was 55 minutes.

Random thoughts

- Shutdown of plant processes must be accomplished as soon as possible by plant operations personnel. This shutdown must be consistent with safety and operational guidelines for the particular plant.

- Coordination between the fire department and plant operations personnel is crucial to the successful control and extinguishment of a fire of this nature.
- Fire department personnel must account for plant operations personnel immediately on arrival at the scene to identify any rescues that may have to be performed.
- The use of portable and fixed monitors to control the spread of fire while equipment is shut down is invaluable. Once these streams are in operation, firefighters can be withdrawn to an area of safety until plant conditions are rendered safe for close-in operations.

NOTES

[1] National Fire Protection Association (NFPA). 2003. *Fire Protection Handbook,* 13th ed., vol. II. Quincy, MA: NFPA, pp. 13–288.

[2] NFPA, *Fire Protection Handbook,* pp. 13–289.

[3] NFPA, *Fire Protection Handbook,* pp. 13–291.

[4] NFPA, *Fire Protection Handbook,* pp. 6–337.

[5] NFPA, *Fire Protection Handbook,* pp. 13–291.

[6] NFPA, *Fire Protection Handbook,* pp. 13–291.

[7] NFPA, *Fire Protection Handbook,* pp. 8–116.

[8] NFPA. 2005. NFPA 850, *Recommended Practice for Fire Protection for Electric Generating Plants and High Voltage Direct Current Converter Stations.* Quincy, MA: NFPA, pp. 850–38.

[9] Douberly, E. B. 2003 (October). Fire Protection Guidelines for Handling and Storing PRB Coal. *Power,* pp. 70–73.

31

Storage Tank Facilities

INTRODUCTION

In the early days of the industry, tank fires were a common experience. As the industry matured, it demanded better design, construction, fire protection, and improvements to the various codes and standards maintained by the American Petroleum Institute (API) and National Fire Protection Association (NFPA). As a result, we see far fewer tank fires today than in the past. It is interesting to note that while the frequency of tank fires has decreased, the size of the tanks has increased, which means a more severe hazard in the event of a fire. As a result of the increase in size, fires involving large aboveground storage tanks can be extremely costly in terms of property damage, business interruption, environmental concerns, and public impact and opinion. Further, the control and extinguishment of full surface tank fires require a big commitment to human logistics and equipment resources (fig. 31–1). Due to the potential of a loss, the fire protection industry has developed ways to effectively control and extinguish fires in large diameter storage tanks. These methods are continuing to be developed and updated.

***Fig. 31–1.* Full surface tank fire** *(Courtesy of Williams Fire and Hazard Control)*

Where are these tanks located? Do you have any in your fire district? These bulk flammable and combustible liquid storage tanks can be found in a multitude of locations, ranging from large refinery and petrochemical plants (chapters 12 and 13) and marine terminals (chapter 24) to intermediate storage facilities such as bulk plants and loading terminals (Fig. 31–2). This chapter will look mainly at atmospheric storage tank types, with some description of how they fit into various operations, types of fires, preplanning, and suggested tactics. Specifics on semifixed and fully fixed foam system design and application are detailed in chapter 8. We will address manual fire suppression from the perspective that the installed protection is not adequate.

Fig. 31–2. **Tank storage facility** *(Courtesy of Paddy Briggs, 2005)*

DESCRIPTION OF PROCESS

Storage tanks

Storage tanks are constructed in several pieces and components. Materials used for the construction of flammable and combustible liquid tanks must conform to various standards such as API 650, *Welded Steel Tanks for Oil Storage*, or the NFPA 30, *Flammable and Combustibles Liquids Code*. The plates and other components used to form the tank must comply with various standards of the American Society of Mechanical Engineers (ASME) and American Standards of Testing and Materials (ASTM). Tanks are highly regulated and are usually checked by state and city inspectors, as well as insurance companies, to ensure compliance with all applicable codes and standards.

Atmospheric storage tanks are used to store or mix flammable and combustible liquids in a variety of ways in many different types of industrial facilities. These tanks can range from a diameter of 10 ft to more than 350 ft, with an average height of about 45 ft. They are capable of holding in excess of 1.5 million barrels (63,000,000 gallons) of flammable or combustible liquids. Some locations might have more than 100 tanks in various sizes and quantities, with many different products. To make things even more interesting, these tanks may be located in close proximity to each other, and/or several tanks may be within a common dike. *Dikes* are physical barriers or dividers used to prevent the spread of tank contents in the event of a tank overflow or tank structural failure. Dikes are also used to segregate and group tanks by the classification of their contents. Dikes constructed of compacted dirt are often called earthen dikes, or they may be made of concrete and similar type materials. Dike height and perimeter are functions of the volume of the tank(s) to be stored in a particular dike area. Many dikes are designed to be able to contain the contents of the tank plus a percentage above this for a safety margin. This safety margin is typically used for the anticipated accumulation of firefighting water during emergencies. If more than one tank is located within a common dike area, the dike should be capable of containing the volume of the largest tank plus a safety margin.

STORAGE TANK AND ROOF TYPES

Tanks are classified or described by their type of roof:

- Fixed roof tanks
- Internal (covered) floating roof tanks
- Open-top (external) floating roof tanks
- Domed external floating roof tanks

Fixed roof tanks

Fixed roof tanks are vertical steel cylinders with a permanently attached roof. In the petroleum and petrochemical industries, these permanent roofs are usually cone shaped and are sometimes referred to as fixed-cone roof tanks. These roofs can be flat or slightly domed to prevent water accumulation. This permits a vapor space between the liquid surface and the bottom side of the roof. These tanks are constructed according to API 650 and have a weak roof-to-shell seam. This design is such that in the event of an incident, such as internal overpressure from an explosion or similar incident, the roof will separate from the vertical shell to prevent failure of the bottom seams causing the tank to rocket. This design is often called a frangible seam roof and requires careful design and installation. Fixed-cone roof tanks are typically used to store Class II and higher combustible liquids (fig. 31–3). Oftentimes, these tanks are insulated and used to store Class IIIB liquids such as asphalt; bunker fuels; and other heavy, viscous liquids. Fixed-cone roof tanks are typically provided with some form of venting capability to allow for the tank to "breathe" during loading, unloading, and extreme changes in temperatures. These vents may be open vents or pressure-vacuum vents, which prevent the release of vapors during the times where there is a change in liquid level or a temperature change. Depending on the location of the tanks with respect to the community, the vents can be equipped with environmental controls and flame arrester/diverters to capture fugitive emissions.[1]

Internal (covered) floating roof tanks

Internal (covered) floating roof tanks have a permanent fixed roof with a floating roof inside. Internal floating roof tanks (referred to as internal floaters) usually have vertical supports within the tank for the fixed roof or have a self-supporting fixed roof. The internal roof floats on the surface of the liquid, rising and falling with the changing level. The internal floating roof, sometimes called a "pan," can also be constructed according to API 650. The pan can be constructed using plastic, aluminum, or steel and sometimes uses pontoons for flotation, as compared to other roofs that actually float directly on the surface of the liquid. The fixed roof above is provided with open-air vents to permit the space above the internal roof to breathe. Fixed roofs are allowed to vent in this manner because their vapor space is considered below the flammable limits. Seals are provided in the rim seal space to prevent the escape of fugitive emissions. The rim seal space is considered to be the area between the tank shell wall and the internal floating roof (the difference in the external tank shell diameter and the internal roof diameter). This distance is usually 1 to 4 feet and may be the origin of some fires. These tanks are typically used to store highly flammable finished products such as gasoline.[2]

Classification of Liquids	Flash Point	Boiling Point
Flammable		
Class IA	Below 73 °F (22.8 °C)	Below 100 °F (37.8 °C)
Class IB	Below 73 °F (22.8 °C)	At or above 100 °F (37.8 °C)
Class IC	At or above 73 °F (37.8 °C) and below 100 °F (37.8 °C)	
Combustible		
Class II	At or above 100 °F (37.8 °C) and below 140 °F (60 °C)	
Class IIIA	At or above 140 °F (60 °C) and below 200 °F (93 °C)	
Class IIIB	At or above 200 °F (93 °C)	

Fig. 31–3. **Classification of flammable and combustible liquids**

Open-top (external) floating roof tanks

Open-top (external) floating roof tanks are vertical steel cylinders with a roof that floats on the surface of the liquid in the tank, but it is open to the atmosphere above, that is, no fixed roof above. The common name used to describe these is external floaters, and the only main difference between internal (covered) floating roof tanks and open-top (external) floating roof tanks is the presence of a fixed roof above to protect it from the environment. As with internal floaters, external floaters, as these are locally referred to, have roofs that float on pontoons or have a double-deck for flotation on the liquid surface. This roof also rises and falls with the changing of the liquid level. These tanks also have rim seals to prevent the vapors from escaping. These tanks are generally used for the storage of crude oil.[3]

Domed external floating roof tanks

Domed external floating roof tanks (or covered floating roof tanks) are similar in functionality to internal floaters and are created by retrofitting a domed covering over an existing external floater. These domed roofs are often referred to as geodesic dome tanks and are really the same as internal floaters. The main purpose of the dome is to serve as a wind or rain barrier, but it also provides environmental control with respect to fugitive emission controls. These tanks are typically permitted to vent freely to the atmosphere. They are typically used to store finished or refined products such as gasoline.[4]

Location of storage tanks

Flammable and combustible liquid storage tanks can be found in many different types of industrial facilities. Some of the more common industries include refineries, petrochemical facilities, bulk storage plants, and marine terminals. Although there are also other facilities that might have some limited bulk storage of flammable and combustible liquids, such as power plants, airports, and large manufacturing facilities like automotive and steel plants, we will focus on those locations that have either a large number of tanks or locations where the individual volume of tanks creates an extreme fire hazard.

Whether storage tanks are located in a refinery or petrochemical plant, their purpose is to store a raw material used for a particular process, mix two or more liquids together, or provide storage for further process at another plant or as a finished product. This is somewhat important because the way the tanks are used indicates the method used to fill or empty the tank contents or the presence of different types of equipment such as mixers, heaters or pumps.

Tanks used for storage of raw material, whether flammable or combustible, for a particular process usually have pump(s) located at a truck or rail unloading rack. These pumps offload the raw material via classified liquid pumps and send the product to the tank. Automatic shutoffs, emergency shutdown devices (ESDs; chapter 9), and fixed protection and detection are usually provided in the area. Some tanks are used for mixing two or more flammable or combustible liquids together. This is common in refineries where mixing of two different gasolines are mixed with varying levels of octane to make a single octane gasoline.

Some plants use tanks with mixers because the combustible liquid being stored may be prone to solidification if not heated and rotated. This is common in the asphalt plant of a refinery or a stand-alone asphalt plant. Some facilities use mixing to create other combinations of flammable or combustible liquids such as motor oil, windshield fluid in concentrate, or other automotive related fluids. The presence of mixers adds additional sources of ignition.

Some tanks are used as simple storage tanks. These can be found at many different facilities—for example, petrochemical, refinery, pulp and paper, and similar process plants—and are simply that, large tanks with flammable or combustible liquids. These locations can have as few as three or four tanks or as many as 100 tanks.

COMMON HAZARDS

Particular with each tank type, there are certain related fire hazards that are common to the tank roof type. These hazards vary in severity from a simple vent fire to a full liquid surface tank fire. The most common of these incidents include overfill ground fire, vent fire, rim seal, obstructed full liquid surface fire, and unobstructed full liquid surface fire.

Overfill ground fires

Overfill ground fires or dike fires result from piping or tank leakage. Many times this is a result of some other cause such as operator error or equipment malfunction. These types of incidents are considered the least severe. If a leak occurs without ignition, caution should be used to ensure that all ignition sources are isolated. If ignition did occur, then this can be simply treated as a large pool fire. Overfill ground fires are common to fixed-cone roof, internal floaters, external floaters, and domed roof tanks.

Vent fires

Vent fires are typically associated with fixed roof tanks such as fixed-cone and internal floating roofs. The main cause is lightning strike. These are a less severe type of fire and can usually be extinguished with a dry-chemical fire extinguisher or reducing the pressure in the tank.

Rim seal fires

Rim seal fires compose the vast majority of fires in external floaters but can occur in internal floaters or domed roof tanks. As with most tank fires, lightning is the primary cause of ignition, although with floating roof tanks, an induced electrical charge without direct lightning hit may occur. These fires, because they are the most common, experience a high rate of extinguishment success, assuming that there is no collateral damage such as a pontoon failure (explosion) or the suppression efforts do not sink the roof. The success of rim seal fire extinguishment can be mostly attributed to the installation of rim seal fire protection systems. These semifixed or fully fixed rim seal fire protection systems (see chapter 8 for details) have a good history of extinguishment, assuming proper design, installation, and maintenance. Rim seal fires for internal floaters are slightly more challenging, especially if semifixed or fully fixed systems are not provided. This means that the only access to the fire area is through the vents or access covers.

Obstructed full liquid surface fires

Obstructed full liquid surface fires can occur in fixed-cone roofs, internal floaters, and external floaters. These tend to be challenging because access to the burning surface is blocked by the roof or pan. The roof or pan can sink for various reasons, such as an increase in vapor of an internal floater, which can cause the pan to tilt. Pontoon failure in external floaters is commonly caused by closed drain valves during rains or mechanical seal failure causing the pan to sink.

Unobstructed full liquid surface fires

Unobstructed full liquid surface fires are relatively easy to extinguish where the tank diameter is relatively small (less than 150 feet) and sufficient resources and trained personnel are available. The most challenging fires involve larger tanks (greater than 150 feet) because of the surface area of the fire and the amount of resources needed to control and extinguish the fire. Unobstructed full surface fires can occur in fixed roof tanks without internal roofs, where the frangible weak seam at the roof-shell joint separated from an explosion or other overpressure event, leaving a full surface tank. External floaters are also prone to unobstructed full surface fires during heavy rain conditions. With closed roof drains, the roof can quickly sink, leaving the exposed liquid surface vulnerable to a lightning strike.

Tank Type	Common Fire Hazard	General Comments
Fixed Roof Tank	· Overfill (dike) ground fire · Vent fire · Obstructed full surface fire · Unobstructed full surface fire	· Atmosphere above UEL/UFL in vapor space · Typically stores Class II and III liquids · Frangible roof-to-seam · Can be insulated · Smaller diameter than other tanks
Internal (covered) Floating Roof Tanks	· Overfill (dike) ground fire · Vent fire · Rim seal fire · Obstructed full surface fire	· Lean atmosphere above floating roof · Usually stores gasoline · No frangible roof-to-shell seam · Larger diameter tanks
Open-Top (external) Floating Roof Tanks	· Overfill (dike) ground fire · Rim seal fire · Obstructed full surface fire · Unobstructed full surface fire	· Lean atmosphere above floating roof · Mainly used for storing crude oil · Caution about sinking the roof during unobstructed surface fires
Domed External Floating Roof Tanks	· Overfill (dike) ground fire · Vent fire · Rim seal fire · Obstructed full surface fire	· Lean atmosphere above floating roof · Usually stores gasoline · No frangible roof-to-shell seam · These are retrofitted internal floaters

Fig. 31–4. **Quick reference guide to common hazards**

Figure 31–4 provides a quick reference guide to common hazards.

FIREFIGHTING TACTICS

As with all firefighting tactics and discussed many times in this book, a well-planned and tested preincident response plan is needed. Preincident response planning should be carried out in bulk storage facilities to identify the hazards, potential to develop into a major incident, and the required resources. In the event of a fire at this type of facility, it is advantageous to have information readily available that relates to the tank and product information, fire safety provisions, access roads, staging areas, and location of water sources (fig. 31–5). Good fire prevention starts with the firesafe design of the tanks, as well as spacing, containment in the form of diking, and drainage. It cannot be stressed enough the importance of proper design and construction of the tanks, as well as the need for detailed analysis for preplanning of all storage tank farms.

Fig. 31–5. **When preplanning storage tanks, make sure that information pertaining to not only the tank but also exposures, dike areas, and associated piping are included in the preplans.** *(Courtesy of G. Noll).*

Tank fires are complex events (fig. 31–6). Fighting them requires implementation of plans, preparation, and proper utilization of resources coordinated by an effective emergency management organization. The following overview of tank fire suppression activities presumes that the planning and preparation stages have been performed by both the plant personnel and fire officials to ensure that proper tank design follows an accepted industry standard and practice.

Experience shows that safe and successful firefighting of tank fires can be achieved when based on this planning and preparation. During an incident is not the time to decide that a plan should be developed (fig. 31–7). However, with a plan in place, suppression activities are only the implementation of that plan. If the plan is not achieving desired results during a fire, the strategy and tactics should be changed to a workable and safe strategy and tactics to achieve such goals.

Fig. 31–6. **Rim seal fire** *(Courtesy of Williams Fire and Hazard Control)*

Fig. 31–7. **Heavy smoke coming from a tank fire** *(Courtesy of Williams Fire and Hazard Control)*

On notifying and activating the fire department, gathering and assessing the incident should begin immediately. Information should be gathered quickly to develop an effective strategy to fight the fires based on the already existing preincident response planning. The following should be considered:

1. Rescue of personnel in the immediate areas
2. Life safety hazard to site personnel
3. Extension
4. Confinement
5. Extinguishment
6. Environmental impact
7. Community impact

After the immediate issues are addressed, the following situation assessment, determining the type of fire, should be performed:

- Vent fire
- Seal fire
- Piping-connection fire
- Full involvement fire
- Overfill fire
- Tank and dike fire
- Multiple tank fires
- Exposures

Determining the type of fire will drive the requirements for resources and will dictate the necessary incident action plan to fight the fire. There are several types of fires that an emergency organization could be faced with and different ways of attacking a fire.

Ground or dike fires caused by tank overfilling or pipe failures can be viewed as simple spill or pool fires. Trying to calculate the area of the oddly shape spill can be challenging, but the best tactic is to establish an adequate water and foam supply and begin to suppress the fire. Exposures such as the tank and associated piping and pumps can be protected with water via ground monitors or those that are installed on hydrants. Firefighters should not attempt to enter the dike area unless safe to do so. This can be verified by atmospheric testing and ensuring that the spill potential does not fill the dike floor. This is especially true for small spills, with or without ignition. For larger spills, where ignition occurred and foam operations are continuing, entry in the dike should not be allowed. Disturbing the foam blanket can have disastrous effects.

For fires involving the rim seal area, these are usually extinguished with the semifixed or fully fixed water/foam system (chapter 8). Responses to these types of fires are similar to fires in sprinklered buildings from the perspective of supporting the fixed systems on arrival. The main difference is that you should not attempt to extinguish the fire until it is confirmed that both the water and foam supply reliability and quantity can be supplied for the entire duration required to extinguish the fire. If semifixed or fully fixed systems are not installed, portable equipment can be used to extinguish these fires. Hoselines and monitors can be used to fill the rim seal area with water/foam solution. Some tools, such as the Daspit™ tool, are specifically designed for rim seal fires (fig. 31–8). This monitor device has a brace or clamp attachment that is designed to secure it to the tank shell at the top lip of the tank. The foam/water solution flow rate for rim seal fires using portable equipment ranges from 250 gpm for small size tanks (diameter up to 90 ft diameter), 550 gpm for medium size tanks (90 to 175 ft diameter), and 950 gpm for larger size tanks (175 to 300 ft in diameter) with an application time of 20 minutes. The actual flow rate for a tank can be determined by multiplying the application rate of 0.50 gpm/ft^2 with the surface area of the rim seal space.

Fig. 31–8. **Daspit tool discharging foam (Courtesy of Williams Fire and Hazard Control)**

In general, the following should be used when fighting tank fires where fixed systems have failed to operate correctly or are not installed. The method of extinguishing tanks fires in manual operations, using portable or mobile monitors, is commonly referred to as *Type III application* or "over-the-top," and includes the following:

1. Minimum application rates
2. Application densities
3. Minimum foam solution application durations
4. Supplemental foam application rate in dike area

These vary based on fuel flash point, water *immiscibility,* type of foam, and application device. For fires involving hydrocarbons such as gasoline or diesel, a 3% concentration is the industry standard. Although there are foams produced that can be proportioned at 1%, these are still considered to be a new technology. For fires involving polar solvents such as alcohols or methyl-tertiary butyl ether (MTBE), most foam should be proportioned at 6% concentration. There are, however, foam concentrates made to be used at 3% concentration on MTBE.

Proportion concentration means the percentage of foam that is proportioned into the water. For example, a 3% foam proportioning means that 3% of the total foam/water solution is foam concentrate, and the remaining 97% of the foam/water solution is water. The application rate of the foam solution is a function of the liquid surface area. As with the rim seal fires, you should not attempt to extinguish the fire until it is confirmed that both the water and foam supply reliability and quantity can be supplied for the entire duration required to extinguish the fire.

The formula is as follows:

foam solution flow rate (gpm of foam/water solution)

$$\text{foam solution flow rate} = \text{tank surface area} \times \text{application rate} \quad (31.1)$$

$$\text{tank surface area} = 3.14 \times \text{radius}^2 \quad (31.2)$$

application rate = minimum of 0.16 gpm/ft^2 of surface area

Application rates will vary depending on the tank diameter (surface area). For larger tank diameters, a higher application rate is required. (fig. 31–9)

Tank Diameter	Application Rate
Up to 150 feet	0.16 gpm/sq.ft
150 to 200	0.18 gpm/sq.ft
200-250	0.20 gpm/sq.ft
250 to 300	0.22 gpm/sq.ft
Above 300	0.24 gpm/sq.ft or higher

Fig. 31–9.* Suggested application rates
*These rates are based on experiences and advice of industry subject matter experts.

foam concentrate flow rate (gpm of foam concentrate)

foam concentrate flow = foam solution flow rate × foam % (31.3)

foam solution flow rate = gpm

foam % = 1%, 3%, or 6%

foam concentrate supply (foam quantity)

foam concentrate supply = foam concentrate flow rate × duration (31.4)

foam concentrate flow rate = gpm

duration = 65 minutes (50 minutes for Class II liquids)

As a sample, the preplan information can be assembled in a form for ease of use. (fig. 31–10):

PRE-INCIDENT FOAM APPLICATION REQUIREMENT

Date of Issue: **Location:**

Revision Due:

Product Information:	
Name:	
Flash Point (F):	
IGN. Temp (F):	
Vapor P (psi):	
Vapor Density:	
Boiling Point (F):	
NFPA 704	

Tank #

Tank Diameter (ft):	
Tank Circumference (ft):	
Surface Area (sq.ft.):	
Tank Height (ft):	
Capacity (Max Safe):	
Dike Surface Area:	

Fire Protection Data:	
Fixed System	
Semi Fixed System	
Subsurface Injection	
Cooling Rings	

Rim Seal Fire

Foam Chambers	
Surface Area of Seal	
Foam Concentrate	
Application Time	
Foam Sol. Rate (gpm)	
Foam Conc. Rate (gpm)	
Foam Quantity (gal)	

Portable Equipment	
Foam Concentrate	
Application Time	
Foam Sol. Rate (gpm)	
Foam Conc. Rate (gpm)	
Foam Quatity (gal)	

Full Surface Fire

Product Type	
Foam Monitors/Nozzles	
Foam Concentrate	
Application Time	
Foam Solution Flow Rate (gpm)	
Foam Concentrate Flow Rate (gpm)	
Foam Concentrate Supply (gal)	

Other Information:

***Fig. 31–10.* Sample preincident foam application requirement worksheet**

Fig. 31–11. Large-flow trailer-mounted monitor *(Courtesy of Williams Fire and Hazard Control)*

Fig. 31–12. Large-flow trailer-mounted monitor in operation *(Courtesy of Williams Fire and Hazard Control)*

Fig. 31–13. Two large-flow trailer-mounted monitors in operation *(Courtesy of Williams Fire and Hazard Control)*

Some of the calculated flows may reach in excess of 10,000 to 18,000 gpm and require large delivery devices such as the trailer-mounted monitors (figs. 31–11 to 31–13) and large portable pumps (fig. 31–14).

Fig. 31–14. Large portable pump *(Courtesy of Williams Fire and Hazard Control)*

Additional information that should be collected includes those associated with the situation of the tank such as the following:

- Position and condition of the roof drains
- Volume of the product
- Depth of water bottoms
- Condition of the structural aspects of the tank
- Product in tank
- Physical properties of the product
- Possible to transload or pump out the contents

Tank spacing (distance from tank shell to tank shell) is very important to know. If tank spacing does not conform to recognized standards, exposure protection becomes a priority. In general, the spacing should be anywhere from 33% to 100% of the tank's diameter. If the tank spacing is too close, and you are not sure if the exposed tank is being heated from the burning tank, set up a ground monitor or hoseline to cool the exposed tank shell. If the application of this cooling stream creates steam on contact, then the tank is getting too hot from the exposure, and cooling should continue until the water does not convert to steam. In general, based on tank size, a flow rate of anywhere from 500 to 1,000 gpm would be sufficient to cool the exposed steel.

It should be noted that transferring product out of the tank immediately might not be the best option. Transferring product from the burning (or exposure tank) increases the amount of tank steel to the fire. With the product inside the tank, the liquid is acting as a heat sink and preventing the tank shell wall from being directly exposed to the fire. Depending on the situation, you might not want to pump product out, or you may even decide to pump product in.

This is especially true for storage tanks containing crude oil where the likelihood of a *slopover, frothover,* or *boilover* may occur. A slopover is a short duration of slopping of froth over the rim of the tank with a minimum of intensity. This can be compared to a frothover, which is similar. Frothover is a steady, slow moving frothing over the tank rim, like a continued slopover. Both slopover and frothover occur without the sudden and violent reaction that is common with boilovers. Frothover occurs if the tank is not on fire; for example, when hot asphalt is loaded into a tank containing water, the product may froth over the top.

A boilover is a sudden and violent ejection of crude oil from the tank due to a reaction of the hot layer and the accumulation of water at the bottom of the tank. A boilover occurs when the residues (heavier particles remaining after combustion) from the burning surface becomes more dense than the surrounding less dense oil, the residue sinks down below the surface level toward the bottom of the tank. As the hot layer of denser, burned oil moves downward, this "heat wave" will eventually reach the water that normally accumulates at the bottom of a tank. When the two meet, the water is superheated and subsequently boils and expands explosively, causing a violent ejection of the tank contents. Water expands at a ratio based on the amount of heat it is absorbing to during firefighting application.[5] (fig. 31–15)

Temperature C	Temperature F	Expansion Rate Steam Ft^3: Water Ft^3
100	212	1,600:1
260	500	2,300:1
537	1,000	3,500:1
815	1,500	4,700:1

Fig. 31–15. **Temperature and related water-to-steam expansion rate**

In general, the heat wave for an average crude oil will progress downward at a rate of 1 to 3 feet per hour. If a boilover is likely, the rule of thumb says that the expelled crude oil may travel up to 10 times the tank diameter around the tank perimeter. So, for example, for a crude oil tank that is 250 feet in diameter, expect burning crude oil to cover an area of 2,500 feet in diameter. Therefore, careful consideration should be given to the location of the incident command post, staging, equipment placement, and medical triage, as well as the location of the safe zone. This is a general rule of thumb, and there is currently no research to verify this or explain the relationship between the diameters of large tanks and boilover radiuses. This is especially true for large tanks (in excess of 300 feet) where the factor of 10 times the tank diameter might need to be increased to 12 or 13.

Firefighting resource characteristics need to be identified. Local plant fire brigade members can provide the much needed support and technical advice. Many of the needed resources will be available at the facility or through industrial and/or municipal mutual-aid agreements. Some facilities have contracts with private, third-party companies that specialize in the extinguishment of large hydrocarbon fires.

Resource allocation should also include:

- Trained fire personnel
- Foam concentrate available based on preincident response plan quantities
- Water and pressure available based on preincident response plan quantities

- Foam proportioning available based on preincident response plan quantities
- Vehicle access and placement
- Delivery devices such as large flow monitors (greater than 6,000 gpm)
- Water transport ability via large diameter/volume hoses (7¼ inch or greater)
- Supplemental pumps for drafting from secondary water supplies

Firefighting strategies and tactics are also important. A discussion of objectives or goals versus the risk should be determined. Strategies include:

- Passive
- Defensive
- Offensive

As with most fires, the benefits must outweigh the risks. If a small diameter tank is burning with no threat to exposures, should we extinguish the fire? If the tank has already lost its contents, is exposure protection more applicable? These considerations are developed as part of the preincident response planning, development of emergency action plans, and the identification of the credible incident scenarios.

In addition, other favorable or unfavorable factors should also be considered, such as the weather-related issues:

- Wind
- Rain
- Temperature/humidity
- Storms

These environmental conditions could create problems with distance/range of the water/foam solution streams, changes in wind directions might cause response changes with respect to changes in staging locations, or an increase in temperature or humidity could force a quicker rotation of firefighters to prevent heat stresses.

Related logistical issues are

- Provision for relief with additional firefighters
- Food and water
- Rehab
- Emergency medical sector

CONCLUSION

Fires involving large aboveground storage tanks can be extremely costly in terms of property damage, environmental concerns, and public impact. The control and extinguishment of full surface tank fires requires a large amount of commitment in human logistics and equipment resources. Tank fires are complex events. Fighting them requires implementation of plans, preparation, and proper utilization of resources coordinated by an effective emergency management organization. Only with training and drills will your department become proficient in these strategies and tactics.

NOTES

[1] American Petroleum Institute (API). 2001. API 2021: *Management of Atmospheric Storage Tank Fires.* Washington, DC: API, p. 49.

[2] API, API 2021, p. 50.

[3] API, API 2021, p. 51.

[4] API, API 2021, p. 51.

[5] Cozad, F. D. 1981. *Water Supply for Fire Protection.* Upper Saddle River, NJ: Prentice Hall, p. 11.

32

Well Drilling Sites

INTRODUCTION

When most firefighters report for duty, they rarely discuss the possibility of responding to an oil or gas well fire (fig. 32–1). However, there are areas in the United States where these wells exist or are being drilled. While oil or gas well fires are rare in today's world, they do exist. In this chapter, we will focus on oil and gas well fires and what the municipal fire department will need to know if it is called to respond to such an incident.

Fig. 32–1. **Fire at an oil well site** *(Courtesy of Williams Fire and Hazard Control)*

Well firefighting is unique with regard to other firefighting operations such as structural, wilderness, marine, or most industrial. In many situations, when the available fuel is contained, the fire can burn itself out. With an oil or gas well fire, if left unchecked, the well can burn for years or decades unless the fire is extinguished and the flow of product stopped. Once a well experiences a *blowout* and the product is ignited, it will require specialists to respond and extinguish the fire and cap the well. This does not mean that the local fire department will not be involved in some capacity.

DESCRIPTION OF PROCESS

The practice of drilling an oil or gas well is a complex and extremely demanding operation, although the mechanics of this operation are quite simple. A rotating bit at the end of a *drillstring* breaks through the soil and rock at the bottom of a hole or well bore, while the loose fragments of soil and rock are carried out of the hole by a liquid drilling fluid known as *mud.* This mud is also used to control the pressure in the well so that a blowout does not occur. Mud is forcibly pushed out of nozzles in the drill bit, which flushes the drill cuttings away from the bit face. This mud then carries the cuttings up the *annulus* and out of the well bore. The mud will also cool and lubricate the bit. Another purpose of the mud column is to exert backpressure so that fluids from the reservoir do not enter the well bore. If liquids or gases from the reservoir enter the well bore, a blowout could occur. The density of the mud depends on the reservoir pressures anticipated.

The purposeful work at a well is performed by the drill bit on the end of the drillstring at the bottom of the well bore, where the bit rotates and drills the rock and soil. The drillstring is a tube of steel pipe that extends from the surface to the drill bit. This pipe consists of 30-foot sections of steel pipe that are screwed together. The drillstring transmits the rotational energy from the surface to the bit as well as conducting the mud stream from the surface to the bit. All other components of the well drilling rig support this operation.

The drilling derrick will be tall enough to pull three sections of the drill pipe as well as strong enough to support the weight of a complete drillstring. Large rotating drums or hoists will be present and are used to raise or lower the drillstring. The drill table is used to spin and rotate the drillstring and, therefore, the drill bit in the well bore. The prime mover for the drill table is the source of power for the entire drilling rig. Most modern rigs are diesel-electric. In this instance, diesel engines drive electric generators that produce electric current to power the drilling rig and its associated equipment.

Small diameter pipelines called flowlines are present within an oil or gas field. These flowlines connect individual oil or gas wells to central treating, storage, or processing facilities within the field. In an oil field, these flowlines may be short and require little pressure to move the product through these lines. In many cases, the natural energy pressure in the reservoir will be enough to cause the oil to flow unaided out of the well bore and through the flowlines. In other cases, bottomhole pumps will be used to lift the oil to the surface and pump it through the flowlines.

Oil flowlines usually connect with a tank battery (see chapter 31), which is the location of central processing, measurement, and storage facilities within the field (fig. 32–2). These tank batteries usually contain a separator that separates oil, gas, and water; a heater to break water/oil emulsions to assist with separation; and storage tanks for the product prior to distribution via pipelines or tanker trucks. Metering equipment is also contained at the tank battery. Other equipment such as electrical desalting equipment may be on-site if the oil contains large quantities of salt. Heated and insulated tanks may be present in some areas to reduce the viscosity of the oil for pumping.

Fig. 32–2. **Pumping station at gathering facilities in the field. Oil is collected from the various wells and pumped through a central pipeline to processing facilities.**

Gas wells and their associated processes are similar to that of crude oil wells, but operating conditions and equipment are different. In general, gas wells and their associated facilities operate at higher pressures than oil wells. In place of pumps to move gas through pipelines, compressors will be used. Gas flowlines will be used to connect individual gas wells with field gas treatment and processing facilities. Gas wells will usually flow with sufficient pressure naturally to force the gas through the flowlines to the processing area. Downhole pumps are not used, and in cases where additional pressures are required such as a well with very low pressure, small compressors located near the well may be used to boost the pressure in the flowline.

These well sites may be common in Indiana, Kansas, Missouri, Colorado, Wyoming, Oklahoma, Texas, and California as well as other states, and can be located near residential areas or highways.

COMMON HAZARDS

Crude oil is a highly volatile mixture of hydrocarbons that is lighter than water and burns twice as hot as coal.[1] The vapors given off by crude oil are easily ignited, so in the event of a blowout, these vapors may ignite. There are instances where drilling crews have intentionally set a blowout on fire to reduce the impact to the environment or limit the spread of toxic gases such as hydrogen sulfide (H_2S) gas. You may hear the term *sour gas.* This means that the gas contains H_2S.

Some wells contain a large percentage of H_2S gas within the oil. H_2S is extremely dangerous and has killed oil field workers and other petrochemical employees during accidental releases. The IDLH (immediately dangerous to life or health) for H_2S is 100 parts per million (ppm). The short-term exposure limit (STEL) for H_2S is 20 ppm, while the permissible exposure limit (PEL) is 10 ppm.[2] Firefighters who may respond to facilities, including oil wells where H_2S may be present, should receive specialized training in H_2S awareness and safety. When operating in and around such facilities, proper monitoring equipment should be available and used. Remember, H_2S is colorless, therefore proper monitoring equipment is essential. Personnel responding to emergencies where H_2S is potentially present must wear positive pressure self-contained breathing apparatus (SCBA).

When ignition of a well occurs, either intentional or accidental, other problems can occur, such as additional wellheads being damaged, contributing to further blowouts. An intentional ignition of a blowout may take place when the well contains high concentrations of H_2S. By igniting the blowout, it will reduce the danger of the H_2S spreading by wind to the surrounding area. However, this creates another problem. H_2S when burning emits a by-product, sulfur dioxide (SO_2). SO_2 is a toxic gas also. The downwind side of the fire still poses a threat from SO_2. Evacuation and constant monitoring for toxic concentrations of downwind areas will be necessary.

Because these sites may be in rural areas, there may be a possibility that the well site may be impacted by wildfires in the area. These fires may affect the well site proper or the pipelines servicing the well site. Area maps used by fire departments should indicate the locations of well sites and associated pipelines.

FIREFIGHTING TACTICS

Fire department operations at oil well fires must begin with preincident response planning. If your department may respond to a fire at an oil well site, then you must be prepared. Contact must be made with the responsible parties at the well site and as much information gathered as possible. Preincident response plans must include the following items at a minimum:

- Location of well
- Prevailing wind direction
- Access routes (minimum of two from opposite wind directions)
- Location of staging areas
- Water supply
- Concentration of all toxic gases such as H_2S (1% concentration equals 10,000 ppm H_2S; ignition can occur at 4.3% concentration, or 43,000 ppm)
- Contact information for well site personnel and company officials
- Community exposures
- Evacuation plans for affected communities
- Information regarding the well sites' emergency response plan so that the fire department's preincident response plan can be integrated with the well site's emergency response plan
- Information regarding chemicals stored on-site.
- Other hazards that may be present.

In addition, a fire department should have a reasonable idea of basic blowout hazards associated with a given site and the likely consequences. Familiarization visits to well sites are essential, so begin strategic alliances with the operating or drilling personnel early on.

Response to an incident at an oil/gas well should be made from the upwind side. As in hazardous materials incidents, a "go slow" approach should be employed. Fire apparatus should be stopped well short of the actual incident and an assessment made of the situation. Company personnel on scene should be consulted during the size-up process as well as continually throughout the operations. The company representative may initially be the operations foreperson or drilling supervisor, but this could vary. As the incident escalates or the time frame of the incident expands, then additional company personnel will arrive on scene. Refer to the glossary for the titles and responsibilities of these people. It may be found that secondary fires have been ignited as a result of the well ignition. If the incident is small and does not involve a well ignition, then immediate fire department actions will be warranted. Ground pool fires from leaks or generator fires are examples. Life hazard should be addressed first in all situations. Remember that the life hazard includes the firefighters as well as the general public and drilling/operating personnel. If there is no known or suspected life hazard, then a nonintervention posture may be called for, such as letting the well burn. Again, consult with the professional well control people on the scene.

The second consideration for the responders should be the prevention of extension. What is the fire exposing? Are there oxygen or acetylene tanks exposed? Are support buildings exposed? At one fire known to the authors, 55 gallon drums of unknown chemicals were exposed. In this instance, the responders opted for a nonintervention strategy until the contents of the drums could be identified. It should be noted here that many well sites are in remote areas with little or no water supply. Other well sites may be in locations with freezing temperatures, causing additional water supply problems. If a nonintervention strategy is employed, then all responders and their equipment should be relocated to a safe distance away from the incident and staged. Even if operations are undertaken initially, all nonessential personnel should be in a staging area at a safe distance.

The route to the scene and the immediate area surrounding the incident should be cleared of unnecessary personnel and vehicles. As the incident and response escalates, the need for this will become apparent. It is easier to start this process earlier in the incident rather than later. Staging areas should be set up in a safe location (upwind), and a staging officer appointed. At least two staging areas allowing for different wind directions should be identified in the preincident response plan.

Other considerations that are important to consider at an oil or gas well fire are as follows:

- *Search and rescue.* At times, the best action for the fire department may be a quick search for casualties with a limited rescue where possible. Protection of rescuers with hose streams may be applicable.
- *Incident/unified command.* Establish incident/unified command at these incidents immediately and implement the incident management system. As the incident continues, a unified command structure shall be used. There will be many organizations with a stake in the operations and the outcomes, such as local emergency management officials, drilling company officials, environmental officials, oil well control specialists, and so forth. Their technical advice and the need for a coordinated solution to the problem justify their participation.
- *Early collapse of the drilling rig.* Drilling rigs and *Christmas tree equipment*, are made of steel. The pressure-fed fire of a blowout will totally destroy the steel structure in minutes. At approximately 600°F a decrease in the strength of steel occurs. At 1,100°F the steel will start to soften and fail, depending on the load and how it is applied. Drilling derricks have been known to fail in less than 30 minutes after blowout ignition.[3]

- *Extended use of SCBA.* Concentrations of H_2S will mandate the use of SCBA during rescue and containment operations. This should be anticipated early. Additional spare cylinders or a portable air-refilling units will be necessary. A breathing air cascade unit with air lines may be beneficial. Well-control specialists use such breathing air cascade systems during their operations to increase the uninterrupted work time in the hot zone.
- *Water supply.* As mentioned previously, water supply may be a problem. At most well incidents, a large pit is dug and water supplied by tankers or by the drilling of water wells nearby is placed in the pit. In other instances, large tanks are brought in, and water provided by the water supplies mentioned earlier. Similar to large foam operations, a main assault on the fire should not begin until enough resources are present on scene. This includes the water supply. Water spray and curtains are used to keep well control specialists cool while they work near the wellhead, and a loss of water at a critical time could mean a loss of life.
- *Accountability.* As soon as possible, accountability must be established. The accountability officer must be established at the entrance to the site. All personnel on the site, including nonfire department personnel must be included in the accountability system. If personnel are operating in the hot zone immediately adjacent to the well, then an additional layer of accountability must be established.
- *Campaign length operations.* These incidents will be extended duration operations sometimes referred to campaign length operations. Considerations must be given to rotation of personnel, as well as assuring the resources are on hand to support the personal needs of the firefighters, for example, toilet facilities, food, and medical.
- *Incident action plans (IAPs).* IAPs must be used at these incidents with operational periods established using an IAP for each period. A planning session will be necessary for this. Briefings should be conducted on a regular basis but at least at the beginning of each operational period.
- *Establishment of a log.* The incident commander (IC) should start a log immediately and all events and actions taken recorded. The names, titles, and telephone numbers of key personnel from the well drilling company as well as other companies and agencies should be listed. Advice given from these individuals should also be recorded.
- *Addressing environmental issues.* As mentioned previously, the smoke from an oil well fire will contain a toxic mixture. Smoke from burning wells contains carcinogens and pollutants such as carbon monoxide, soot, and sulfur dioxide, the prime component in acid rain. The well fires of Kuwait are known to have released 500 million tons of carbon dioxide into the atmosphere.[4] Water runoff will also be a concern. At one well fire in the United States, the runoff was entering a ecologically sensitive area, and a lined containment pit was developed on-site to contain the runoff.
- *Extinguishing agents.* The use of foam during well fire extinguishment has limited applicability; however, it has been used to control ground fires and has applications for other fires in and around well sites, for example, fuel fires from generators. Dry-chemical extinguishing agents will also have some limited use at oil well sites, but for an ignited blowout, they are not the method of choice. The amounts needed and the one-shot characteristics preclude their use.
- *HazMat.* Hazardous materials and decontamination issues will be present. Decontamination should be established because of the various chemicals including acids as well as the well fluids that are present. Hazardous materials teams should be called to address these issues.

WELL CONTROL COMPANY OPERATIONS

Well control specialists will most likely be called in to assist with well fire incidents. Major oil companies will not have the equipment or full-time experienced blowout control and associated personnel on their payroll due to costs. Most oil companies employ a strategic partnership with other companies and specialist contractors.

The local fire department may be asked to assist when one of the companies is on-site with support activities. These companies will usually bring their own high-capacity firefighting pumps with its associated piping. These pumps typically pump 5,000 gallons per minute at 150 psi. Personnel assistance will be necessary, as well as some other logistical support.

Generally, the well control company operations during an ignited blowout will consist of the following:

- *Clearing the debris from the wellhead.* Large bulldozers and other such equipment are used for clearing debris (fig. 32–3). Debris is cleared to have unimpeded access to the wellhead. This is why the access routes and area around the well site needs to be clear. There will be much heavy equipment being brought in to support such operations.
- *Cooling of surrounding areas.* Cooling will reduce the chance of the well being reignited after the fire is extinguished and before the capping takes place.
- *Extinguishing and maintaining cooling.* Extinguishment can be one of many methods. Water alone has extinguished some of the world's largest oil well fires. Water can extinguish by one of the following methods:
    - o Cooling the fire below its auto-ignition temperature by absorbing heat as it is flashed to steam.
    - o Water flashed to steam displaces oxygen and smothers the fire.
    - o Powerful water streams displace the fuel from the fire.

  Chemicals have been used in limited roles. Dry-chemical extinguishment agents have been used where lateral flow has led to a large fire surface area. Foam has been used for ground fires to allow well control specialists to work near the flow source. Explosives are still used to temporarily drive fuel away from the point where the flame develops and deprives that immediate area of oxygen to support instant reignition.[5]
- *Capping the well.* Capping consists of placing equipment including valves and rams on the well piping and closing of these valves to stop the flow of fluid.

Fig. 32–3. **Large backhoe clearing debris at well fire**
(Courtesy of Williams Fire and Hazard Control)

If the pressures in the well do not allow the operations to take place, then it is sometimes advisable to drill a relief well a distance from the well on fire (fig. 32–4). This process could take weeks to complete. The well connects with the well on fire hundreds or thousands of feet below the surface and requires pumping in drilling mud to control the flow from the initial well. This may

***Fig. 32–4.* High pressures from a well blowout may require the drilling of a relief well to relieve pressures to cap the initial well.** *(Courtesy of Williams Fire and Hazard Control)*

completely "kill" the fluid flow from the well or reduce the pressures to a point where capping operations can take place.

CONCLUSION

While oil well fires are rare, there have been instances where fire departments have responded. The fire service must be prepared if there are wells in their response district and be familiar with the sites, operations, and hazards present if a successful response is to be made. At any fire incident the loss of a firefighter's life is a possibility. With unfamiliar surroundings and operations, the chance of firefighter injury or death is increased. We can reduce the risks by preincident response planning, training, and strict adherence to guidelines, standard operating procedures (SOPs), and the incident management system.

CASE STUDY

One night during October 1998 an explosion and fire occurred at an oil well location in Louisiana. The local fire department arrived on scene and was confronted with an oil rig fully engulfed in flames. Workers on the oil rig at the time of the explosion sustained burns, and there were six missing workers who were presumed to have sustained fatal injuries. The oil well was located in a remote location, and there was an absence of exposures. Initial efforts of the fire department were directed at treating the injured workers. The fire department was able to eventually extinguish some of the fire that was located remotely from the actual well by closing a pipeline valve. During initial survey, drums of hazardous chemicals were observed by responders, and a hazardous materials team was dispatched. Eventually, well control specialists that had been contracted were able to control the well and extinguish the fire.

Random thoughts

- Establishing staging at the scene and staging nonessential vehicles and equipment will keep roads clear for ambulances and other key apparatus.
- Preincident response planning is invaluable for the successful mitigation of incidents and will help define roles and responsibilities for responders as well as specialized resources that may be needed.
- Use company employees with knowledge of the operations as well as specialized technical experts to assist incident commanders with decision-making responsibilities.
- A go-slow approach to well fires is essential for the safety of all personnel operating on the scene.[6]

NOTES

[1] Tony Peterson quoted by Matthews, C., and Flak, L. (n.d.). *Part 9: Firefighting. The Mechanics of Oil/Gas Fires, Meltdown and Secondary Damage, Water/Chemical/Explosive Extinguishing Methods and Voluntary Ignition.* Retrieved January 13, 2005 from John Wright Company, www.jwco.com/technical-literature

[2] Bevelacqua, A. S., Noll, G. G., and Hildebrand, M. S. 2005. *Hazardous Materials Managing the Incident Field Operations Guide.* Chester, MD: Red Hat, p. 130.

[3] Matthews and Flak, *Part 9: Firefighting.*

[4] Imster, E. 2003. *Oil Well Fires.* Retrieved January 13, 2005 from Earth and Sky, http://earthsky.com/shows

[5] Matthews and Flak, *Part 9: Firefighting.*

[6] The information contained in this case study is based on Taylor, C., and Waits, M. 2003 (May). Oil Well Explosion. *Fire Engineering.*

33

Pipelines

INTRODUCTION

Pipelines have been described as the energy lifelines of our society. They carry everything from home-heating oil to aviation jet fuel. Pipeline systems will differ in size, purpose, complexity, and operating environments. Pipelines are found in every state and in almost every community in one form or another, whether they are largely aboveground pipes, or smaller pipelines buried beneath the surface (fig. 33–1).

***Fig. 33–1.* Pipelines may be above- or below-ground.**
(Courtesy of G. Noll)

DESCRIPTION OF PROCESS

Pipeline gathering and distribution systems may vary in size from 2 to 60 inches in diameter. Crude oil pipelines, for example, are constructed of steel and range in size from 8 to 48 inches in diameter. Smaller gathering lines are 2 to 8 inches in diameter. Refined products may typically be carried in pipelines of 8 to 42 inches in diameter. Natural gas pipelines may be in the size range of 2 to 42 inches. Natural gas gathering and transmission lines are constructed of steel while a natural gas distribution system may use both steel and plastic piping.

Gathering lines are those pipelines that transport crude oil or natural gas from the wellhead to a processing facility where the oil gas and water are separated and processed. The size required will vary based on the capacity of the well being served, the length of the line, and the pressure at the well. For crude oil lines, the pressure may be below 100 pounds per square inch (psi), but for gas lines the pressures will be much higher, in some cases up to 2,000 psi, but many will be operating at several hundred psi. Transmission lines are pipelines used to transport natural gas from a gathering or storage facility to a processing or storage facility, large volume customer, or a distribution system. When used to describe a hazardous liquid pipeline, a transmission pipeline is one that is used to transport crude oil from a gathering line to a refinery and refined products from a refinery to a distribution center. Distribution lines are lines used to supply natural gas to a customer. They are located downstream of a natural gas transmission line. Compressors are used to create the pressures needed to move the natural gas through the system (fig. 33–2). The pressures in gas transmission lines may range from 500 to 1,000 psi. The compressor stations are spaced 50 to 100 miles apart along the pipeline. The distance between compressor stations vary based on the volume of gas, pipeline size, and other factors. The capacity of the system may be increased by adding compressors at existing compressor stations, or by adding additional compressor stations.

Fig. 33–2. **Compressor** *(Courtesy of G. Dodson)*

Much of the pipeline systems that exist in the United States are not visible. What are visible are the pumping stations and other support facilities required for pipeline operations (fig. 33–3).

Fig. 33–3. **Pipeline manifold at pipeline station** *(Courtesy of G. Noll)*

We normally associate oil and gas as the products contained and transported in pipelines, but other liquids and solids may also be transported. Products that are transported in pipelines include the following:

- Gasoline
- Diesel fuel
- Kerosene
- Natural gas
- Heating oil
- Propane
- Aviation gasoline
- Jet fuel
- Carbon dioxide (CO_2)
- Ethane
- Crude oil
- Coal
- Liquefied natural gas (LNG)
- Coal slurry

While most of these pipeline products are familiar to us, some are not as familiar as others. Some of those that may not be as familiar when we think of products carried in pipelines are described in the following sections.

Carbon dioxide

Carbon dioxide (CO_2) pipelines are used where CO_2 is used to inject the product into an oil field's reservoir for enhanced oil recovery operations. Pipelines are used to move the CO_2 to the oil production field, and smaller distribution lines will then bring the CO_2 to the individual injection well. Overall mileage of CO_2 pipelines is relatively small, but increased growth is possible as more CO_2 is used to enhance oil recovery.

Liquefied natural gas (LNG)

LNG is natural gas that has been cooled and compressed to a temperature where it becomes a liquid. The demand for LNG has increased substantially because of the increased energy needs of the United States. Much is known about the product, but there are many who would like to prevent its appearance in some areas of the country. As the demand for additional energy sources increases and the LNG infrastructure grows, it is probable that more fire departments and fire chiefs will come into contact with this infrastructure. In recent years, eight communities have said no to LNG facilities.[1] Normally, large volumes of LNG are transported by ocean tankers rather than pipelines, but there are LNG pipelines throughout the United States. To maintain the gas as a liquid in a pipeline, the gas must be kept at a low temperature. This requires the insulation of the pipeline and cooling stations that remove the heat caused by the pumping of the product. Special steel is also required for the pipeline due to low operating temperatures. Short pipelines are present at gas liquefaction and vaporization plants as well as terminals for loading and unloading the tankers.

Coal slurry

Coal slurry pipelines carry finely ground solids in water and have been in operation for many years. These pipelines require large volumes of water, and because of this, these pipelines are not often used.

Most pipelines are owned by pipeline companies whose sole function is to operate the pipeline system. Pipeline corridors may contain three or more pipelines running parallel to each other. These corridors may also be known as pipeline rights of way. This area may be 25 to 150 feet wide and it provides access to the pipeline for repairs or maintenance. There are also building restrictions close to these rights of ways. The rights of way may be owned by the pipeline or acquired for use from the owner of the land. In addition to pipelines, other subsurface utilities such as communications lines may also use the pipeline corridor.

BATCHING

Batching is a process in which multiple products can be transported using the same pipeline. Refined product pipelines can transport several products in a pipeline simultaneously. These products can be separated with a physical barrier, or they may have no physical barrier between products. Where no physical barrier is present, the difference in density between products maintains the product separation. Dozens of products or grades of a single product can be transported this way. If a rupture were to occur in a pipeline carrying batched liquids, it is possible that first responders would encounter more than one product flowing with different properties and hazard characteristics.

PIPELINE MARKINGS

Pipelines are required by law to be clearly identified using markers and warning signs. The markers and warning signs are in high visibility colors such as yellow or orange and indicate the approximate route of the pipeline. They are placed at frequent intervals along the pipeline rights of way (figs. 33–4 and 33–5). They are located where a pipeline intersects a street, highway, railway, waterway, and other prominent points along the rights of way such as near buildings and structures.[2] They are also found at locations where the pipeline is aboveground in areas of public access. The marker will identify the material being carried in the pipeline, the name of the company operating the pipeline, and a contact number where the operator can be reached in case of emergency. The markers will also carry the words "warning," "danger," or "caution." The markers or signs only indicate the general location of the pipeline. In actuality, the markers do not indicate the exact location because the pipelines do not always follow a straight course between markers. Also, markers do not give information on the depth of pipelines present. The absence of a pipeline marker is no indication that a pipeline is not present. In urban high-density areas where the placement may be impractical, no pipeline markers may be present. An example may be areas with hard surfaces and parking lots. In some instances, vandalism or damage by outside forces may have removed the pipeline markers.

Fig. 33–4. **Pipeline marker** *(Courtesy of G. Noll)*

Fig. 33–5. **Pipeline markers** *(Courtesy of G. Noll)*

COMMON HAZARDS

Causes of pipeline accidents and incidents

Historically, pipelines are normally a very safe method of transporting products when compared to rail or truck transportation. However, pipeline ruptures do occur. The United States Department of Transportation, Office of Pipeline Safety, maintains data on the frequency, causes, and resulting consequences of pipeline failures. The Office of Pipeline Safety indicates that "outside force" damage contributes to a larger number of pipeline accidents and incidents than any other category of causes when the data from all accidents involving hazardous liquid, natural gas transmission, and natural gas distribution lines are considered together.[3] Outside force damage can be the result of a variety of factors, including:

- Extreme temperatures
- High winds
- Heavy rains and flooding
- Fires or explosions external to the pipeline
- Vehicles striking the pipeline
- Lightning
- Earth movement
- Vandalism
- Excavation damage

The largest portion of outside force damage is the result of excavation damage.[4] Another leading cause of pipeline failures is corrosion of the pipeline, both internal and external. To prevent outside force damage to pipeline as well as other essential buried utilities such as telephone, fiber optics, water, sewer, and underground power lines during excavations, each state in the United States has a one-call program that provides a single toll-free telephone number, so that contractors and homeowners can notify pipeline operators of excavation plans. Telephone notification to the one-call center is usually given 48 to 72 hours in advance. The center then notifies the owners and operators of the underground equipment so that they can go to the area and mark their cables, piping, and so forth.

Under normal operating conditions, there would be no hazards to first responders associated with pipelines. However, if a rupture or leak takes place, various hazards may be present. These hazards may include the flammability of the product, high pressures of the product, explosion potential, fires, product toxicity, or skin irritation. Carbon dioxide, for example, poses no threat of toxicity, but on release from the pipeline, it may create an asphyxiation hazard. Carbon dioxide is heavier than air, so it will stay near the ground. If the leak is in a depressed area, then the area may be oxygen deficient. When responding to a pipeline incident, it should be treated as a hazardous materials incident. Responders should be protected from inhalation, absorption, and ingestion of the product. Removal of ignition sources should be undertaken when it is safe to do so.

Natural gas gathering pipelines, leading from well to processing facilities, may contain hydrogen sulfide (H_2S), also known in the oil industry as "sour gas." Breathing hydrogen sulfide gas can cause death. The immediately dangerous to life or health (IDLH) for H_2S is 100 parts per million (ppm), and the short-term exposure limit (STEL) for H_2S is 20 ppm.[5] The permissible exposure limit (PEL) for hydrogen sulfide as set by the Occupational Safety and Health Administration (OSHA) is 10 parts per million (ppm). Sour gas is found in gathering pipeline systems that move the crude oil from the well to the field processing plant, where it is removed so the gas entering the transmission line for long distance movement does not contain toxic amounts of hydrogen sulfide. Associated sour gas may also be found in crude oil pipelines, and responders must be aware of this and plan accordingly for the possibility.

FIREFIGHTING TACTICS

Preincident response planning for pipeline emergencies

As with other hazards within a fire department's response jurisdiction, if pipelines are present, they should be preplanned. Preincident response planning is essential for a safe and effective response. The response to the event actually begins well before notification is made to a department that an emergency has taken place. Preincident response planning must involve representatives of the pipeline operating company as well as local emergency management officials. Pipeline incidents may be multijurisdictional due to a large geographical area and involve many agencies. A unified command structure is required.

Preincident response planning should contain the following information at a minimum:

- Location of the pipelines within your response area. Fire departments that may be called on mutual or automatic aid to a jurisdiction with pipelines should be familiar with those pipelines as well as any preincident plans developed by the initial responding agency.
- Maps of the pipeline routes
- Product(s) being transported in the pipeline and their hazard.
- Construction of the pipeline
- Location of valves associated with the pipeline
- Location of pumping and compressor stations
- Direction of flow of the products
- Emergency contact numbers
- Amount of product carried in the pipeline on a daily basis, as well as the amount of product that would be in the various sections of pipe should a leak occur and the pipeline section was isolated. Once the valves are closed isolating a section, there still remains a considerable amount of product in the line that will flow. Departments must be able to control this situation.
- Any specialized equipment that may be required. Large caliber portable monitors to control vapors, large quantities of foam for spill coverage, specialized foam, for example, high-expansion for vapor control of unignited LNG.
- Predesignated staging areas for pipeline sections. It is recommended that the pipeline be separated into sections for response protocols. This will allow for annexes to the overall preincident response plan, identifying specific hazards within a designated section. Primary and secondary staging areas should be selected depending on weather and topographical considerations.

- Target hazards that would require special handling in case of a pipeline incident, for example, schools, hospitals, nursing homes, places of assembly. Shelter-in-place or evacuation procedures should also be established for the target hazards identified.
- Other exposures that could be affected including the jurisdiction's infrastructure such as bridges, tunnels, and so on.
- General firefighting guidelines for responding units
- Evacuation and isolation distances
- Water supplies
- Material safety data sheets (MSDS) for the products in the pipeline
- Prevailing wind direction.

A major metropolitan fire department that has a major pipeline running through the city has developed a comprehensive preincident response plan and standard operating procedure (SOP). The pipeline is actually two pipelines running together. One contains gasoline, and the other, a kerosene-based aviation fuel. The pressures in this pipeline range from 200 to 1,200 psig. The preincident response plan contains a complete description of the pipeline and its associated facilities, including the failsafe systems designed to minimize damage in case of an incident. Control valves will shut the system down on any excessive pressure in the system. If these valves fail, then pressure switches will shut down the pumps. In the event of a pressure rise due to excessive heating of the product, then pressure relief valves will operate. The system is subdivided using automatic valves for isolation of the system into sections. Each section can then be shut down using manual valves.

The fire department has an SOP that was developed in conjunction with the pipeline operating company whereby the fire department will shut the manually operated valves located on each side of the confirmed leak. Normally, emergency response personnel should never isolate or close in any pipeline valves on transmission or distribution lines unless directed by pipeline operations personnel (fig. 33–6).[6] In this department's case, it has been agreed to in advance that the department will isolate the pipe section where the leak has been found. Included in the preplan is the number of turns required to shut down a valve. The number of turns is considerable, so by knowing this information, the firefighter turning the wrench or valve wheel will not think the valve is broken, or stop too soon if resistance is met.

Fig. 33–6. **Normally, emergency response personnel should never isolate or close in any pipeline valves on transmission or distribution lines unless directed by pipeline operations personnel.** *(Courtesy of G. Noll)*

Other areas covered by this department's preincident response plan are the role of supporting agencies and other department divisions such as the communications division. Drill frequencies are established and a mechanism for evaluation and modification has been incorporated.

NOTIFICATION OF A PIPELINE INCIDENT

The response to a pipeline incident will be usually initiated in one of three ways:

1. Notification by the pipeline company to the local emergency response system, that is, 911. Pipeline companies monitor their pipelines using instrumentation and monitoring systems. These monitoring systems are known as SCADA (supervisory control and data) systems. These systems will provide pipeline operators at the control station with up-to-date readings on pressures, flows, volume, alarms, temperatures, and other conditions regarding the pipeline. Alarms are generated when preset levels are exceeded. Pipeline operators will be able to detect a leak by comparing total product being delivered by the system to total product being pumped into the system and activate emergency procedures to shut down the affected section of the pipeline, or in some cases, the entire pipeline. Pipeline operators also use aerial and surface inspection of their pipeline rights of way. When a leak is detected, emergency procedures will call for the notification of responders such as local fire departments.
2. Notification by nonpipeline employees is the second way that a fire department can be notified of a pipeline incident. These persons may be present or nearby when the pipeline ruptures and call 911. They may also see the effects of a pipeline leak such as an oil spill or strong odor and report it to 911. In this case, the reason for the response may not be made clear to the dispatcher or call receiver. In one case, civilians reported a strong odor in the area to the emergency call center on numerous occasions due to a pipeline leak, but when the fire department responded, they assumed the odor was from an asphalt plant nearby. The resultant fire of the leak caused three deaths.
3. The third method is a verbal alarm from a fire department unit passing by. There are times where firefighters out of the station on routine assignments have spotted pipeline or other emergencies.

As already described, the response to a pipeline incident may have been initiated by a caller stating a little defined problem, that is, unknown hazardous materials spill or an unknown odor. Therefore, when responding to such vague reports, certain information must be considered by the crew, and certain clues must be recognized if we are to have a safe and effective response. When receiving reports such as the one above, we must be familiar with the district and determine if there are any pipelines in the area that could give such indicators if there were a leak or rupture. The units must respond with full protective equipment. Remember, how we respond to the small or unknown incident is how we will respond to the major incidents. Are we disciplined, prepared to work, and in the correct frame of mind? When firefighters approach the area, we must be cognizant of our surroundings. Observe the area as you approach. Is there dead or discolored vegetation? Is there a strong odor present? What are the sounds that you hear? Is there a hissing noise? On a hot day, you may see the vapors from a spilled product. If you are near a body of water, do you see evidence of product spillage such as sheen on the surface or pooling or discoloration of the shoreline? With some gases, a vapor cloud may be present. Dirt being blown into the air may indicate a buried pipeline leak. The pooling of liquid may also be an indication of a leak.[7] Remember, these signs may not be clearly visible during the day, but especially at night.

If we are responding to a reported pipeline leak we must treat it just as we would a hazardous materials incident and respond with caution. The use of combustible gas indicating equipment and establishment of hot zones are required. Another consideration when responding is that there may be more involved than a leak or spill. A pipeline response may be instituted for a confined space or other technical rescue scenario. Respond and position the apparatus from the upwind and uphill side of the incident. Refer to your preincident response plan for the product information and evacuation and isolation distances. Never place your apparatus closer than the recommended evacuation distances. Apparatus must not become an ignition source. The authors know of instances where the fire department apparatus was the ignition source. One was a natural gas leak where the resulting fire destroyed a number of city blocks, and the other was an industrial incident resulting in the shutdown of a plant for an extended period of time. Also, try to place the apparatus facing the direction of evacuation, in case a hasty retreat is required. Evacuation and accountability procedures should be practiced during incident response drills.

Apparatus operators should locate and test their water source, but refrain from connecting the pumper until the incident commander (IC) determines the incident objectives and strategies. The position of the apparatus may change when the strategies are defined. Ensure that apparatus and hoselines do not block evacuation routes.

The fire department shall establish isolation distances and prevent entry by unauthorized persons. Persons allowed into the hot zone should be equipped with the proper personal protective equipment including self-contained breathing apparatus (SCBA), and monitoring for flammable, toxic, and oxygen-deficient atmospheres should be performed. SCBA is required for all pipeline firefighting and control operations.

The IC should establish command and open lines of communications with the pipeline company. The IC must verify the product and its hazards and attempt to verify the quantity of the product leaked. Liquids leaking under pressure may give the appearance of a harmless geyser of water, but do not assume that it is not a hazardous material. The ignition of the product will be easier when it is in vapor or leaking under pressure. In addition, the IC must notify other response and support agencies. Because this a pipeline incident, it will involve more than one agency and may involve more than one jurisdiction, so a unified command system should be established. Some of the key players will be pipeline company representatives, the Environmental Protection Agency, environmental cleanup contractors, support agencies, the US Coast Guard (if the spill is located near a waterway), the National Transportation Safety Board, the US Department of Transportation, law enforcement agencies, and other fire departments.

As with any response, the life hazard must be the first priority. This includes the lives of our firefighters. Perform a risk-benefit analysis before committing personnel to a rescue attempt. Remember, there may be times where casualties must be left in place until the area is deemed safe to enter. In addition to any life hazard in the hazard zone, the life hazard to of those persons in the surrounding area needs to be addressed, also. Is evacuation necessary, or will shelter in place be appropriate? Remember to address those in the area most likely to be affected first. This would normally be the downwind direction. When addressing the life hazard, it should be noted that during these times where terrorism is a threat to our society, we should always consider that the incident we have responded to may be the result of a terrorist act. ICs should ensure that the area is secure for fire department operations. Be alert for unexploded improvised explosive devices (IEDs) that may be in the area. Heavy sandbags or other objects may be used to hold such devices to the pipeline for maximum effect. Be observant of all suspicious objects and take the necessary precautions. Staging areas should be established so that nonessential vehicles, equipment, and personnel are staged at a safe distance, ready for assignment and deployment as ordered by the IC.

If no life hazard is present, the next priority will be incident stabilization. Incident stabilization may require the following:

- Coordination with the pipeline company to verify that the pipeline has been shut down, the flow redirected, or the section has been isolated.
- Containing the spilled product. Damming or diking may be required to control the product runoff.

- Elimination of ignition sources.
- Vapor disbursement using fog streams. In this instance, it is desirable to use unattended monitors, to avoid placing personnel in the hazard zone. When using a fog stream to disperse vapors, water runoff should be monitored. It is possible that the water runoff will weaken the trench walls if present, collapsing the trench onto the pipeline.
- Control of flammable vapors using fog patterns or a foam blanket application. The vapors of an LNG spill can be controlled using high-expansion foam. Additional foam supplies may be required other than what is carried on the responding apparatus. This should be identified in the preincident response plan and resources identified.
- Extinguish secondary fires. Secondary fires may have been started by the primary fire or explosion. These should be addressed as soon as practical.
- Protection of exposures. In the case of a pipeline fire, it may be appropriate to let the product continue burning until the fuel can be shut down. This is essential where a flammable gas is involved. Cooling of exposures, including other pipeline or support facilities, is warranted where fire is involved.
- Fire extinguishment where applicable. Remember, gas fed fires should not be extinguished unless the fuel source has been isolated and the pipeline operator advises you to do so. When a pool of LNG is on fire, the use of water on the fire will increase the vaporization rate of the LNG and will intensify the fire. Dry-chemical would be the preferred extinguishing agent for LNG fires. For large fires involving LNG, high-expansion foam may be used. High-expansion foam, while not completely extinguishing the fire, will provide a controlled burn and significantly reduce the radiant heat.

  When a high-expansion foam blanket is applied to the surface of a burning LNG pool, it reduces the vaporization rate, resulting in the fire reaching a steady-state condition.[8] Dry-chemical can then be used to extinguish the remaining fire. The use of foams for hydrocarbon fires may require additional foam supplies to be delivered on-site. This should be identified in the preincident response plan and resources identified.
- Establishment of water supplies. Due to the potential of having the incident location in a remote area, water supply may be scarce or nonexistent. A water resource officer will be needed and plans made to establish water supplies as needed. This may include water tanker shuttles or long hose lays with pumper relays. Practice these relays during drills and exercises.

During or after the incident, decontamination may be required. This would include personnel as well as tools and equipment. Because of the hazardous properties of many of the products being transported via the pipeline system, a careful analysis of the product and determination of the decontamination required should be performed at all pipeline incidents.

As with all incidents, we must protect the environment. Close liaison with Environmental Protection Agency (EPA) representatives during the incident will allow discussion concerning our actions during the incident. At times, our actions may actually hurt the environment, and we should be aware of this and prevent it as much as possible. Control water runoff from cooling and extinguishment streams as much as you can. Documentation of all actions taken at the incident is an essential part of the recordkeeping and subsequent investigation of the incident.

CONCLUSION

Responding to pipeline incidents may be rare for most fire departments, but statistics highlight that they do occur. Fire departments must be prepared to respond to these incidents when called on just as they are prepared to respond to the bread-and-butter calls to structural fires and emergencies. As with all responses, preincident response planning and drills are the keys to success. Remember that pipelines may contain almost any substance. Do not assume that the pipeline that has been in reliable service for 20 years does not pose a potential threat to your community.

CASE STUDY

A large county fire department was dispatched to a reported jet fuel spill at approximately 5 p.m. on a sunny April afternoon. The pipeline ruptured due to unknown reasons, and on arrival rescue crews began evacuating residents. Twenty homes were evacuated, and fire department hazardous materials crews set up booms in a nearby lake that was contaminated with the jet fuel. The pipeline was 12 inches in diameter and contained pressures of 1,000 psi.[9]

Random thoughts

- Pipeline incidents will involve more than just the immediate area of the leak. In this case, the environmental impact was considerable. Trees, shrubbery, and even structures were covered with the fuel. Much of the fuel seeped into the ground soil and also entered a nearby lake.
- Evacuation of surrounding structures and eliminating sources of ignition are essential. Combustible gas indicators should be used to determine the hazard zones and areas that could ignite.
- Pools of unignited flammable or combustible fuel will have to be covered with foam to reduce vapors and the possibility of ignition.
- Private cleanup companies may be required to respond to clean up the spilled fuel.
- The pipeline should be isolated to stem the flow of fuel to the leak area.

NOTES

[1] National Association of State Fire Marshals (NASFM). 2005. *Liquefied Natural Gas: An Overview of the LNG Industry for Fire Marshals and Emergency Responders.* Washington, DC: NASFM, p. 5.

[2] United States Department of Transportation (DOT), Office of Pipeline Safety. Retrieved August 12, 2005, from http://primis.phmsa.dot.gov.

[3] DOT, Office of Pipeline Safety.

[4] DOT, Office of Pipeline Safety.

[5] Bevelacqua, A. S., Noll, G. G., and Hildebrand, M. S. 2005. *Hazardous Materials Managing the Incident Field Operations Guide,* Chester, MD: Red Hat, p. 130.

[6] Noll, G. G., and Hildebrand, M. S. 2004. *Pipeline Emergencies.* Chester, MD: Red Hat, p. 105.

[7] Noll and Hildebrand, *Pipeline Emergencies.* pp. 79–80.

[8] Texas A & M University, Emergency Services Training Institute. *Liquefied Natural Gas (LNG) Spill Control and Fire Suppression Training,* Student Notebook. (College Station, TX: Texas A & M, n.d.) p. 85.

[9] Information for the case study was based on the article Virginia Haz-Mat Crews Face Jet Fuel Pipeline Rupture in *Firehouse.* Retrieved April 12, 2005, from http://cms.firehouse.com.

34

Automotive Manufacturing

INTRODUCTION

Automotive production consists of nearly every known type of manufacturing process from metal working, plastics, painting, utility, and assembly. Building a car or truck involves large structures that can exceed 90 acres under a single roof and contain a small city in itself. These small cities can include police, fire, and EMS services, and may contain a clinic, offices, cafeteria, warehouses, railroad, and many other services commonly found in cities. The plants also contain a large number of people, exceeding a few thousand per shift. Because of the intricacy of this type of facility, preincident response planning and emergency response can be very challenging. This chapter will discuss some of the basic manufacturing processes commonly found in different automotive facilities and some suggested hazards and firefighting tactics to address those hazards. This chapter will not discuss a specific manufacturing process but a more generalist approach.

DESCRIPTION OF PROCESS

The manufacture of a car or truck begins with a simple roll of coiled steel and ends with a finished product in the form of a car or truck. The process is broken down into the following process areas:

- Stamping shop
- Body weld shop
- Paint shop
- Plastics shop
- Powertrain
- Assembly
- Miscellaneous operations (inspection, test track, and support)

Stamping shop

Stamping is where the entire automotive manufacturing process starts. Here, rolled coils of steel are delivered from the steel manufacturing plant (chapter 26). The coiled steel comes in different lengths, widths, and thicknesses depending on the specific use intended and is segregated into different areas. Large overhead cranes, usually controlled from the ground with a remote control, pick up these coils of steel to begin the process. The coiled steel is then sent to the first process, cleaning. The coils are cleaned in a bath to remove any excess oil, dirt, or rust that may have accumulated during shipment. From the cleaning process, the coils of steel are straightened and ready for stamping. During the stamping process, large hydraulically operated presses stamp out the designated vehicle part depending on the inserted die in the machine. Some 200 parts begin their life in this manner, ranging from hoods, trunk lids, body panels, engine cradles, and floor boards. The large stamping machines can be more than 10 feet tall and may extend below the finished floor to the basement. These large pieces of machinery contain large quantities of combustible liquids. After the specific part is stamped, it is once again cleaned to remove oils that have accumulated in the stamping process. From the stamping process, the next stop is the body weld shop.

Body weld shop

The body weld shop is an intricate network of conveyors and robotic welders. Each part, whether it is a door or vehicle hood, proceeds down a designated path toward the welding operations. The welding operations consist mainly of robotic welding units, each specifically designed for a series of welds on a specific part. These are highly accurate and expensive machines that can complete a multitude of welds in a fraction of a minute. Human welders are still present in some of the different processes, where current robotic technology has not yet developed the ability to complete all types of welds. This building, although under a single roof, is very noisy and dangerous with respect to the amount of conveyor and welding activity. After this stage of the process, the separate pieces of stamped metal are beginning to take the form of a vehicle. After the welding is complete, the vehicle now heads toward the paint shop.

Paint shop

The paint shop is another complicated area containing large dip tanks, paint booths, and ovens for curing the vehicles. The first stop for the vehicle in the paint shop is the preparation area. This area begins the prepping of the vehicle for painting. During this phase, the vehicle undergoes priming and sanding to ensure the vehicles surface is ready to be painted. This is completed by robots, similar to those used in the body weld shop. As the vehicle progresses down the conveyor system, it enters the paint booth. The paint booth is a fully enclosed structure, within the paint building. Some of the newer designed paint booths have large positive pressure fans to pressurize the paint booth so that breathing air is not required for entry. There, large robotic arms surround the vehicle, applying the paint in a sweeping motion. After the vehicle is painted, it then heads toward the curing ovens. The curing ovens, are large, multiple level ovens that operate at about 600°F to 700°F. These ovens can be 50 to 80 ft in height and cover half the size of a football field. The vehicle continues along this conveyor path through the curing ovens and cools down area until it is ready for assembly.

Simultaneously during the stamping, welding, and painting operations, two other processes are also producing vehicle components, plastics, and powertrain.

Plastics shop

In the plastics shop, the process of vehicle components such as instrument panels, glove box covers, bumpers, and other finishing details are produced. This is typically done via injection molding or casting (chapter 20). These areas also contain a small selection of robots, and parts are transported via conveyor to the assembly plant.

Powertrain

Powertrain is another process that does not begin in the stamping building. Powertrain operations are located where the engine and transmissions are manufactured. Although the block and other large metal components such as the transmission housing are generally not manufactured here, they are assembled and tested in the powertrain building. The process begins with a simple engine block or transmission housing, and the remaining components are added in a smaller scale but just as long conveyor system. This area typically contains a large amount of robotic transporters or carts, which carry the engine, transmission, and their components through the process. The assembly of the engines and transmissions consist of installing the cams, cylinders, fuel injection system, cooling system, oil system, gears, and other components. After the assembly is complete, the engines and transmissions are tested in a piece of machinery called a test stand. The engines and transmissions are loaded onto the test stand, and fuel, coolant, and oil are pumped through the engine and back out again through special drains. This permits the engine and transmission to be tested without having to fill them with fluids, only to empty them again for final assembly. After the test stand tests are completed satisfactorily, the units are sent to the assembly area to be installed in the vehicle.

Assembly

The assembly, sometimes called final assembly, is the center of the most activity. At the beginning of the assembly process, vehicles and their components are beginning to come together. The painted vehicle is now exiting the paint shop's curing oven, and engines and transmissions are arriving from the powertrain building. Bumpers, dash boards, and other vehicle trim components from the plastics shop are also arriving in the assembly building. Other vehicle components that are not manufactured on-site such as radios, steering wheels, tires, seats, windshields, wiring harnesses, and brakes are delivered via truck and sent to a specific area of the final assembly line for installation into the vehicle.

The assembly building is a very complex and highly complicated process. Although no production of components takes place here, many assembly processes rely on a philosophy of just-in-time assembly. As the vehicle travels down a series of conveyor systems, sometimes three conveyor systems stacked high, the various vehicle components are arriving via another conveyor for installation. For example, as the vehicle approaches the location on the assembly line for installation of the seats, the seats for that specific vehicle are arriving on a separate conveyor system. Another example would be that as the vehicle arrives at a designated point, the engine and transmission for that specific vehicle arrives from the powertrain building. This just-in-time assembly process ensures that only the needed parts are used and limits the amount of combustibles on the assembly floor. This entire assembly process continues until the final stage where the vehicle fluids (gasoline, oil, coolant, windshield fluid) are added, and the vehicle is started. At this point, the vehicle is driven to the marshalling yard to be loaded onto rail or truck for delivery to the dealers or staged until those services are available.

Miscellaneous operations

There are several other areas and operations that take place at an automotive manufacturing facility. These can include

- Offices
- Inspection
- Research and development
- Utilities

Offices

As with most manufacturing operations, offices are used to house support operations ranging from payroll, engineering, design, safety, and other office-related business. Some automotive facilities also have clinics and cafeterias, as well as day care and banking services. These do not present unusual hazards.

Inspection

Automotive manufacturing facilities also have some form of inspection department. This department handles the quality assurance (QA) and quality control (QC) of the manufacturing process, as well as approval of the newly assembled vehicle for sale to the dealers. This operation usually involves both office and laboratory work (chapter 23), as well as field work in either the test track or with the research and development teams. It is common for the inspection team to drive a vehicle off the assembly line (perhaps a few times per shift) and straight to an inspection area, for QA and QC audits.

Research and development

Not all, but most automotive manufacturing facilities have some type of research and development (R&D). Some facilities have a large and involved R&D department, while other have only the R&D facilities to support everyday manufacturing operations. These R&D operations, similar to the inspection department, consist of both office and laboratory work and field work. The R&D department is involved in various aspects of the manufacturing process such as new processes, materials, and assembly procedures, to investigating recalls and the creation and development of new designs. Additional field work could include testing vehicles with the inspection department on a test track or in a controlled environmental laboratory.

Utilities

Large manufacturing plants have large demands for power, water, steam, and other needed process services. Automotive plants, particularly larger and older plants might even have their own power plants. Other facilities might just have large transformers that are supplied from the local utility company. It is typical to find large boilers for steam generation and heating for the winter, wastewater treatment plants for cleaning process water used during production, and storage tanks (chapter 31) to store flammable and combustible liquids such as gasoline, oil, transmission oil, windshield washer fluid, and engine coolant.

COMMON HAZARDS

Automotive manufacturing facilities contain a potpourri of hazards that span a wide range and create challenging situations. Some hazards are inherent with the process itself or are a result of a failure of a process or process containment. Some of the more common hazards include

- Flammable and combustible liquids
- Flammable gas
- Electrical
- Equipment hazards
- Hazardous materials/chemicals

Flammable and combustible liquids

The presence of flammable and combustible liquids is a recurring theme in industrial facilities. More apparent is the magnitude at which these are present. Automotive-manufacturing facilities have both flammable and combustible liquids that are used in either to process (e.g., hydraulic oil in a stamping press) or as a component of the finished product, in this case gasoline or motor oil for the vehicle. Flammable and combustible liquids can be commonly found in all parts of the manufacturing process. This includes areas such as the stamping presses, which use hydraulic fluid; conveyor systems; mechanical devices such as robotic arms; paint booth where the vehicles are painted; plastics shop, where large quantities are plastics and hydraulic fluid are present; powertrain operations, where the engines are filled with different vehicle fluids such as gasoline and motor oil; and the assembly building when the vehicle is again filled with various flammable and combustible liquids. With the use of these fluids comes the hazard of storing such liquids. This is typically done using storage tanks (chapter 31) located on the property. These storage tanks could range in size from small to medium size tanks.

Flammable gas

The presence of flammable gas (chapter 25 and 28) is always a hazard. This hazard is typically confined to the curing ovens in the paint shop. Although it can be found in the utilities and support services used to power boilers or other support services, as well as in the body weld shop, where welders are used to weld metal car parts together. As with all flammable gases, escape from containment is always a concern. Careful review of combustion controls and procedures is warranted here to ensure that proper lighting of ovens, furnaces, and boilers.

Electrical

Automotive manufacturing uses large amounts of energy in the production of vehicles, which includes operating the normal lighting and similar power users to operating enormous quantities of robotic welders in the body weld shop. Many of these facilities will have large power substations or even their own power plant. The biggest user of electricity is the robotic welders. These units, costing over $1 million each, use electrical power to move and create an arc for welding.

Equipment hazards

The level of process equipment in this facility is incredible. Large hydraulic presses, multilevel conveyors, ovens, welders, and injection-molding machines, all with open trenches, hanging hazards, electrical and flammable or combustible liquids present, creates a hazard itself. These large pieces of process equipment, most of which are automatic and/or robotic, generate a dangerous environment. This equipment is subject to activating itself on a preprogrammed operation and could trap firefighters. Additional equipment hazards to consider include overhead cranes, open trenches leading to electrical or support equipment, overhead and ground conveyors, recovery pits, confined spaces created by large pieces of equipment, and limited access, egress, and dead ends created by equipment.

Hazardous materials and chemicals

Hazardous materials and chemicals (chapter 11) are also present in most of the manufacturing process. These are especially concentrated in the plastic shop, paint shop, wastewater treatment plant, and laboratories. As with flammable and combustible liquids and gases, the loss of containment creates the hazard. Some of the chemicals present include chlorine, solvents, natural gas, resin, carbon monoxide, caustics, acids (sulfuric and others), and not otherwise classified substances such as waste products and sludge.

FIREFIGHTING TACTICS

As with all firefighting tactics and strategies, whether for a house or business, a well-planned and tested emergency response plan is needed. Preincident response planning should be carried out in automotive manufacturing facilities to identify the hazards, which have the potential to develop into a major incident and the required resources. In the event of a fire at this type of facility, it is advantageous to have information readily available that relates to the many common hazards, fire safety provisions, access roads, staging areas, and location of water sources. Good loss prevention starts with the firesafe design of the facility such as the installation of fixed suppression and detection systems and construction features. It cannot be stressed enough the importance of proper prevention measures needed to ensure a firesafe facility.

Arriving at the scene of an incident at a large manufacturing facility such as an automotive plant can be very confusing, with multiple gates, and buildings, as well as the magnitude and size of the plant. Proper preincident response planning and several site visits will help eliminate some of this confusion. Depending on the type of incident, most often establishing a reliable water supply and supporting the fixed systems are the priorities.

After arriving at the scene, establish a command post, if one has not already been set up. Try to first determine the specific building that is involved in the incident. By doing this, you can begin to list the possible hazards common to that part of the plant. For example, if there is a fire or other incident in the weld shop, actions could include mechanically and electronically isolating the robotic welders. An incident in the stamping shop would require the same procedure. This type of action will help ensure that firefighters don't crawl into an area prone to automatic equipment activation.

Before firefighters enter these facilities, they should be wearing full personal protective equipment (PPE) and self-contained breathing apparatus (SCBA). Also, they must have communication equipment or the ability to communicate to the command post, hand lights and forcible entry/exit equipment. Complex interiors created by equipment and conveyors should be reviewed on the preincident response plan before entry. When firefighters enter large facilities such as these with high ceilings, even a serious fire may not appear dangerous. Products of combustion such smoke and carbon monoxide conditions can quickly build up and incapacitate unprepared firefighters.

Because of the large areas covered by these plants, locations of standpipes and hose cabinets are important. Similar to warehouse firefighting (chapter 17), the use of high-rise or hotel packs is beneficial. Connecting into the standpipe (preferably a Class I) will help reduce, although not necessarily eliminate, extremely long hose lays and expedite manual firefighting efforts. These buildings can be rather large and long (several hundred to several thousand feet), it may be possible to drive inside and through an uninvolved section of the plant and set up operations on the unexposed side of a fire wall. It then may become necessary to set up attack lines from the truck and gated down to smaller handlines to extend the effective range and attack the fire from there.

Unless your department owns truck-mounted positive-pressure fans, (those that exceed about 80,000 CFM) roof operations should begin after the initial support of the fixed systems. Because of the large, open-areas (more than 60 to 90 acres under a single roof) of these plants, venting is important so that smoke, heat, and other products of combustion can be removed for interior firefighters. This is especially important because of the number of open trenches and pits, as well as overhead and ground conveyors. Overhead conveyors, holding in process vehicles can become weak in a fire and be a cause of failure. Ground level conveyors have open pits and create obstacles for crawling firefighters. Further, firefighters could crawl into a paint booth or curing oven, causing them to be disoriented and become lost. Many large, open areas such as these facilities, because of code-related issues, have automatic or manual smoke vents. These are ideal for rapid ventilation of a building. Many roof vents simply require the push of a button to activate several large vents for easy ventilation. If manual vents are present, simple removing the lock and pulling the lever will operate the spring-loading vent. These methods are preferred over cutting holes in the roof because they are easier to use and do not cause unnecessary damage to the building.

Staging of equipment is also important. In the event of a fire deep inside the facility or even on a different level, staging of spare air cylinders, nozzles, and hoses prevent the unnecessary trips back outside. The large openness of these structures permits the staging of these materials in a relatively safe area, away from the origin of the fire.

Confinement of the fire in one portion of the structure or piece of equipment may be possible. In many cases these facilities are built with proper fire barriers and fire walls. The fire may burn in a section of the facility, but automatic and manual efforts to confine the fire to the immediate area of origin may be the best tactic. Once confined using fire doors, all power sources to the affected area should be shut down. Confinement of the fire to a single piece of equipment may further be possible if the equipment has a local fixed system. Many times single pieces of production equipment like a dip tank, plastic machine, or similar might have a local flooding fixed system like carbon dioxide (chapter 8) installed. Activation of this system, combined with manual firefighting efforts, could control the fire.

Because of the nature of these facilities, a wide range of fires can be expected. These fires might involve

- Flammable and combustible liquids
  - o Bulk storage inside or outside
  - o Paint operations
  - o Final assembly line
  - o Hydraulic spray fires in stamping operations
  - o Plastic operations
  - o Powertrain operations
- Flammable gases
  - o Paint-curing ovens
  - o Welding operations
  - o Boilers and other utility areas
- Electrical fires
  - o Electrical panels throughout facility
  - o Cable trays
  - o Cable tunnels
  - o Manufacturing equipment

- Rubber fires
    - o Tires on the assembly line
    - o Storage of tires
- Plastic fires
    - o Bumpers and other car components
    - o Plastic operations
    - o Assembly operations
- Poly-foam fires
    - o Seats and car components
    - o Plastic operations
    - o Assembly operations
- Vehicle fires
    - o In-process vehicles
    - o Final product vehicles
    - o Marshalling/storage yard
- Ordinary combustibles fires
    - o Office
    - o Warehouse
    - o Cafeteria

Most of these areas are protected by an overhead sprinkler system. A single building might have 20 or 30 risers, depending on the building footprint. These risers are located along the outside wall and are equipped with fire department connections (FDC). Your response plan should include the location of these fixed systems as well as the specific area each riser covers. Support of these systems is important, but remember that industrial water supplies are different, and the capacity is designed to handle large flows from both fixed systems and manual firefighting efforts.

CONCLUSION

Preincident response planning should be carried out in automotive manufacturing facilities to help identify some of the common hazards and different firefighting strategies and tactics needed to combat such fires. The shear size and magnitude of these facilities necessitate the need for frequent site visits. On-site training with local plant brigade members and frequent facility tours are also encouraged. Remember that these facilities contain several different types of hazards, based on where in the property the incident occurs.

Glossary

ablate To remove by cutting, abrading or evaporating.

accountability systems (See personnel accountability systems.)

aft Toward the rear of the vessel.

alkylate A mixture of several chemicals, all of which have high octane rating, burn relatively cleanly, and have good antiknock properties.

alkylation A process in which straight chain hydrocarbons are added to or substituted in a compound to make high-octane fuels.

annealing A heat treatment process that is used to heat metal close to its melting point and cool it over a long period of time. The result of this causes billions of tiny atoms to align themselves in such a way that it produces tough and hardened structures in some types of metals.

annulus In drilling, the space between the drillstring and the hole. Also, the annular space between two strings of pipe, either casing in casing or tubing in casing.

anode Negative electrode.

aqueous film-forming foam (AFFF) A common firefighting foam used in fire suppression systems, aircraft firefighting and other Class B fire situations. It is a combination of fluorocarbon surfactants and synthetic foam agents that produces a thin film that spreads across the surface of the Class B liquid.

authority having jurisdiction (AHJ) The organization, office, or individual responsible for approving equipment, materials, an installation, or procedure. It is usually the fire chief or fire marshal.

backdraft An explosion resulting from the sudden introduction of air (oxygen) into a confined space containing oxygen-deficient superheated products of incomplete combustion.

ballast Weight put into a vessel to improve stability. This weight may be steel, concrete, or water.

barrels A measure of volume. One barrel of oil equals 42 US gallons.

basic oxygen steelmaking (BOS) A method of steelmaking, that is, converting carbon-rich molten iron to steel.

battery limit The physical boundary or separation of a process unit from other processes or separation of processes within a process unit. Normally are identified by the valves required to isolate the unit from the rest of the refinery.

bauxite An aluminum ore that consists largely of the aluminum minerals.

billets Long square bars of metal (usually steel or aluminum) that can measure 4 to 5 inches square and up to 30 feet in length.

blowout An uncontrolled flow of fluids from a wellhead or well bore. The flow might include hydrogen sulfide (H_2S), which is fatal at low concentrations if breathed.

blowout preventer A device used to control formation pressures in a well by sealing the area around a drill pipe while the pipe is suspended in a hole, or by alternatively sealing across the entire hole if no pipe is in it.

boiling liquid expanding vapor explosion (BLEVE) A catastrophic container failure with a release of energy, often rapidly and violently, which is accompanied by a release of vapor to the atmosphere. The main cause is usually related to external heating via flame impingement on a shell of a pressure vessel above its liquid level, thus weakening the container and leading to a sudden shell rupture.

boiling point The temperature of a liquid when the vapor pressure equals or exceeds atmospheric pressure.

boilover A condition that occurs during a fire in an open-top tank containing certain types of crude oils. After a time of quiescent burning, a sudden eruption of the tank occurs with an increased volume of fire. This is caused by a heat wave inside the tank that comes into contact with residual water in the tank, causing the water to boil, resulting in a rapid expansion of steam-oil froth.

boot pit The bottom component of the bucket elevator.

bow The forward part of the vessel.

bubble cap Inside a vertical vessel, each opening on the horizontal trays is fitted with small openings and a short vertical pipe, which is in turn is fitted with a cap to allow for collection of product on the horizontal trays, depending on the product's chemical characteristics.

burrow To penetrate a surface as if by digging.

butylene Typically used as a monomer in making synthetic rubber, it is one of the LP gases.

camber The deviation of a side edge from a straight edge.

casing Steel pipe set in the hole as drilling progresses to line the hole wall, preventing caving in and providing a passage to the surface for drilling fluid and for hydrocarbons if the well is proved productive.

catalyst Chemicals that promote chemical reactions in other materials without being affected themselves. There are many common materials that are used as industrial catalysts such as platinum and silver.

cathode Positive electrode.

caustic The opposite of an acid on the pH scale, often called a base or alkali. These compounds contain a hydroxide ion.

central processing facilities (CPF) A more complex facility, similar to a GOSP, where other processes are taking place. This may include degassing, stabilization, dewatering, and other similar crude treatment processes.

Christmas tree A high-pressure assembly of valves, pipes, and fittings installed on a wellhead after completion of drilling to control the flow of oil and gas from the casing.

Class I laser Low-powered devices that are considered safe from all potential hazards. Some examples of Class I laser uses are laser printers, CD players, CD ROM devices, geological survey equipment, and laboratory analytical equipment.

Class II laser Low-power, visible light lasers that could possibly cause damage to a person's eyes. Some examples of Class II laser use are classroom demonstrations, laser pointers, aiming devices, and range-finding equipment.

Class III laser Intermediate power devices. Some examples of Class III laser uses are spectrometry, stereolithography, and entertainment light shows. Direct viewing of the Class III laser beam is hazardous to the eye, and diffuse reflections of the beam can also be hazardous to the eye.

COAL TWAS WEALTHS Acronym used in size-up and preincident response planning. The letters of the acronym stand for construction, occupancy, apparatus and staffing, life hazard, terrain, water supply, auxiliary appliances and aids, street conditions, weather, exposures, area, location and extent, time, height, and special considerations.

coker Converts the heavy hydrocarbon remains (asphalts) from the crude distillation column to more useful products.

collating A process of assembling sheets of paper in order for binding.

commodities The basic product itself (e.g., can of soda drink), the packaging (e.g., 12 soda drink cans in a cardboard box or case), and its container (e.g., 50 cases of soda drink, wrapped in plastic on a wood or plastic pallet).

communications security (COMSEC) Measures and controls taken to deny unauthorized persons information derived from radio and telecommunications.

continuity of combustibles The result of poor housekeeping, storage arrangements, or other related habits that permit combustibles to spread fire to uninvolved sections of a plant or building.

cracking (cracked) process Thermal decomposition, sometimes with catalysis, of a complex substance, especially the breaking of petroleum molecules into shorter molecules to extract low-boiling fractions such as gasoline.

cryogenic Gases that have been transformed into extremely cold liquids and are stored at temperatures below –130°F.

cryolite An uncommon mineral of very limited natural distribution. It is typically used as an ore of aluminum and later in the electrolytic processing of the aluminum.

crystalline Like a crystal. It implies a uniform structure of molecules in all dimensions.

dangerous cargo/goods manifest Listing of hazardous cargo carried on a vessel.

dangerous goods Commodities that seriously endanger the safety of the vessel in which they are carried.

day tanks Tanks used in a process in which only a single day's worth of storage is provided in an attempt to reduce fuel loading or other hazardous conditions.

debutanizer The process equipment used to remove butane from the hydrocarbon feed stream.

decantation A process for the separation of mixtures, carefully pouring a solution from a container, leaving the precipitate (sediments) in the bottom of the container. Usually a small amount of solution must be left in the container, and care must be taken to prevent a small amount of precipitate from flowing with the solution out of the container.

deckhouse The superstructure on a ship's deck.

deethanizer The process equipment used to remove ethane from the hydrocarbon feed stream.

defensive strategy Tactics where certain areas may be conceded to the incident and actions are limited to protecting exposures and limiting spread of the incident.

deflagration Burning which takes place at a flame speed below the velocity of sound in the unburned medium.

demethanizer The process equipment used to remove methane from the hydrocarbon feed stream.

depropanizer The process equipment used to remove propane from the hydrocarbon feed stream.

detonation Propagation of combustion from the ignition point at greater than the speed of sound in the unreacted medium.

dielectric electrical insulator, is a substance that is highly resistant to electric current. Although a vacuum is also an excellent dielectric, the following discussion applies primarily to physical substances.

distillation column A large vertical vessel used to distill raw materials and compounds to remove water, particulates, and other impurities.

dolomite The name of both a carbonate rock and a mineral consisting of calcium magnesium carbonate found in crystals. Dolomite rock (also dolostone) is composed predominantly of the mineral dolomite.

downcomer As the condensing liquids in a vertical vessel accumulate on each tray, some of the liquid overflows to lower trays through this opening.

downstream The next step in the petrochemical manufacturing process in the production of a finished commodity.

draft marks Numerals on the ends of the vessel indicating the depth of the vessel in the water.

drilling engineer Responsible for technical work on drilling programs, technical evaluations, and engineering studies.

drilling manager The top drilling position in most oil companies. The drilling manager reports to the operations manager.

drilling supervisor Often called the "company man"; the most senior representative of the operator at the well site. Coordinates all drilling-related activities.

drillstring The pipes and tools that are run into the well hole to drill.

dunnage Any material, permanent or temporary, that is used to ensure good stowage; and protect cargo during transport.

electric arc furnace (EAF) A method of steelmaking that uses a system that heats charged material by means of an electric arc.

electrolytic The use of electrolysis industrially to refine metals or compounds at a high purity and low cost such as those common to aluminum smelting.

electromagnetic induction The production of an electromotive force either by motion of a conductor through a magnetic field so as to cut across the magnetic flux or by a change in the magnetic flux that threads a conductor.

electrostatic precipitators A particulate collection device that removes particles from a flowing gas (such as air) using the force of an induced electrostatic charge.

emergency operation center (EOC) The physical location at which the coordination of information and resources to support emergency incidents takes place.

emergency response team (ERT) (see plant emergency response team) Teams of specially trained personnel used within industrial facilities for the control and mitigation of emergency incidents such as fires, hazardous materials spills, or rescues.

emergency shutdown system (ESD) Control the plant's capability to shut down in the event of any type of emergency or process abnormality. It is intended to stop various operations of a process and isolate them from the overall process to limit or reduce the likelihood of an uncontrollable event escalating.

encapsulation Commodities are wrapped in plastic for weatherproofing and transportation.

ethylene Unsaturated hydrocarbon.

exfoliate To come off in thin layers or scales.

exothermic A process or reaction that releases energy in the form of heat.

ferromanganese A ferroalloy with high content of manganese, made by heating a mixture of the oxide's manganese oxide (MnO_2) and iron oxide (Fe_2O_3), with carbon in a furnace.

fire control plan Set of general arrangement plans for each deck that illustrate fire stations, fire resisting bulkheads, and fire-retarding bulkheads, together with particulars of fire detection systems, manual alarm systems, fire extinguishing systems, fire doors, means of access to different compartments, and ventilation systems.

fire detection systems Designed to allow for building and facility occupants to be notified prior to a building becoming untenable due to smoke, heat, and gases.

fission The process when the nucleus of an atom splits into two or more smaller nuclei.

flare The area where controlled burning of a high-vapor-pressure liquid or compressed gas takes place to reduce or control the pressure and/or dispose of the product.

flaring The actual event of venting gases from a process or emergency condition.

flexography A process of rotary letterpress printing using flexible plates and fast-drying inks.

fluoroprotein (FP) foam A protein-based foam with fluorinate surfactants. It has some qualities of protein foam (such as resistance to heat) and other qualities of synthetic foam (such as the ability to flow across fuel).

foam tenders Large-capacity tankers used for delivering large volumes of foam concentrate to the scene of an incident.

foam tote Large containers used for the storage and transport of foam concentrate. These containers can be easily moved by forklifts. Totes usually have 275- or 330-gallon capacity.

fore Toward the forward end of the vessel.

Fourdrinier process Putting a single ply of paper on top of a single-layer, moving wire mesh. Paper is formed and pressed by the machine, then air-dried.

fractional distillation The separation of a mixture into its component parts, or fractions, such as in separating chemical compounds by their boiling point by heating them to a temperature at which several fractions of the compound will evaporate.

free surface effect Tendency of a liquid within a compartment to remain level as a vessel moves, which allows the liquid to move unimpeded from side to side.

freeboard (1) Height that the outboard edge of the deck is above the water level. (2) Vertical distance between a vessel's lowest open deck and the water surface.

frothover The overflowing of a container not on fire when water boils under the surface of a viscous hot oil.

furnish The pulp mixture is further diluted with water, resulting in very thin slurry. This is then run onto the forming fabric or wire of the paper machine under pressure in the headbox.

garners To gather or store grains in bins, similar to day tanks.

gas detection systems When activated by the presence of a fugitive gas (e.g., H_2S CO, NH_3), can shut down process equipment, activate fans, initiate fixed systems, and notify first responders.

gas oil separation plant (GOSP) Facilities that process crude oil from the wells. The purpose is to remove water and gas from the raw crude oil.

gathering pipeline systems Smaller pipelines, usually 2 to 8 inches in diameter, that move natural gas or crude oil mixtures from individual wellheads and production locations to a central facility where the oil, gas, and water are separated and processed.

generator A machine by which mechanical energy is changed into electrical energy.

geothermal Relating to or using the heat of the earth's interior.

global 360 A 360° visual size-up of an incident. Where possible, this visualization should include above as well as below the incident. If we picture the incident inside of a clear plastic globe, our visual of the scene should include the entire surface of the clear globe.

headhouse The top component of the bucket elevator.

heat of combustion The energy released as heat when a compound undergoes complete combustion with oxygen. The chemical reaction is typically a hydrocarbon reacting with oxygen to form carbon dioxide, water, and heat.

heat release rate The amount of energy released by burning materials, recorded in kilowatts (kW) or megawatts per square meter (MW/m^2)

hydrocarbons Chemical compounds containing hydrogen and carbon.

hydrogenation A class of chemical reactions in which the net result is the addition of hydrogen (H_2). The usual targets of hydrogenation are unsaturated organic compounds, such as alkenes, alkynes, ketones, nitriles, and imine. Most hydrogenations involve the direct addition of diatomic hydrogen under pressure in the presence of catalysts.

hydrotreater A common process unit in an oil refinery that is used to treat products such as gasoline, kerosene, diesel, and intermediates such as gasoil. A hydrotreater uses hydrogen to saturate aromatics and olefins to remove undesirable compounds of elements such as sulfur and nitrogen.

immediately dangerous to life of health (IDLH) 1. An atmospheric concentration of any toxic, corrosive, or asphyxiant substance that poses an immediate threat to life or would cause irreversible or delayed adverse health effects or would interfere with an individual's ability to escape from a dangerous atmosphere. 2. The maximum level to which a healthy worker can be exposed for 30 minutes to a chemical and escape without suffering irreversible health effects or escape-impairing symptoms.

incident action plan (IAP) An oral or written plan containing general objectives reflecting the overall strategy for managing an incident. It may include the identification of operational resources and assignments. It may also include attachments that provide direction and important information for management of the incident during one or more operational periods.

incident command post (ICP) The field location at which the primary tactical-level, on-scene incident command functions are performed.

incident command system (ICS) A system that defines the operating characteristics, interactive management components, and structure of incident management and emergency management organizations engaged throughout the life cycle of an emergency incident. ICS is the combination of facilities, equipment, personnel, procedures, and communications operating within a common organizational structure.

incident management system (IMS) (See incident command system.)

industrial fire brigade An organized group of personnel within an industrial occupancy assigned to perform some level of fire protection, usually within their particular industrial facility. These personnel are trained and skilled in at least basic firefighting operations. (See emergency response team and plant emergency response team.)

initiating device circuits (IDC) A nonintelligent circuit that monitors electrical activity through the circuit by use of an end-of-line resistor. When the resistance changes, this signifies an alarm to the fire alarm control panel (FACP).

inorganic Not containing carbon. An example of an inorganic substance is phosphorus or chlorine. Inorganic compounds come principally from mineral sources of nonbiological origin.

inside battery limits (ISBL) Inside the process area, normally identified by the valves required to isolate the unit from the rest of the refinery or plant.

intermodal container Shipping container having standard dimensions so that it can be secured for shipment by truck, rail, or vessel.

International Fire Service Accreditation Congress (IFSAC) A peer-driven, self-governing system that accredits both fire service certification programs and higher education fire-related degree programs. IFSAC is a nonprofit organization authorized by the Board of Regents of Oklahoma State University as a part of the College of Engineering, Architecture and Technology.

international shore connection A pipe flange that consists of a standardized pattern, including bolt position and size that allows land-based firefighters to connect to and charge with water a vessel's fire main system.

intrinsically safe A device or circuit in which a spark or thermal effect in normal or abnormal conditions is incapable of causing the ignition of a flammable mixture in air.

intumescent The process of swelling or enlarging.

iron reduction process A process used to reduce the amount of oxide in iron oxide to produce and transform iron ore to directed reduced iron (DRI) through heating and chemical reduction by natural gas.

isobutene A flammable, colorless, polymerized gas with a faint petroleum odor.

jargon The technical terminology or characteristic idiom of a special activity or group.

keel Principal member of a ship's construction. It lies fore and aft along the centerline of bottom. It forms the backbone of the vessel.

large-diameter hose (LDH). A hose with a minimum diameter of 3½ inches and a minimum designed test pressure of 200 psi. When responding to industrial incidents, it is good practice to consider LDH to be any hose with a diameter of 5 inches or greater.

LCES The acronym for lookouts, communications, escape routes, and safety zones. While this acronym was developed for use in wildland fire incidents, it has application for industrial fires.

legs Vertical bucket elevators.

lignin A gluelike substance in wood to hold small cellulose fibers together.

liquefied natural gas (LNG) Methane liquefied by cryogenic temperatures.

liquefied petroleum gas (LPG) A gas predominantly comprising light hydrocarbons such as propane and butane, whose vapor pressure exceeds 40 psia at 100°F.

list Continuous transverse inclination of a vessel to one side due to an imbalance of weight within the vessel.

live load A transient or moveable load, such as a building's content, the occupants, the weight of firefighters, or the water discharge from hose streams.

local emergency planning committee (LEPC) A committee appointed by a state emergency response commission as required by the Superfund Amendments and Reauthorization Act of 1986 (SARA) Title III, to formulate a comprehensive emergency plan for its corresponding local government or mutual aid region.

lockout/tagout A specific set of safety-related practices and procedures that reduce the risk of harm from the unexpected activation of machinery or energization of electrical systems during maintenance activities. These procedures require workers engaged in service activities to disconnect the affected systems and either lock or tag the systems to prevent them from being reactivated by other workers who are unaware of the maintenance.

longitudinal flue Flue spaces are perpendicular to the direction of the loading (space in between racks), and transverse flue spaces are parallel to the direction of loading (between loads or racks side by side).

loom A frame or machine for interlacing at right angles two or more sets of threads or yarns to form a cloth.

lower explosive limit (LEL) Minimum concentration of vapor or gas in air below which it is not possible to ignite the vapors with a proper ignition source.

magnetography The magnetographic print process consists of an array of electromagnets that forms a magnetic image on the surface of a metal drum. The image is exposed to toner for developing, and the developed image is then transferred to the substrate, where the toner particles are permanently flash-fused.

material safety data sheets (MSDS) Documents that contain information regarding the chemical composition, physical and chemical properties, health and safety hazards, emergency response, and waste disposal of a material.

meltdown In the case of uranium, splitting into fragments may release more than 100 million volts of energy. The free neutrons may trigger additional breakups in the form of a chain reaction. If this is the case, then the chain reaction may be controlled, releasing energy that can be used to boil water and produce steam to drive a turbine. If this chain reaction cannot be controlled or begins to run away, a meltdown occurs.

Military Sealift Command Provides ocean transportation of equipment, fuel, supplies, and ammunition to sustain US forces worldwide.

minimum explosion concentration (MEC) The minimum concentration of dust required to generate an explosion.

monomer Low molecular weight formations of organic elements; small molecules. These are usually chemically bonded together to form polymers. Examples of monomers are hydrocarbons such as styrene and ethene.

moulaged To have simulated accident victims wear costumes and makeup to simulate wounds and add realism for the drill responders.

mud Liquid drilling fluid used to control the pressure in a well bore. The rams are activated by hydraulic pressure when a blowout threatens and can be locked shut.

multisecular Governmental and geographical response agencies that may be called to respond to an emergency situation.

nacelle The cover around an aircraft engine.

National Board on Fire Service Professional Qualifications (Pro Board) An organization whose goal is to establish an internationally recognized means of acknowledging professional achievement in the fire service and related fields. The accreditation of organizations that certify uniformed members of public fire departments, both career and volunteer, is the primary goal. However, other organizations with fire protection interests may also be considered for participation. The Pro Board accredits fire service training agencies that use the professional qualification standards of the National Fire Protection Association (NFPA).

National Incident Management System (NIMS) A system to provide a consistent approach for federal, state, local, and tribal governments to work effectively and efficiently together to prepare for, prevent, respond to, and recover from domestic incidents.

notification appliance circuit (NAC) A circuit that contains a group of appliances whose intended purpose is to alert building and facility occupants of a potential problem in the building. Notification equipment can include audible appliances (such as bells, horns, speakers, or chimes), visible appliances (such as strobes, lamps, or printer), vibrating appliances, or appliances that combine some or all of these different types.

offensive strategy Aggressive tactics used to control an incident.

olefins Unsaturated hydrocarbons containing one double bond, such as polyethylene or polypropylene.

operational security (OPSEC) A process of identifying critical information and operating procedures and selecting and executing measures to protect this information from being used by other than authorized persons, for reasons that may be detrimental to facilities or persons.

operations manager (drilling) Generally responsible for all operational matters to do with drilling and production-related activities. Will usually take over command of company incident response team in the event of a major incident such as a blowout.

outside battery limits (OSBL) Everything outside of the inside battery limits (ISBL).

outside screw (or stem) and yoke valve (OS&Y) A valve used to control fluid flow through piping. The position of the screw indicates whether it is open or closed. When extended so the screw is visible, the valve is in the full or partially open position.

overhead gases The gases rise in the column and are drawn off through a large pipe at the top of the column.

passive strategy This strategy in firefighting is essentially a nonaction mode when the risks associated with intervening are unacceptable. All personnel are withdrawn to safe areas.

pass-on log Pertinent information that is recorded for oncoming fire officers or members to review as they report to duty at shift change.

permissible exposure limit (PEL) The maximum time-weighted concentration at which 95% of exposed, healthy adults suffer no adverse effects over a 40-hour workweek.

permit-to-work system (See work permit.)

personnel accountability system Systems designed to account for the location and welfare of personnel during an emergency incident or daily work activities.

plant emergency response team (PERT) (See emergency response team.)

pneumonitis Inflammation of the lungs; caused by a virus or an allergic reaction.

polymer Chains of monomers that form plastics like polystyrene and polyethene.

polymerization A process of reacting monomer molecules together in a chemical reaction to form three-dimensional networks or polymer chains.

pooling During a major fire or spill, the entire curbed area may fill with product and firefighting water. During pooling, flammable liquids that float on the water surface are spread as the curbed area is filled with firefighting or storm water.

post indicator valve (PIV) A valve used to control the water for fire suppression systems. The post indicates the open or closed position of an underground valve on a water distribution system. Through an window opening in the post, the word *open* or *shut* will be visible.

private-mode notification where the signal is transmitted to a person responsible for taking action or who might notify occupants when to evacuate, such as in a hospital or assisted-care facility.

Pro Board (See National Board on Fire Service Professional Qualifications)

propylene A double-bond hydrocarbon.

pyrophoric Ignites upon contact with air.

public-mode notification The occupants of the building receiving evacuation signals.

quenching The rapid cooling of a solid to lock it into a metastable crystal structure rather than allow it to cool slowly and revert to a softer structure. It is most commonly used to harden steel.

radiological safety officer (RSO) A person who is qualified by training, experience, and certification in radiation protection, and who is available for advice and assistance on radiological safety matters.

rapid intervention crew (RIC) See rapid intervention team.

rapid intervention team (RIT) A team of firefighters used to assist other firefighters in need of emergency assistance.

ratio controller Installed in fire protection systems to control the proportion of water to foam induction.

recausticizing plant A plant where the green liquor is clarified to remove insoluble impurities, then causticized in the lime slaker to convert the sodium carbonate to sodium hydroxide.

regeneration The process to reactivate a catalyst by removing buildups or impurities that have accumulated during its use.

rotogravure 1. The printing process that transfers an image from a photogravure plate to a cylinder in a rotary press. 2. Using photography to produce a plate for printing.

salt cake Chemical ash collected from the bottom boiler hoppers at paper mils.

self-contained breathing apparatus (SCBA) A positive-pressure, self-contained breathing apparatus certified by an appropriate approval agency for use in atmospheres that are immediately dangerous to life or health (IDLH).

settler A tank where the dense liquids are separated from the lighter liquids.

short-term exposure limit (STEL) The 15-minute, time-weighted average exposure that should not be exceeded at any time, nor repeated more than four times daily, with a 60-minute rest period required between each STEL exposure.

signaling line circuits (SLC) A circuit or path between any combination of circuit interfaces, control units, or transmitters, over which multiple system input signal or output signal or both are carried.

silicomanganese An alloy of manganese and iron with a high percentage of silicon.

Sked® A specific brand of lightweight stretcher used in confined space, hazardous materials, and other applications.

slopover Results when a water stream is applied to the hot surface of burning oil, causing the burning oil to slop over the tank sides.

sour gas Gas containing high concentrations of hydrogen sulfide (H_2S).

sparging Commonly used as a method of removing a liquid from a solution by bubbling dry air through the solution. Liquids with higher vapor pressures will tend to evaporate faster than liquids with lower vapor pressures.

spheroidization Annealing or tempering process used to soften and make more ductile, as well as reducing any internal stresses.

spudding the well Starting to drill the well.

stack effect The updraft caused by hot air or gases rising in a confined area. The stack effect occurs in stairways of tall buildings and can affect the smoke travel. In very tall buildings, smoke may spread from the stairways into the floor areas because of this stack effect.

standard operating guideline (SOG) (See standard operating procedure.)

standard operating procedure (SOP) A set of organizational directives that establish a course of actions both administratively and operationally.

stern After end of the vessel.

stoichiometric limits Composition limits (maximum and minimum) of an ignited oxidizer-fuel mixture what will burn indefinitely at given conditions of temperature and pressure without further ignition.

strike team A set number of resources of the same kind and type that have an established minimum number of personnel.

surfactant Wetting agents that lower the surface tension of a liquid, allowing easier spreading and lowering the interfacial tension between two liquids.

taconite Low-grade iron ore.

target hazard A building, complex of buildings, or any potential problem areas that, because of construction, occupancy use, or size, presents serious and/or unusual operational problems for firefighters. These identified hazards require preincident response planning and on-site drills.

target occupancy An occupancy, structure, or system that, because of its use or local/national symbolic or historical significance, has been identified as a potential target for terrorist attacks. These may include government buildings, industrial facilities, controversial businesses and organizations, public buildings, and infrastructure systems.

task force Any combination of resources assembled to support a specific mission or operational need. All resource elements within a task force must have common communications and a designated leader.

thermoplastic Capable of becoming soft when heated and hardened when cooled.

thermoset Set into a permanent shape from heat and applied pressure. Reheating will not soften these materials.

three-dimensional fires Fires where burning liquids cascade down through a process structure, showering burning liquids onto platforms and equipment below the source of the fuel. Three-dimensional fires spread out horizontally from their source as burning liquids splash off of vessels, equipment, and structural components located below the fuel source.

tide Periodic vertical rise and fall of the water surface level of the oceans, bays, gulfs, inlets, and tidal regions of rivers caused by the gravitational attraction of the sun and moon.

toolpusher The person in overall command of the drilling rig.

tramp elements Chemical elements that unintentionally find their way into steel melts during the production process. They accumulate further as the steel is subsequently recycled.

tundish car A piece of equipment used in steel manufacturing. It is usually located between the ladle and the caster mold and allows the molten steel to enter the mold.

upper explosive limit (UEL) Maximum proportion of vapors or gas in the air above which it is not possible to cause ignition with a proper ignition source.

vapor cloud explosion (VCE) An explosion occurring outside that produces damaging overpressures. It is initiated by the unplanned release of a large quantity of flammable vaporizing liquid or high-pressure gas from a storage tank or system, process vessel, pipeline, or transport vessel.

veneer Thin slices of wood, usually thinner than 1/8 inch. Veneer layers are usually glued and pressed onto core panels of different materials (such as wood, particle board, or medium density fiberboard) to obtain doors, tops, and side panels for cabinets, parquet floors, and pieces of furniture.

venturi The creation of a slight negative pressure by water flowing past an opening in a pipe. This slight negative pressure is used to draw foam concentrate into foam-proportioning equipment.

vermiculite A natural, nontoxic mineral that expands with the application of heat. The expansion process is called exfoliation, and it is routinely accomplished in purpose-designed commercial furnaces.

weapons of mass destruction (WMD) A term used to describe munitions with the capacity to indiscriminately kill large numbers of human beings. The phrase broadly encompasses several areas of weapon synthesis, including nuclear, biological, chemical, and, more recently, radiological weapons.

weir A small overflow dam installed on the top of a vessel such as a liquefied petroleum gas vessel used to spread cooling water evenly down the sides.

wetting agents A substance that, by becoming adsorbed, prevents a surface from being repellant to a wetting liquid; used in spreading liquids on surfaces.

windshield survey A quick visual survey of an industrial facility while driving a vehicle, as observed through the windshield.

work permit system The system of written permits/certificates by which work at industrial installations can be authorized. Before commencement of repairs, maintenance, or other nonroutine work in an industrial installation, full precautions must be taken to safeguard the persons performing the work and others in the vicinity, and prevent material damage. The work permit provides a record showing that the conditions of work have been checked by the parties responsible, and it identifies the risks present and precautions to be taken.

work permits (See work permit system.)

workhouse Contains storage bins used to hold in-process grain for loading and shipping, as well as additional cleaning and bagging for local packaging.

zeolite A type of catalyst.

Acronyms and Abbreviations

AED automated external defibrillator

AFFF aqueous film-forming foam

AHJ authority having jurisdiction

API American Petroleum Institute

ARFF aircraft rescue and fire fighting

BLEVE boiling liquid expanding vapor explosion

COAL TWAS WEALTHS construction, occupancy, apparatus and staffing, life hazard, terrain, water supply, auxiliary appliances and aids, street conditions, weather, exposures, area, location and extent, time, height, and special considerations

CFR Code of Federal Regulations

COMSEC communications security

CPF central processing facility

DOT Department of Transportation

EAP emergency action plan

EOC emergency operations center

EOP emergency operations plan

EPA Environmental Protection Agency

EPCRA Emergency Planning and Community Right-to-Know Act

ERT emergency response team

ESD emergency shutdown device

FEMA Federal Emergency Management Agency

FFFP film-forming fluoroprotein foam

FP fluoroprotein foam

GOSP gas oil separation plant

gpm gallons per minute

HazMat hazardous materials

HAZWOPER hazardous waste operations and emergency response

HFC halocarbon agent

IAP incident action plan

ICP incident command post

ICS incident command system

IDC initiating device circuits

IDLH immediately dangerous to life and health

IFSAC International Fire Service Accreditation Congress

ILBP in-line balanced proportioning

IMS incident management system

ISBL inside battery limits

LDH large-diameter hose

LEPC local emergency planning committee

LPG liquefied petroleum gas

LNG liquefied natural gas

MEC minimum explosion concentration

MSDS material safety data sheet

NAC notification appliance circuit

NASA National Aeronautical and Space Administration

NFPA National Fire Protection Association

NIMS National Incident Management System

NST national standard thread

OPSEC operations security

OSBL outside battery limits

OSCP on-scene command post

OS&Y outside screw and yoke or outside stem and yoke

OSHA Occupational Health and Safety Administration

PEL permissible exposure limit

PERT plant emergency response team

PKP potassium bicarbonate

PIV post indicator valve

psi pounds per square inch

psig pounds per square inch gauge

psia pounds per square inch absolute

RIC rapid intervention crew

RIT rapid intervention team

RSO radiological safety officer

SCBA self-contained breathing apparatus

SLC signaling line circuit

SOG standard operating guide

SOP standard operating plan

STEL short-term exposure limit

T.O. technical order

VCE vapor cloud explosion

WMD weapons of mass destruction

Bibliography

Abel, W. L., Bowden Sr., J. R., and Campbell, P. J. *Firefighting and Blowout Control.* Spring, TX: Wild Well Control, 1994.

Adams, N. *Terrorism and Oil.* Tulsa, OK: PennWell, 2003.

American Bureau of Shipping (ABS). *Guide for Building and Classing Fire Fighting Vessels.* New York: ABS, 1981.

American Chemical Society. "Production by the US Chemical Industry." *Chemical and Engineering News,* June 1996.

American Forest and Paper Association, Environmental and Recycling. Retrieved September 23, 2006. from www.afandpa.org

American Petroleum Institute, (API). Recommended Practice (RP) 2021, *Management of Atmospheric Storage Tank Fires,* 4th ed. Washington, DC: API, 2001.

American Petroleum Institute (API). API 2021a: *Interim Study—Prevention and Suppression of Fires in Large Aboveground Atmospheric Storage Tanks.* Washington, DC: API, 1998.

American Petroleum Institute (API). API 2510a: *Fire Protection Considerations for the Design and Operation of Liquefied Petroleum Gas (LPG) Storage Facilities.* Washington, DC: API, 1996.

Ansul. "A Comparison of Foam Agents to Emulsifying/Wetting Agents." Retrieved June 6, 2006, from http://www.tychotech.com.

Associated Press. *Sprinkler System Caused Huge Fire at Magnesium Plant.* January 16, 2005, Andersen, Indiana. ABC channel 7 Chicago. Retrieved September 23, 2006, from http://abclocal.go.com/wls/story?section=News&id=2623415&ft=lg.

Avillo, A. L. *Fireground Strategies.* Tulsa, OK: PennWell, 2002.

Bevelacqua, A., and Stilp, R. *Terrorism Handbook for Operational Responders.* Albany, NY: Delmar, 1998.

Bevelacqua, A. S., Noll, G. G., and Hildebrand, M. S. *Hazardous Materials, Managing the Incident, Field Operations Guide.* Chester, MD: Red Hat, 2005.

Brown, G. J., and Crist, G. S. *Confined Space Rescue.* Albany, NY: Delmar, 1999.

Burdick, D. L., and Leffler, W. L. *Petrochemicals in Non Technical Language,* 3rd ed. Tulsa, OK: PennWell, 2001.

Calhoun, T. "Disaster Averted at Sunny Point." *The State Port Pilot,* July 18, 2001. Retrieved July 5, 2006, from http://www.sppilot.com/front/shipfire.html.

Center for Chemical Process Safety (CCPS) of the American Institute of Chemical Engineers. *Guidelines for Analyzing and Managing the Security Vulnerabilities of Fixed Chemical Sites.* New York: CCPS, 2003, p.10.

Chemistry.org. *Production by the US Chemical Industry*. Retrieved August 2006 from http://pubs.acs.org/hotartcl/cenear/960624/prod.html.

Colletti, D. *Class A Foam, Best Practice of Structure Firefighters*. Royersford, PA: Lyon, 1998.

Conaway, C. F. *The Petroleum Industry: A Nontechnical Guide*. Tulsa, OK: PennWell, 1999.

Copenhaver, J. "Emerging Risk: Are We Prepared?" *Continuity Insights*. Retrieved July 1, 2006, from http://www.continuityinsights.com.

Cozad, F. D. *Water Supply for Fire Protection*. Upper Saddle River, NJ: Prentice Hall, 1981.

Devereux, S. *Drilling Technology in Nontechnical Language*. Tulsa, OK: PennWell, 1999.

Douberly, E. B. "Fire Protection Guidelines for Handling and Storing PRB Coal." *Power*, 147(9). Retrieved June 19, 2006, from http://www.powermag.com/archive_search.asp?a+20x0e410810132Y1ax7b.

Downey, R. *The Rescue Company*. Saddle Brook, NJ: Fire Engineering Books and Videos, 1992.

Dunn, V. *Collapse of Burning Buildings, A Guide to Fireground Safety*. New York: Fire Engineering, 1998.

Dunn, V. *Command and Control of Fires and Emergencies*. Saddle Brook, NJ: Fire Engineering Books and Videos, 1999.

Dunn, V. *Safety and Survival on the Fireground*. Saddle Brook, NJ: Fire Engineering Books and Videos, 1992.

Eckman, W. F. *The Fire Department Water Supply Handbook*. Saddle Brook, NJ: Fire Engineering Books and Videos, 1994.

Electricity Generation. Retrieved June 19, 2006, from http://en.wikipedia.org/wiki/Electricity_generation.

Fire, F. L., *The Common Sense Approach to Hazardous Materials*. Saddle Brook, NJ: Fire Engineering Books and Videos, 1996.

FM Global. *Guidelines for Evaluating the Effects of Vapor Cloud Explosions Using a TNT Equivalency Method*. Johnston, RI : Factory Mutual Insurance Company, 2001.

FM Global. Property Loss Prevention Data Sheets, *7–74, Distilleries*. Norwood, MA: FM Global, 2000.

FM Global Property Loss Prevention Data Sheets. *8–1 Commodity Classification*. Norwood, MA: FM Global, 2000.

Food Processing Technology, American Italian Noodle Production Plant, Retrieved on September 23, 2006, from http://www.foodprocessing-technology.com/projects/aipc.

Free Software Foundation. *Nuclear Power*. Retrieved July 2, 2006, from http://en.wikipedia.org/wiki/nuclear_power.

Friedman, Raymond, *Principles of Fire Protection Chemistry and Physics*, 3rd ed. Quincy, MA: National Fire Protection Association, 1998.

Gagnon, R., and Kirby, R., *A Designers Guide to Fire Alarm Systems*. Quincy, MA: National Fire Protection Association, 2003.

Grace, R. D. *Blowout and Well Control Handbook*. London: Gulf Professional, 2003.

Habashi, F. A. *Textbook of Hydrometallurgy*. Quebec, Canada: Metallurgie Extractive, 1993.

Hawley, C., Noll, G. G., and Hildebrand, M. S. *Special Operations for Terrorism and HazMat Crimes*. Chester, MD: Red Hat, 2002.

Hildebrand, M. S., and Noll, G. G. *Storage Tank Emergencies, Guidelines and Procedures*. Annapolis, MD: Red Hat, 1997.

HM Fire Service Inspectorate Publications Section. *Fire Service Manual, Vol. 2: Fire Service Operations, Marine Incidents*. London: The Stationary Office, 1999.

Imster, E. *Oil Well Fires*, 2003. Retrieved January 13, 2005, from Earth and Sky on http://earthsky.com/shows.

International Fire Service Training Association (IFSTA). *Industrial Emergency Services Training: Incipient Level,* 2nd ed. Stillwater, OK: Fire Protection Publications, 2003.

International Fire Service Training Association (IFSTA). *Marine Firefighting for Land-Based Firefighters.* Stillwater, OK: Fire Protection Publications, 2001.

International Maritime Organization (IMO). *International Maritime Dangerous Goods (IMDG) Code.* London: IMO, 2004.

International Maritime Organization. *International Ship and Port Facility Security Code (ISPS) and SOLAS Amendments 2002.* London: IMO, 2003, sect. 3.1.

International Maritime Organization (IMO). *SOLAS Consolidated Edition, 1992.* London: IMO, 1992.

Invitesite.com. "Descriptions of Printing Processes." Retrieved April 6, 2006, from http://www.invitesite.com/white_pages/printing_processes.html.

Jakubowski, G. "Unfamiliar Territory: Industrial Fires Pose Unique, Dangerous Challenges." *FireRescue Magazine,* June 2006. Retrieved July 11, 2006, from http://www.firerescue1.com/firerescue-magazine/24-2/16969.

Kanterman, R. "MAC-SICS: Industrial Mutual Aid in New Jersey." *Fire Engineering,* 2005, November, pp. 97–98.

Kennedy, J. L. *Oil and Gas Pipeline Fundamentals,* 2nd ed. Tulsa, OK: PennWell, 1993.

Kennedy, P. M., and Kennedy, J., *Explosion Investigation and Analysis, Kennedy on Explosions.* Investigations Institute, 1990.

Linstrom, J. "Pipeline Plans Improve Info Flow." *Fire Chief,* August 4, 2004. Retrieved July 6, 2006, from http://firechief.com/mag/firefighting_pipeline_preplans_improve/index.html.

Matthews, C., and Flak, L. *Part 9: Firefighting. the Mechanics of Oil/Gas Fires, Meltdown and Secondary Damage, Water/Chemical/Explosive Extinguishing Methods and Considerations for Voluntary Ignition.* Retrieved January 13, 2005, from http://www.jwco.com/technical-litterature,p09.htm.

Merriam-Webster. *Merriam Webster's Collegiate Dictionary,* 10th ed. Springfield, MA: Merriam-Webster, 2002.

Miller, A. G. W. *Dictionary of Nautical Terms.* Glasgow: Brown, Son & Ferguson, 1994.

National Association of State Fire Marshals (NASFM). *Liquefied Natural Gas: An Overview of the LNG Industry for Fire Marshals and Emergency Responders.* Washington, DC: NASFM, 2005.

National Emergency Training Center (NETC). *Initial Response to Hazardous Materials Incidents, Course I: Basic Concepts.* Emmitsburg, MD: NETC, 1992.

National Fire Protection Association (NFPA). *Fire Protection Handbook,* 19th ed., vols. I and II. Quincy, MA: NFPA, 2003.

National Fire Protection Association (NFPA). *Fire Protection Handbook, Business and Industry Edition.* Quincy, MA: NFPA, 2003.

National Fire Protection Association (NFPA). NFPA 921, *Guide for Fire and Explosion Investigation.* Quincy, MA: NFPA, 2004.

National Fire Protection Association (NFPA). *Industrial Fire Hazards Handbook,* 3rd ed. Quincy, MA: NFPA, 1990.

National Fire Protection Association (NFPA). NFPA 72, *National Fire Alarm Code.* Quincy, MA : NFPA, 2002.

National Fire Protection Association (NFPA). NFPA 291, *Recommended Practice for Fire Flow Testing and Marking of Hydrants.* Quincy, MA: NFPA, 2002.

National Fire Protection Association (NFPA). NFPA 850, *Recommended Practice for Fire Protection for Electric Generating Plants and High Voltage Direct Current Converter Stations.* Quincy, MA: NFPA, 2005.

National Fire Protection Association (NFPA). NFPA 851, *Recommended Practice for Fire Protection for Hydroelectric Generating Plants.* Quincy, MA: NFPA, 2005.

National Fire Protection Association (NFPA). NFPA 471, *Recommended Practices for Responding to Hazardous Materials Incidents.* Quincy, MA: NFPA, 2002.

National Fire Protection Association (NFPA). NFPA 1620, *Recommended Practice for Pre-Incident Planning.* Quincy, MA: NFPA, 2003.

National Fire Protection Association (NFPA). NFPA 15, *Standard for Water Spray Fixed Systems for Fire Protection.* Quincy, MA: NFPA, 2007.

National Fire Protection Association (NFPA). NFPA 17, *Standard for Dry Chemical Extinguishing Systems.* Quincy, MA: NFPA, 2002.

National Fire Protection Association (NFPA). NFPA 20, *Standard for Installation of Stationary Pumps for Fire Protection.* Quincy, MA: NFPA, 2003.

National Fire Protection Association (NFPA). NFPA 403, *Standard for Aircraft Rescue and Fire-Fighting Services at Airports.* Quincy, MA: NFPA, 2003.

National Fire Protection Association (NFPA). NFPA 801, *Standard for Fire Protection for Facilities Handling Radioactive Materials.* Quincy, MA: NFPA, 2003.

National Fire Protection Association (NFPA). NFPA 1081, *Standard for Industrial Fire Brigade Member Professional Qualifications.* Quincy, MA: NFPA, 2001.

National Fire Protection Association (NFPA). NFPA 11, *Standard Low-, Medium-, and High Expansion Foam.* Quincy, MA: NFPA, 2005.

National Fire Protection Association (NFPA). NFPA 12, *Standard on Carbon Dioxide Extinguishing Systems.* Quincy, MA: NFPA, 2005.

National Fire Protection Association (NFPA). NFPA 2001, *Standard on Clean Agent Extinguishing Systems.* Quincy, MA: NFPA, 2004.

National Fire Protection Association (NFPA). NFPA 1600, *Standard on Disaster/Emergency Management and Business Continuity Programs.* Quincy, MA: NFPA, 2004.

National Fire Protection Association (NFPA). NFPA 1561, *Standard on Emergency Services Incident Management System.* Quincy, MA: NFPA, 2005.

National Fire Protection Association (NFPA). NFPA 1500, *Standard on Fire Department Occupational Safety and Health Program.* Quincy, MA: NFPA, 2002.

National Fire Protection Association (NFPA). NFPA 600, *Standard on Industrial Fire Brigades.* Quincy, MA: NFPA, 2002.

National Fire Protection Association (NFPA). NFPA 750, *Standard on Water Mist Fire Protection Systems.* Quincy, MA: NFPA, 2003.

National Fire Protection Association (NFPA). NFPA 704, *Standard System for the Identification of the Hazards of Materials for Emergency Response.* Quincy, MA: NFPA, 2001.

Naylor, C. "HazMat on the High Seas." *Industrial Fire World Magazine,* 13(3). Retrieved July 12, 2005, from http://www.fireworld.com/site/articles/Hazmat_on_the_High_Seas.html.

Nolan, D. P. *Encyclopedia of Fire Protection.* Clifton Park, NY: Thomson Delmar Learning, 2006.

Noll, G. G., Hildebrand, M. S., and Yvorra, J. *Hazardous Materials, Managing the Incident,* 3rd ed. Chester, MD: Red Hat, 2005.

Noll, G. G., and Hildebrand, M. S. *Pipeline Emergencies.* Chester, MD: Red Hat, 2004.

Noll G. G., and Hildebrand M. S., *Storage Tank Emergencies, Guidelines and Procedures.* Annapolis, MD, Red Hat, 1997

Norman, J. *Fire Officer's Handbook of Tactics.* Saddle Brook, NJ: Fire Engineering, 1991.

Office of Pipeline Safety. "Frequently Asked Questions." Retrieved August 12, 2005, from http://primis.phmsa.dot.gov/comm/FAQs.htm.

Oil Well Fires. Retrieved January 13, 2005, from http://www.earthsky.com/shows/shows.php?t=20030624.

Pipeline Safety 101. American Heat. VHS. Carrollton, TX: Fire and Emergency Training Network, 2004.

Printers' National Environmental Assistance Center. "Print Process Descriptions: Printing Industry Overview." Retrieved August 3, 2006, from http://www.pneac.org/printprocesses/screen.

Society of Fire Protection Engineers (SFPE). *The SFPE Handbook of Fire Protection Engineering,* 3rd ed. Quincy, MA: National Fire Protection Association, 2002.

Stubbles, J. *The Basic Oxygen Steelmaking (BOS) Process.* Mason, Ohio: Author, 2000.

Taylor, C., and Waits, M. C. "Oil Well Explosion." *Fire Engineering,* May 1999: 75–82.

Terpak, M. A. "Size-Up: Updating an Old Acronym. *Fire Engineering,* August 2002: 69–80.

Texas A & M University, Emergency Services Training Institute. *Liquefied Natural Gas (LNG) Spill Control and Fire Suppression Training,* Student Notebook. College Station, TX: Texas A & M, n.d.

The State Port Pilot. "Disaster Averted and Sunny Point." Retrieved July 20, 2001, from http://www.sppilot.com/front/shipfire.html.

Thomson, N. *Fire Hazards in Industry.* Oxford: Butterworth-Heinemann, 2002.

Tremblay, K. J. (2005, September/October). Fire Destroys Cold Storage Fruit Packing Buildings. *NFPA Journal.* Retrieved September 23, 2006, from http://www.nfpa.org/publicColumn.asp?categoryID=&itemID=25645&src=NFPAJournal.

Turbines, Generators, and Power Plants. *The Energy Story.* Retrieved June 19, 2006, from http://www.energyquest.ca.gov/story/chapter06.html.

United States Congress, HR 2237, *Chemical Security Act of 2005,* sect. 2. Retrieved June 1, 2006, from http://www.theorator.com/bills109/hr2237.html.

United States Enrichment Corporation (USEC). Retrieved July 1, 2006, from http://www.usec.com/v2001_2/html/aboutusec_quickfacts.asp.

United States Environmental Protection Agency, *AP 42,* 5th ed., vol. I, chap. 9: Food and Agricultural Industries.

United States Environmental Protection Agency, Understanding Radiation, Gamma Rays. Retrieved on September 23, 2006from http://www.epa.gov/radiation/understand/gamma.htm.

United States Department of Homeland Security (DHS). *National Response Plan.* Washington, DC: 2004.

United States Department of Homeland Security (DHS). *National Incident Management System.* Washington, DC: DHS, 2004.

United States. Department of Labor. *Hazardous Waste Operations and Emergency Response, 1910.120.* Retrieved July 28, 2005, from http://ww.osha.gov/pls/oshaweb/owadisp.show_document?p_table=STANDARDS&p_id=9765.

United States Department of Transportation. *Emergency Response Guide.* Washington, DC: 2004.

United States Department of Transportation, Office of Hazardous Materials Safety, About HMIS. Retrieved on September 23, 2006, http://hazmat.dot.gov/enforce/spills/abhmis.htm.

United States Department of Transportation, Office of Pipeline Safety. Retrieved August 12, 2005, from http://primis.phmsa.dot.gov.

United States Enrichment Corporation (USEC). Retrieved July 1, 2006, from http://www.usec.com/v2001_02/html/aboutusec_quickfacts.asp.

United States Environmental Protection Agency (EPA), Rhinehart Tire Fire. Retrieved September 23, 2006, from http://epa.gov/reg3hwmd/npl/VAD980831796.htm.

United States Environmental Protection Agency, (EPA). *Title 42—The Public Health and Welfare,* chap. 116: Emergency Planning and Community Right to Know, 1986.

United States National Transportation Safety Board (NTSB). *Natural Gas Pipeline Rupture and Fire Near Carlsbad, New Mexico, August 19, 2000.* Washington, DC: NTSB, 2003.

United States National Transportation Safety Board (NTSB). *Pipeline Rupture and Subsequent Fire in Bellingham, Washington, June 10, 1999.* Washington, DC: NTSB, 2002.

United States Nuclear Regulatory Commission (NRC). Retrieved June 4, 2006, from http://www.nrc.gov/reactors.html.

United States Occupational Safety and Health Administration (OSHA). *Title 29: PART 1910—Labor-Occupational Safety and Health Standards, Subpart H—Hazardous Materials. 1910.120 Hazardous waste operations and emergency response.*

United States Occupational Safety and Health Administration (OSHA). *Title 29: PART 1910—Labor-Occupational Safety and Health Standards, Subpart L—Fire Protection. 1910.156 Fire Brigades.*

White, D. (Ed.). Racked Up. *Industrial Fire World,* January/February 2003: 6–9, 24.

Wright, B. R. Virginia HazMat Crews Face Jet Fuel Pipeline Rupture. *Firehouse.com.* Retrieved April 21, 2006, from http://cms.firehouse.com/content/article/printer.jsp?id+48889.

Zalosh, R. G. *Industrial Fire Protection Engineering.* West Sussex, UK: Wiley, 2003.

Index

See also glossary of key terms, 331–341.

A

B

C

D

E

F

G

I

J

K

L

M

N

O

P

Q

R

S

T

U

V

W

X–Y

Z

About the Authors

Craig H. Shelley, EFO, CFO, MIFireE is a 40-year veteran of the fire service. He has served in volunteer fire/EMS, career municipal and career industrial fire departments. Mr. Shelley served for 26 years with the City of New York Fire Department (FDNY) retiring as the Chief of Marine Operations. In addition he has served as the Fire Chief for the City of Rutland (VT) Fire Department, Deputy Chief with the West Hamilton Beach Fire Department, and as a Task Force Leader with the Federal Emergency Management Agency's (FEMA) Urban Search and Rescue Task Force NY TF-1. In this capacity Mr. Shelley responded to major incidents including the Oklahoma City bombing. He has served on the National Fire Protection Association's (NFPA) technical committee on training and on the marine firefighting vessels committee.

Mr. Shelley is an adjunct associate professor for the University of Maryland University College teaching their Managerial Issues in Hazardous Materials course and also serves as an adjunct associate professor for Charter Oak State College teaching strategic planning. Mr. Shelley currently serves as a fire protection advisor with a major oil company operating in the Middle East and is a partner with the World Safe Group providing international consulting services. Mr. Shelley holds a Bachelor of Science degree in Fire Service Administration and a Master of Science degree in Executive Fire Service Leadership. He is a frequent contributor to industry trade publications and speaks at conferences both nationally and internationally. He may be contacted at MARDIV4715@aol.com.

Anthony Cole, CFPS, CFEI, MIFireE has 20 years of experience in fire protection ranging from fire fighting to engineering and design. Currently, Mr. Cole is a Fire Protection Engineer with the Saudi Arabian Oil Company (Saudi Aramco) in Dhahran, Saudi Arabia. His responsibilities include project plan review, technical standards and specifications development, plant fire risk analysis, special studies and projects, and large-scale emergency response.

Prior to his work at Saudi Aramco, Mr. Cole was a Fire Protection Engineer with Rolf Jensen and Associates in the Denver, Colorado office. His responsibilities included project management, plan and code reviews, plant surveys, and consulting. A former Fire Chief/Senior Fire Engineer for the Saudi Iron and Steel Company in Jubail, Saudi Arabia, Mr. Cole is also a former Loss Prevention Engineer with FM Global. He started his fire career as both a paid and volunteer firefighter in Ohio, he continued on several volunteer fire departments in Kentucky, Mississippi, and Missouri.

Mr. Cole earned a B.S. in Fire Protection Engineering Technology from Eastern Kentucky University and a M.S. in Fire Protection Engineering from Worcester Polytechnic Institute. Serving on three NFPA technical committees (NFPA 12, NFPA 15, NFPA 59A), he currently serves on two SFPE committees and is the Vice President/Co-founder of the Saudi Arabian Chapter of SFPE. Mr. Cole is a Certified Fire Protection Specialist and a Certified Fire and Explosion Investigator and is a member of SFPE, NFPA, NAFI, IFE, and IAFC.

Timothy E. Markley, CFEI, CFEII, CFPS, started his career in the fire service in 1975 at Friendship Volunteer Fire Department in Winchester, Virginia. After serving at Friendship VFD, he joined the United States Air Force as a Fire Protection Specialist. He completed assignments within the United States, England, Korea, Honduras, Germany, Egypt, and Iraq. While in the Air Force, he completed his Associate of Science and Bachelor of Science Degrees. At Sembach AB, Germany he served as Fire Chief and in1994 was assigned to Headquarters Air Combat Command Inspector General at Langley AFB. Mr. Markley inspected active duty, National Guard and Reserve fire departments. He retired from the USAF in 1997 and worked for TRW as a Fire Protection Advisor to the Royal Saudi Air Force in Dhahran Saudi Arabia until 2001.

Mr. Markley also worked for Kellogg Brown and Root in Taszar, Hungary to support U. S. military operations. He moved back to Saudi Arabia in 2002 as a Fire Protection Advisor. He is currently a Fire Protection Advisor to a major oil company in Saudi Arabia. Responsibilities encompass advising on fire protection matters both on and offshore, including facilities such as airstrips and heli-ports, gas-oil separation plants, hydrocarbon refineries, marine terminals, gas plants, industrial support facilities and residential communities.

Selected certifications include International Fire Service Accreditation Congress Fire Officer IV, Hazardous Materials Technician Trainer and Incident Command Trainer. In addition, Mr. Markley is a certified Fire and Explosives Investigator Trainer through the National Association of Fire Investigators. He has also taught fire science for City Colleges of Chicago and Central Texas College.